Sensory Evaluation Practices

Fourth Edition

Herbert Stone

Rebecca N. Bleibaum

Heather A. Thomas

AMSTERDAM • BOSTON • HEIDELBERG • LONDON
NEW YORK • OXFORD • PARIS • SAN DIEGO
SAN FRANCISCO • SINGAPORE • SYDNEY • TOKYO
Academic Press is an imprint of Elsevier

Academic Press is an imprint of Elsevier
32 Jamestown Road, London NW1 7BY, UK
225 Wyman Street, Waltham, MA 02451, USA
525 B Street, Suite 1800, San Diego, CA 92101-4495, USA

First edition 1985
Second edition 1993
Third edition 2004
Fourth edition 2012

Notice
No responsibility is assumed by the publisher for any injury and/or damage to persons or property as
a matter of products liability, negligence or otherwise, or from any use or operation of any methods,
products, instructions or ideas contained in the material herein. Because of rapid advances in the medical
sciences, in particular, independent verification of diagnoses and drug dosages should be made

British Library Cataloguing-in-Publication Data
A catalogue record for this book is available from the British Library

Library of Congress Cataloging-in-Publication Data
A catalog record for this book is available from the Library of Congress

ISBN: 978-0-12-382086-0

For information on all Academic Press publications
visit our website at elsevierdirect.com

Typeset by MPS Limited, Chennai, India
www.adi-mps.com

Printed and bound by CPI Group (UK) Ltd, Croydon, CR0 4YY

Contents

Foreword

Since the publication of the third edition eight years ago, sensory science has grown exponentially and is now recognized as a strategic information source for many Fortune 500 companies. This growth in sensory science programs within corporations has been reflected in the formation and increasing membership of sensory professional organizations and in sensory related presentations at national and international meetings, as well as in the increase in submissions for sensory related research journals. This growth in the field has also been accompanied by an increase in the number of individuals who are employed as sensory professionals. Many scientists working in this field depend on the core textbooks such as this one to enhance their working knowledge base with practical applications that can be applied to business objectives.

Stone, Bleibaum, and Thomas have worked together in sensory evaluation, along with the previous co-author, Sidel, for over 25 years. They are recognized as visionary leaders in the fields of sensory evaluation and consumer research. They have been tireless advocates for the rigorous design of sensory experiments and statistical analysis of the data for actionable outcomes. As the visionary developers of Quantitative Descriptive Analysis (QDA) over 30 years ago, they pioneered the methodology of descriptive analysis, now a staple of most sensory evaluation programs in the consumer product industry worldwide. As a team, they have consulted on and conducted thousands of sensory research projects around the world. The authors also are actively engaged in teaching sensory evaluation principles in workshops, University certificate programs, and other professional organizations.

Sensory Evaluation Practices contributes admirably to help fill this knowledge gap both as a reference for sensory professionals and as a textbook in educational institutions. This text was specifically designed to give practical guidance on sensory procedures that could be directly applied to the strategic issues encountered in a business and research environment.

This fourth edition draws even more heavily on their practical experience in partnering with business associates in marketing and development teams to bring creativity and innovation to consumer driven product development. This new edition provides updates on sensory evaluation methods and provides additional discussion on applications in today's global business environment. The information contained in this textbook can further enhance the sensory professional in performing their responsibilities at a higher level of professionalism and competence.

We appreciate the invitation to write the foreword to this book and are confident that *Sensory Evaluation Practices, Fourth Edition* will contribute in a meaningful way to the reader's development as a sensory professional.

Howard G. Schutz
Professor Emeritus of Consumer Sciences
University of California, Davis

Jean-Xavier Guinard
Professor of Sensory Science
Department of Food Science & Technology
University of California, Davis

Preface

In this edition, readers will find that there have been changes in content as well as in authorship. After three editions, my co-author, Joel L. Sidel, has chosen not to continue, and I acknowledge his invaluable contributions. Two new co-authors are introduced. Both have decades of experience in sensory evaluation, both were students of Professor Rose Marie Pangborn, and both bring new perspectives and additional experiences to this edition.

It has been 8 years since the publication of the third edition of this book, and developments continue to be made, offering new approaches to the product evaluation process, to the analysis of the data, and to the applications of sensory resources in a wide range of product-related topics. One of these new topics is involvement in early stage or front-end research focused on techniques such as qualitative research, imagery, and related social psychological methods such as laddering and emotional profiling. These techniques provide insight into how consumers think about a product. In various publications, the senior author has called attention to the importance of sensory professional involvement in these activities as part of any product development effort. Awareness of these activities is important because it enables the sensory staff to decide how best to test a particular product, to become familiar with the language used by consumers, and to create more synergy within the research team.

Such efforts serve as a reminder of the high level of interest in the sensory sciences and specifically sensory testing activities. As noted in the previous edition, much remains to be done to improve the quality of the information obtained. Some concerns expressed previously remain concerns, such as an almost blind reliance on software that enables a wide range of multivariate analyses to be done without regard for whether the original data set yielded significant differences or whether the results make sense (i.e. whether it has face validity). The power of today's PCs and access to cloud computing enable the sensory professional to provide information with minimal delay anywhere in the world on a 24-hour basis. It is a blessing and a curse; marketing and technology want to be able to respond quickly to reformulate or to proceed with a market test with minimal delay. This process makes sense, but the downside is the pressure on sensory professionals to make decisions without adequate time to determine if the results make sense. As Groopman (2009) noted, "Statistical analysis is not a substitute for thinking"—a statement that is essential for all sensory professionals to keep in mind when examining test results.

Despite these impediments, the field continues to grow and to attract interest. Whereas two decades ago there were very few texts on the topic, today there are literally dozens available and more in press. New journals devoted to sensory evaluation have appeared, and all are available in hard copy and/or via the Internet. Professional societies continue to grow. In addition, the number of conferences and short courses has increased dramatically. In any one year, there are at least 25 courses and as many as six conferences throughout the world. Perhaps most gratifying has been the increase in academic programs offering course work and degrees in sensory evaluation. This has occurred in both the United States and Europe, and some universities in China have offered courses in sensory evaluation. In addition, the University of California at Davis has been offering a yearlong web-based distance learning class since 2001 attended by more than 400 students throughout the world. These programs are welcome because they will eventually lead to a more scientific approach to the testing process. In this edition, we have reviewed the organizational issues and where necessary we have made changes that we believe will help maintain programs despite the many changes taking place in the consumer products industries. Consideration has also been given to the reviews of earlier editions and comments provided by numerous individuals who have written and/or talked with us about what is missing in the book.

As in previous editions, we provide detailed information about test methods, in part because there continues to be confusion about these methods. We encounter individuals describing a method that bears little resemblance to the method of the same name as described in the literature. Failure to understand the methods and how they were developed leads to their misuse, and recommended practices no longer appear to be practiced as rigorously as in the past. One important example is the use of analytical methods (i.e. discrimination and descriptive) without replication, which is akin to a chemical analysis done only once. Some have advocated the use of large numbers of subjects to avoid qualifying subjects or to compensate for subject variability. On the one hand, there is use of sophisticated statistical procedures, and on the other hand, there is use of data obtained randomly and/or without evidence that the individual was qualified to be in the test. Finally, the search continues for the universal scale and the invariant subject despite evidence that such a search will not be successful.

As stated in the first edition, and restated here, this book is not a review of the literature. We do, however, discuss literature relevant to specific issues and cite what we consider to be pertinent to the applications of sensory resources to provide actionable information. In this edition, we continue to emphasize the importance of planning, and the importance of learning how to manage and communicate product sensory information in a way that is actionable, that is understood by others.

Finally, we acknowledge the important early contributions of our longtime friend and associate, the late Professor Rose Marie Pangborn of the University of California at Davis, who worked tirelessly to educate students, encouraged them to pursue a career in sensory evaluation, and worked with a total commitment for the betterment of the science of sensory evaluation.

Herbert Stone
Rebecca N. Bleibaum
Heather A. Thomas

Introduction to Sensory Evaluation

1.1 Introduction and objective

Since publication of the third edition in 2004, developments continue to be made in the sensory sciences and their application in the product evaluation process. In today's highly competitive, global business environment, company executives acknowledge the need for actionable product sensory information to supplement what they already believe they know about consumer behavior. This is especially the case for foods, beverages, and most other consumer products, including the ubiquitous electronic devices that have such a large impact on our daily lives. As noted elsewhere (see, for example, Stone and Sidel, 2007, 2009), brand managers and marketing researchers (also known as consumer insights professionals), like sensory scientists, look for ways to increase their knowledge about consumer responses to products before and after purchase. Of major interest is identifying variables besides those that are typically measured (e.g. preference) to gain a competitive advantage. Product sensory information has long been one of those variables often overlooked or not well understood, but it is now becoming a more integral part of a product's business strategy. This has led to increased participation of sensory professionals as part of project teams versus being on the outside, waiting for test requests to arrive, if at all, without knowing the background or the basis for a request. Where successful, it has yielded benefits for the sensory professional in the form of improved status, increased compensation, and, for some, a stronger and earlier voice in a company's product decision-making process. Direct and indirect rewards also were realized, including willingness to support university research, more course offerings, and an increasing demand for newly trained sensory professionals.

H. Stone, R. Bleibaum, H. A. Thomas: Sensory Evaluation Practices, fourth edition.
DOI: http://dx.doi.org/10.1016/B978-0-12-382086-0.00001-7

Another development has been a media connection. In the past, the likelihood of media mention of sensory evaluation was rare; today, there are many books, blogs, and other media outlets describing results from "sensory tests" or interviews with individuals possessing unique sensory skills. Although one may question some of these claims/curiosities, this exposure has brought more attention to the field and that is a positive effect. Changes within the profession also continue to take place, such as more professional meetings and symposiums discussing research results. Other changes are more focused on the ways in which professionals design, analyze, and report results. There has been a major increase in use of direct data capture and a concomitant reduction in use of paper ballots—a more rapid turnaround from data collection to reporting results. The availability of relatively inexpensive software has enabled professionals to more easily undertake design studies. However, none of this has come without some costs—for example, use of software not appropriate for behavioral data or using data that are not appropriate for a particular analysis. This makes it relatively easy for the inexperienced professional to satisfy a request but significantly increases the likelihood of decisions errors. As Groopman (2009) noted as part of a discussion about the state of clinical trials but equally relevant to sensory tests, "Statistical analysis is not a substitute for thinking." Just because one obtains a graphical display or a series of tables with associated statistical significances does not mean it has any meaning or external validity. Considering the major increases in computing power and the trivial cost, one can undertake a wide range of analyses with little or no knowledge as to their relevancy or if the data set is appropriate for that analysis (May, 2004). One needs to appreciate that output will always be obtained regardless of the quality and quantity of the original data. When unexpected product changes are obtained and not understood, it is the responsibility of the sensory staff to explain the result and minimize confusion. We explore these issues in more detail later in this book.

As mentioned previously, using sensory information as a part of a product marketing strategy has given it unprecedented attention. Identifying specific sensory and chemical measures that have a significant effect on preference and purchase intent has important consequences for a company. In those instances in which this information has been used, its significance has been recognized and the product success in the marketplace is appreciated. Clearly, it is a powerful approach to enhancing product preference versus using a simple trial-and-error process. However, this has only been possible with the use of quantitative descriptive analysis to identify the sensory differences and similarities among competitive products and the availability of user-friendly software. The next logical step in this process has been to incorporate imagery into the process. Exploiting this information to the fullest extent possible has enabled companies to grow their market share as well as implement cost savings through better use of technology, etc. (Stone and Sidel, 2007, 2009). All this has been possible as a direct result of using sensory resources effectively, a better understanding of the measurement process, combined with a more systematic approach to the testing. Much of this progress has been achieved within the technical and marketing structures of companies that recognized the unique contributions of sensory evaluation. In the past, such developments were the exception; today, it is a more common occurrence, again reflecting the increased awareness of sensory information. For a summary of these developments, the reader is directed to Schutz (1998). However, much more needs to be done, in part because the links between sensory, marketing, and production are not strong and in part because there is a lack of appreciation for the principles on which the science is based. For some, sensory evaluation is not considered a science capable of providing reliable and valid information. This is not so surprising, given that this perception is fostered in part by the seemingly simplistic notion that anyone can provide a sensory judgment. We are born with our senses and barring some genetic defects, we are all capable of seeing, smelling, tasting, etc. It certainly seems simple enough, so why should a technologist or a brand manager believe results from a test that are inconsistent with their expectations and their own evaluations? As a result, demonstrating that there is a scientific basis to the discipline continues to be a challenge. Further challenges develop when tests are fielded without qualified and sufficient numbers of subjects, again leading to incorrect recommendations. It is no longer a surprise to hear statements such as "We don't have the time or money to do it right, but we will be able to do it over again later." It takes a lot of effort to overcome this kind of thinking. Since the previous editions of this book, advances continue to be made, albeit at a slow pace, not because test procedures are inadequate but, rather, as noted previously, because the science is not readily acknowledged as such. In all fairness, it should be noted that sensory professionals have not been effective spokespeople for their work or for the science. In one company, sensory evaluation will be used successfully, but in another it will be misused or the information will be ignored because it is inconsistent with expectation. Unfortunately, this latter situation has encouraged use of other information sources or the development of competing test capabilities in the hopes of obtaining acceptable information without fully appreciating the consequences.

Throughout the years, numerous efforts have been made and continue to be made to develop a more permanent role for sensory evaluation within a company. Reviewing the technical and trade literature shows that progress in the development and use of sensory resources continues. There has been a noticeable increase, and much of the impetus continues to come from selected sectors of the economy, notably foods and beverages, and their suppliers (Jackson, 2002; Lawless and Heymann, 2010; Meiselman and MacFie, 1996; Piggott, 1988). In their seminal textbook on sensory evaluation published more than 45 years ago, Amerine *et al.* (1965) correctly called attention to three key issues: the importance of flavor to the acceptance of foods and other products, the use of flavor-related words in advertising, and the extent to which everyday use of the senses was largely unappreciated, at that time. Perhaps a secondary benefit of today's concerns about food safety has been awareness by consumers of the sensory aspects of the foods they purchase.

Current interest in sensory evaluation reflects a more basic concern than simply being able to claim use of sound sensory practices. A paper published more than three decades ago (Brandt and Arnold, 1977) described the results of a survey on the uses of sensory tests by food product development groups. The survey provided insight into some of the basic issues facing sensory evaluation then and now. Of 62 companies contacted, 56 responded that they were utilizing sensory evaluation. If this question were asked today, one would expect the number to be close to 100%. Since publication of that survey, there have been many similar surveys with additional results reflecting greater interest in and use of sensory evaluation. However, descriptions of tests being used revealed then, as now, that confusion existed about the various methods; for example, it was found that the triangle test (a type of discrimination test) was the most popular, followed by hedonic scaling (a type of acceptance test) and paired comparison (either an acceptance test or a discrimination test). Because these and other methods mentioned in the survey provide different kinds of information, it is not possible to evaluate the listing other than to assume that most companies use a variety of methods.

Not surprisingly, there was confusion about acceptance test methods and the information that each provides. For example, single-sample presentation is not a test method, but 25 of the 56 companies responded that it was one of the test methods in use. It is in fact a serving procedure, and such responses may reflect poorly worded or a misunderstood question in the survey. Single-sample testing would be referred to as a monadic test. Another example of the confusion is "scoring," which was said to be in use by only 7 of the 56 companies contacted. However, all sensory tests involve some kind of scoring. Statistical tests included in the survey elicited similarly confusing responses. The failure to clearly define the terms confirms that the consumer packaged goods industry uses sensory evaluation but there is confusion as to what methods are used and for what applications. As previously mentioned, sensory evaluation is still not well understood, methods are not always used in appropriate ways, and results are easily misused. There continues to be a lack of qualified professionals, but more are being trained. Books on sensory evaluation continue to be published; however, the science still has not achieved a status commensurate with its potential. Similarly, its role within individual companies is far from clear. One of the goals of this text is to provide a perspective on all these issues and how one should develop resources and integrate them into the business environment. This is not to imply that the scientific basis of sensory evaluation is overlooked. Rather, it is not as well understood as it should be considering its widespread use and strategic potential.

This edition continues to emphasize the importance of a systematic approach to the organization, development, and operation of a sensory program. Primary emphasis is directed toward the more practical aspects of sensory evaluation and approaches to implementing a respected and credible program, but we do not neglect the fundamental, underlying issues, including but not limited to experimental design, the reliability and validity of results, and related topics. From a sensory science perspective, reliability and validity are essential to developing a credible program and providing actionable recommendations that enable a company to build a strong brand consistent with its brand strategy (Aaker, 1996). From a business perspective, it is these latter issues that loom as most important because they build trust that others will place on these recommendations and the extent to which managers will act on them. Sensory professionals must communicate information clearly to ensure that superiors, peers, and subordinates understand what was done and what the results mean in terms of meeting that test's specific objective and how the research facilitates the product link with the business strategy.

The importance of the relationship between reliability and validity of results and the credibility assigned to sensory recommendations and, in a global sense, the credibility of a sensory program are at stake. Although it is logical to assume that a test result should speak for itself, and that it will be understood, this is not the usual situation. It takes a conscious effort to gain the confidence of a brand manager and consumer insights advisors so that recommendations will be acted on sufficiently. When it succeeds, it is reassuring, but more often it does not, which raises questions as to why sensory information is not better understood and used more effectively. There is no single or simple answer. However, it is clear that the ease with which one can evaluate a product makes it easy for a product manager to disregard information that does not satisfy expectations and substitute one's own judgment. That and an unwillingness to appreciate the complexity of the judgmental process leads to decision problems and product failures. However, it is the responsibility of the sensory professional to be sure that tests are organized and fielded using best practices, that replication is an integral part of every sensory analytical test, that subjects are qualified and understand the task, that the scorecard is relevant to the test objective, etc. Product evaluation is a complex multistep process in which a group of individuals respond to stimuli (a set of products) by marking a scorecard (electronically or by paper ballot) according to a specified set of instructions. There are many steps in the process where errors can occur, and not all of them are obvious. Of particular importance is the matter of whether the test objective is understood and the requester can explain how the results will be used; that is, what questions will be answered? Failure to obtain such information is a clear sign of problems when reporting results. Once there is agreement on the objective, a test plan can be established, a design prepared, and actual fielding initiated. As already emphasized, there are many challenges faced by sensory professionals; some are self-induced, whereas others come from requesters specifying the test method, the type of scale, or output that yields a single numerical value. The concept of number biases impacting responses seems to have been lost or not at all understood. Measurement or scaling is another area of confusion. Requests to use only a company's "standard scale" so that it is possible to compare results reflect a poor understanding of the measurement process, as is the request to use a "universal scale" for which there is no evidence. These measurement issues are discussed in detail in Chapter 3.

Subjects have a major impact on a program's credibility. How they were selected and what kind of training, if any, they received are important considerations, as are their responses, in terms of both their sensitivity and their reliability. To a degree, a panel of subjects can be considered as functioning like a null instrument, recording what is perceived.

Within the sensory community, there is the implication of a human (or group of humans) functioning like an instrument, which has obvious appeal in a technical sense. However, there is an equally large body of scientific evidence that recognizes individual variability in perception based on genetic differences at the receptor level, differences in sensory acuity, and experiential differences that all impact perception. The human instrument position within sensory science is sensitive, particularly to those who mistakenly envision an invariant system yielding the same numerical values time and time again. It has particular appeal among those that lack an understanding of psychology, human behavior, and the perceptual process. The realities of the situation are, of course, quite different. Subjects, no matter what their level of skill or number of years of training and practice, exhibit differences in sensitivity from one another, as well as differences in variability that are unique (to the individual). Some training programs (e.g. see Spectrum analysis in Meilgaard *et al.*, 2006) imply that this sensitivity and accompanying variability can be overcome through training and use of appropriate reference standards that represent absolute intensities. Such training, as much as 10 hours per week and often lasting 4 or more months, has considerable appeal; however, there is no evidence that such an approach has scientific merit. Lawless and Heymann (2010) stated, "We are somewhat skeptical of this claim since there are no published data to support it" (p. 360). It is inconsistent with our knowledge of human perception and the physiology of the senses. There are no peer-reviewed results reported in the literature, and one should not expect to obtain invariance. "Theoretically, if panelists were in complete agreement one would expect the standard deviation for any specific product-attribute combination to be close to zero. However, most Spectrum studies have attributes with non-zero standard deviations indicating that the panel is not absolutely calibrated" (Lawless & Heymann, 2010, p. 361). Such an approach is a form of behavior modification rather than a means of capturing responses as a function of a stimulus (whether that stimulus is a purified chemical or a consumer-ready beverage). It ignores the fact that subjects themselves are changing within a trial and across time. The stimulus also changes, which is why one uses replication to capture sufficient data to partition the different sources of variation before concluding that the difference of interest was detected. It is misguided to assume that a human can be invariant, just as it is misguided to assume that an instrument yields the same value every time it is used and also to assume that product is always the same. In each instance, the end result is to overcome what appear to be limitations of the sensory process in the mistaken belief that they will make results more acceptable. Short of directly telling an individual what should be an answer, there will always be variability. Nonetheless, the requestor of a test expecting that test to yield an invariant result (the same response or numerical value every time) is disappointed and concerned about this unique information source. This disappointment is also accompanied by reservations as to any conclusions and recommendations; that is, the issue of credibility arises. Alternatively, when results are not in agreement with information obtained elsewhere (and often not sensory information) and no attempt is made to understand and explain the basis for any disagreement, then further erosion of program credibility will occur. After all, products are not the same despite a documented formulation. The precision of weighing and volumetric devices, flow through pumps, and heating and cooling equipment varies so that the finished product will exhibit this variation. To somehow expect a subject to provide the same judgment each time the "product" is presented is naive, at best.

Sensory professionals must not ignore fundamental scientific truths in our data collection instrument: the human. If sensory is indeed a science, we must embrace the knowledge of genetic differences at the receptor level, perceptual differences, and behavioral differences and provide scientifically derived quantitative data to measure and understand perception.

The success of a sensory program, and particularly its credibility, begins with having a plan and an organized effort and making sure that the testing process follows accepted procedures and practices—what method was used, who the subjects were and how they were selected, what test design was used, how the data were analyzed, including the evidence of reliability, and so forth. In a business sense, it begins with an explanation of what sensory information is and is not, how results are communicated and whether they are understood, and ends with actionable recommendations being implemented to everyone's satisfaction. Although these issues are discussed in detail in succeeding chapters, their inclusion here is to provide an initial perspective to the issue of business credibility and direct involvement of sensory evaluation in product decisions. Without an organized product evaluation effort and demonstrated reliable and valid results that are communicated in a way that is readily understood, one is returning to reliance on product experts, the "N of 1," who made product decisions by virtue of their expertness and not because there were data to support their judgments. Decisions derived in this manner are neither new nor unusual; however, they make it very difficult for individuals trying to organize and operate a credible sensory test program. As noted by Eggert (1989) and by Stone and Sidel (1995), sensory evaluation must develop a strategy for success. It must reach out to its customers; it must educate them about the benefits that can be realized from using sensory information. At the same time, it must gain management support through an active program of selling its services and how the company can benefit from those services.

This book is not intended as an introduction to the topic of sensory evaluation. Nonetheless, for some aspects of sensory evaluation, considerable detail is provided in an almost stepwise manner. Readers, however, will probably derive more from this book if they have a basic understanding of sensory evaluation, experimental design and statistics, and especially the perceptual process.

Where appropriate, background information sources are cited in this text and should be considered recommended reading. In addition to its benefit to the sensory professional, this book is intended to serve as a guide for the research and development executive seeking to have a capability in sensory evaluation and to develop a more efficient and cost-effective product development program. It should also be of interest to marketing, market research/consumer insights, and technical managers, all of whom have an interest in their company's products and their quality as measured by consumer responses and through sales, market share, and profitability.

1.2 **Historical background**

Of the many sectors of the consumer products industries (food and beverage, cosmetics, personal care products, fabrics and clothing, pharmaceutical, etc.), the food and beverage sectors provided much early support for and interest in sensory evaluation. During the 1940s and through the mid-1950s, sensory evaluation received additional impetus through the U.S. Army Quartermaster Food and Container Institute, which supported research in food acceptance for the armed forces (Peryam *et al.*, 1954). It became apparent to the military that adequate nutrition, as measured by analysis of diets or preparation of elaborate menus, did not guarantee food acceptance by military personnel. The importance of flavor and the degree of acceptability for a particular product were acknowledged. Resources were allocated to studies of the problem of identifying what foods were more or less preferred as well as the more basic issue of the measurement of food acceptance. These problems and their solutions were overlooked or forgotten during the 1960s and early 1970s when the federal government initiated its "War on Hunger" and "Food from the Sea" programs. The government's desire to feed the starving and malnourished met with frustration when product after product was rejected by the recipients primarily because no one bothered to determine whether the sensory properties of these products were acceptable to the targeted groups. This is not to suggest that each country's ethnic and regional food habits and taboos were not important but, rather, in the context of these government programs, there was scant attention given to the sensory evaluation of the products as they were being developed. Unfortunately, this situation continues to exist because there remains a fundamental lack of appreciation for the importance of the sensory properties of the consumer packaged goods being developed. There is no question that providing nutrients to those in need is the first priority, and in extreme situations that is about all there is. However, large giveaway programs often yield less than satisfactory results that are well below everyone's expectations, in part because the sensory properties fail to come close to the expectations of a culture or intended consumer group.

The food and beverage industry, possibly taking a cue from the government's successes and failures in sensory evaluation (Peryam *et al.*, 1954), provided support for this emerging science. Although many industries have since recognized its value in formulating and evaluating products, general appreciation for sensory evaluation as a distinct function within a company remained relatively minimal until approximately the past two decades. In general, there is agreement on the role of sensory evaluation in industry but not necessarily how sensory evaluation should be organized and how it should operate within a company. As with any discipline, divergent opinions and philosophies on sensory evaluation exist both within and outside the field. It is not necessary that we examine all these opinions and philosophies in detail; however, some discussion is appropriate to enable the reader to gain a greater appreciation for the challenges involved in the organization and operation of a sensory program.

The food and beverage industry traditionally viewed sensory evaluation in the context of the company "expert" (the N of 1) who through years of accumulated experience was able to describe company products and set standards of quality by which raw materials would be purchased and each product manufactured and marketed. Examples of such "experts" include the perfumer, flavorist, brewmaster, winemaker, and coffee and tea tasters. In the food industry, experts provided the basis for developing the "cutting sessions" and "canning bees" (Hinreiner, 1956). In the canning industry, products usually were evaluated on a daily basis and in comparison with the previous day's production, competitive products, new products, etc. In addition, there were industrywide cutting bees to assess general product quality. These sessions enabled managers and experts to assess product quality at their own plants as well as to maintain a familiarity with all other companies' products. This process continues today in most companies as well as in trade associations seeking to solve common problems that are usually related to product quality. In recognizing the purpose of the cutting bee and its overall function, Hinreiner described the efforts undertaken to improve the quality of the information derived from one group, the olive industry. The Processed Products Branch of the Fruit and Vegetable Division of the U.S. Department of Agriculture (File Code 131A-31, 1994) has updated its guidelines for approved illumination for cutting rooms, an action that recognizes the importance of providing a more standardized environment for product evaluations. In addition to the evaluation, record-keeping was formalized, making it possible to compare results from one year with those of another and thus provide for a great degree of continuity. It is important to remember that the industry recognized a problem, and with assistance from sensory evaluation, it took some action to improve its product quality information. This activity continues today, especially in industries that rely on basic agricultural products that do not experience substantial value-added processing—for example, the wine industry, fresh fruits and vegetables, and olive oil. The enormous attention given to the Mediterranean diet and emphasis on olive oil has resulted in a major increase in consumption of olive oil, specifically extra virgin olive oil. This in turn has resulted in renewed focus on how olive oil quality is determined, along with conferences about olive oil, its nutritional benefits, and the criteria by which quality is determined. As such, extra virgin olive oil is an interesting case study to illustrate new challenges with combining expert evaluations, U.S. Department of Agriculture (USDA) grading, and sensory sciences. This topic is discussed in more detail later in this book.

Cutting sessions and the work of product experts have continued; however, the impact of both has been reduced but not entirely eliminated. In retrospect, the results accomplished (and failed to be accomplished) by these experts and the cutting sessions were quite remarkable. By and large, experts determined which alternatives—from among many alternative ingredients, products, and so forth—were appropriate for sale to the consumer. Their success reinforced their role for establishing quality standards for particular products, such as canned fruits and vegetables, and these standards in turn received additional support through official USDA standards that referenced these results. As long as the food industry was involved solely in the preserving of a basic agricultural crop (e.g. frozen peas, canned fruits and vegetables, juice, or oil such as extra virgin olive oil), then it was relatively easy (or uncomplicated) for the product expert to understand a particular product category and to make reasonable sound recommendations.

In the early stages of the growth of the food-processing industry and where competition was primarily regional in character, such standards and evaluative procedures by experts were extremely useful. In most industries, experts also created an array of scorecards and unique terminologies to serve as a basis for maintaining records and presenting a more scientific process. Subsequently, numerical values were assigned to the various attributes being measured, as described by Hinreiner (1956). These scores soon became targets or standards; for example, the 100-point butter scorecard, the 10-point oil quality scale, and the 20-point wine scorecard all had specific numbers that connoted levels of product acceptance (equated with quality). All of these and others continue to be used within their respective industries. Certain values became fixed in people's minds, and they were transposed inappropriately into measures of consumer acceptance, creating a multitude of new problems. That some of these scorecards have survived virtually intact after 50+ years is remarkable, considering their weaknesses. Where they have not survived one can usually find the concept still alive, particularly that of the single number equated with quality and the associated belief of the invariance of the expert. Although it is more common in quality control, the reemergence of experts in sensory evaluation is somewhat surprising and discouraging. Not only does it reflect a basic lack of understanding of human behavior and the perceptual process but also it may reflect a wistful desire of some to reduce behavior to some simplistic level. As noted previously, humans are not invariant and their responses to products are not invariant; they should not be as no two products are the same. Sensory professionals do an injustice to themselves and to the science when they embrace these simplistic notions about human behavior without fully appreciating the consequences. They also do a disservice when they participate as subjects, thereby perpetuating the notion of the expert, the N of 1 who can make these absolute judgments. In one sense, the activities associated with product grading, merit awards in product competitions, obtaining the right to label a product as coming from a particular growing region, and so on are linked with sensory evaluation but not in the mainstream of how sensory professionals define the work that they do. These systems provide a basis for assuring the buyer that the particular product meets the previously established criteria. They also have value when buying and selling goods; however, they do not reflect the marketplace, which is dynamic, whereas grading systems are static. The problem arises when the leap is made to assume that highest grade equates with highest quality and with product best liked by the consumer. There is more to product success than simply a designated emblem of best quality.

With the growth of the economy and competition and the evolution of processed and formulated foods, experts faced increasing difficulty in maintaining an awareness of all developments concerning their own product interests. As a further complication, product lines expanded to the extent that it was virtually impossible for an expert to have detailed knowledge about all products, let alone the impact of different technologies. While the expert was required to continue making finite decisions about product quality, consumer attitudes were changing in ways that were not fully appreciated. With the development of contemporary measurement techniques and their application to sensory evaluation, it became evident that reliance on a few experts was questionable. To deal with this problem, some companies turned to sensory evaluation (which was often referred to as "organoleptic analysis" in the early literature). In truth, companies did not turn directly to sensory evaluation as a solution to the inadequacies of experts; rather, the marketplace created opportunities. As competition increased and became more national (and eventually international) in scope, the need for more extensive product information became evident. Managers were disappointed with results from some types of consumer tests and/or costs became increasingly difficult to justify to management, and now they were more willing to consider alternative sources of product information. For those companies in which there were sensory resources, opportunities developed, and in some instances considerable success was achieved. To that extent, sensory evaluation represented a new and as yet untried resource. Before discussing this contemporary view, it is necessary to further explore the earlier developments of sensory evaluation.

As noted previously, sensory evaluation was of considerable interest in the late 1940s and into the 1950s, prompted in part by the government's effort to provide more acceptable food for the military (Peryam *et al.*, 1954), as well as by developments in the private sector. For example, the Arthur D. Little Company introduced the Flavor Profile Method (Caul, 1957), a qualitative form of descriptive analysis that minimized dependence on the technical expert. Although the concept of a technical expert was and continues to be of concern, the Flavor Profile procedure replaced the individual with a group of approximately six experts (whom the technical expert trained) responsible for yielding a consensus decision. This approach provoked controversy among experimental psychologists who were concerned with the concept of a group decision and the potential influence of an individual (in the group) on this consensus decision (Jones, 1958). Nonetheless, at that time and continuing to the present, the method provided a focal point for sensory evaluation, creating new interest in the discipline, which stimulated more research and development into all aspects of the sensory process. This topic is covered in more detail in the discussion on descriptive methods in Chapter 6.

By the late 1950s, the University of California at Davis was offering a series of courses on sensory evaluation, providing one of the few academic sources for training of sensory evaluation professionals. Other universities, including Oregon State University, the University of Massachusetts, and Rutgers University, offered course work in sensory evaluation but not to the extent as offered by the University of California. Subsequently, many other universities initiated independent courses in sensory evaluation, producing more professionals in the discipline. These developments are reflected in the food science literature of this same period, which includes many interesting studies on sensory evaluation by Boggs and Hansen (1949), Giradot *et al.* (1952), Baker *et al.* (1954), Harper (1950), Pangborn (1964), and Anonymous (1968). These studies stimulated and facilitated the use of sensory evaluation in the industrial environment. The early research was especially thorough in its development and evaluation of specific test methods. Discrimination test procedures were evaluated by Boggs and Hansen (1949), Giradot *et al.* (1952), and Peryam *et al.* (1954). In addition to discrimination testing, other measurement techniques also were used as a means for assessing product acceptance. Although scoring procedures were used as early as the 1940s (Baten, 1946), primary emphasis was given to use of various paired procedures for assessing product differences and preferences. Rank-order procedures and hedonic scales became more common in the mid- to late 1950s. During this time period, various technical and scientific societies, such as Committee E-18 of the American Society for Testing and Materials (now ASTM International), the Food and Agriculture Section of the American Chemical Society, the European Chemoreception Organization, and the Sensory Evaluation Division (now the Sensory and Consumer Sciences Division) of the Institute of Food Technologists, organized activities focusing on sensory evaluation and the measurement of flavor. For a review of the activities of ASTM, see Peryam (1991) or any of the more recent documents on the topic.

1.3 Development of sensory evaluation

It would be difficult to identify any one or two developments that were directly responsible for the emergence of sensory evaluation as a unique discipline and its acceptance (albeit, on a limited basis) in product business decisions. Certainly the international focus on food and agriculture in the mid-1960s and into the 1970s (and that continues today), the energy crisis, food fabrication, the cost of raw materials (Stone, 1972), competition, and the internationalization of the marketplace have, directly or indirectly, created opportunities for sensory evaluation. For example, the search for substitute sweeteners stimulated new interest in the measurement of perceived sweetness along with an interest in time-intensity measures. This, in turn, stimulated development of new measurement techniques (Inglett, 1974) and indirectly stimulated interest in development and use of direct data entry systems as a means for evaluating the sweetness intensity of various ingredients (for more information about the latter topic, see Anonymous, 1984; Gordin, 1987; Guinard *et al.*, 1985; Winn, 1988). Today, this situation has changed with respect to capturing responses and real-time data analysis; other opportunities remain to be satisfied. Whether companies are developing new products, attempting to enter new markets, or attempting to compete more effectively in existing markets, the need for sensory information remains (Stone, 2002). Although much more could be written on and speculated about these opportunities and their antecedents, it is more important that our attention be focused on how this sensory resource should be structured so it can function more effectively in the future.

After a long and somewhat difficult gestation, sensory evaluation has emerged as a distinct, recognized scientific specialty (Sidel *et al.*, 1975, see also Sidel and Stone, 2006; Stone and Sidel, 1995). Although the focus of the former article was on the use of sensory evaluation in the development of fabricated foods, there were implications for sensory evaluation in general. As a unique source of product information, it had important marketing consequences, providing direct, actionable information quickly and at low cost. It was proposed that organizing of sensory evaluation test services along well-defined lines (e.g. formal test requests, selection of a test method based on an objective, and selection of subjects based on sensory skill) would increase the likelihood of such services being accepted as an integral part of the research and development process or other business units within a company. It has become clearer that without an organized approach, a management-approved plan, and an operational strategy, sensory resources are rarely used effectively and are less likely to have a significant, long-term impact.

In a short course given several decades ago, Pangborn (1979) called attention to misadventures that have occurred in sensory evaluation. The article was one of several by this author as part of her continuing efforts to improve the quality of the research being done and indirectly enhance the skill level of sensory professionals in companies. The three issues of particular concern were the lack of test objective, adherence to a test method regardless of application, and improper subject selection procedures. These three issues remain even now. These are not the sole property of the sensory literature (and by default many in teaching roles) but also are quite commonplace in the business environment. It is clear that much more needs to be done to improve the quality of sensory information.

An interesting development for sensory evaluation has been the continued growth in the number of short courses and workshops being offered. When there were few university offerings, such programs served a useful purpose for individuals with responsibility for their company's sensory program. In the past decade, there has been a quantum increase in the number of courses being offered, including distance learning certificate programs in sensory and consumer sciences, which suggests that university offerings are still insufficient for industry's needs. Our own experience in offering courses during the past four decades reflects a continued interest in sensory evaluation, especially the more practical issues of developing a program within a business environment. Some of the material presented in this book evolved from workshop material that has proven especially beneficial to participants. Newspapers and other public information sources present articles about sensory evaluation (not regularly, but often enough to be noticeable). These articles usually include some impressive revelations (to attract the reader) about the special "tongue" or "nose" of certain individuals who are claimed to have almost mystical powers. These individuals are generally associated with such products as wine, beer, coffee, and fragrance, or they function as wine and food writers. Still other approaches convey an impression that the subconscious mind is being tapped by a new sensory technique, with the end result being the ideal consumer product. A current example is the development of methodology described as "emotional profiling" about which more can be found in Chapter 6. Although sensory professionals should welcome new methods, such developments need to be examined before their adoption. After all, it is very easy to obtain a response from a consumer/subject; it does not mean that the question was understood or that the responses will yield actionable information and eventually a superior product versus competition. Despite this lack of consistency, continued discussion about sensory evaluation is helpful to the field, if only because it reaches key people who might not otherwise read about it. These changes, and a greater awareness of sensory evaluation, appear to have coincided with a dramatic shift by the consumer packaged goods industry toward a stated consumer-oriented environment and away from the more traditional manufacturing/production-oriented environment. By that we mean a recognition that understanding consumer attitudes and behavior is essential information and ought to be known before one formulates a product rather than manufacturing a product and looking to others (e.g. marketing) to convince the consumer to purchase that product.

Opportunities for sensory evaluation continue to develop primarily as a result of significant changes in the marketplace and to a much greater extent than changes in sensory evaluation methodology. Mergers, leveraged buyouts, and other financial restructuring activities and the internationalization of the marketplace have created even greater challenges in the consumer products industry. There are numerous choices in terms of brands, flavor alternatives, convenience, pricing, new products, and combinations not thought of a decade ago (e.g. yogurt beverages). Many companies have determined that new products at competitive prices are essential for long-term growth and success. However, this has presented its own unique challenges and risks (Meyer, 1984). New product development and the proliferation of choices within a product category rapidly accelerated in the 1980s at a rate neither appreciated nor believed possible in the past. This acceleration was accompanied by considerable financial risk (Anonymous, 1989). In separate 1977 publications on the topic, Crawford and Carlson determined that the failure rate of new products has, at times, been as high as 98% for all new products. In the time between then and now, 30+ years, one can find similar reports in the trade literature; clearly, this situation has not changed very much and certainly not for the better. From a business perspective, this risk severely challenges creative skills and available technical resources and has provided renewed interest in resources such as sensory evaluation (Stone, 2002; Stone and Sidel, 2007) as well as alternative approaches to the development process (Smith, 2007; Smith and Oltmann, 2010). Companies are now more receptive to new approaches and to new ways of anticipating and measuring the potential for a product's success in the marketplace. Of course, some companies may choose to not introduce new products and thereby minimize much of that risk but to rely on brand and line extensions (Lieb, 1989). Here too, the need for sensory information is essential if such products are to be brought into the market within reasonable time and budgetary considerations. These changes should have accelerated the acceptance of sensory evaluation; however, this has not occurred to any great extent until recently. Companies are now more aware of sensory evaluation; however, the organization and operation of sensory business units with full management support still lag other related activities such as consumer insights (a successor to marketing research). Nonetheless, the fact that some programs are fully operational bodes well for the future.

Although much progress has been made, considerably more remains to be achieved, particularly within the business environment. In the next chapter, organizational issues are more fully explored, with particular emphasis on structural issues and their integration with the other product information resources—that is, how methods and subjects are developed and used to solve specific problems and maximize sensory's benefits to a company.

1.4 Defining sensory evaluation

To more fully appreciate the situation, it is helpful if we first consider the manner in which the two words *sensory evaluation* are defined. The paucity of definitions is surprising: A perusal of various texts and technical sources reveals one prepared by the Sensory Evaluation Division of the Institute of Food Technologists (Anonymous, 1975), quoted here, that provides insight into the subject:

> *Sensory evaluation is a scientific discipline used to evoke, measure, analyze, and interpret reactions to those characteristics of foods and materials as they are perceived by the senses of sight, smell, taste, touch, and hearing.*

This definition represented a conscious effort to be as inclusive as is possible within the framework of food evaluation, with the word "food" considered global; that is, an ingredient is a food, a beverage is a food, and so forth. Similarly, materials can be products for the home such as furniture polish, a product for personal care such as a shampoo, a hair colorant, or a lipstick, etc. It can be argued that this definition is too narrow or the converse, it is too broad, based on one's perspective. The main purposes for the definition were to make clear that sensory evaluation encompasses all the senses and is not limited to food. The former concern about all the senses is especially important because the non-sensory staff often approach sensory information in a univariate way when the information is multivariate. One of the outcomes of this view is the request to field a test in which only flavor is measured and the subjects are told to ignore the aroma and the color as if the brain works in some kind of compartmentalized way. Although the subjects may agree to this plan, it is unlikely that their responses will reflect this instruction. If there is no place on the scorecard to record responses to these other attributes, they are likely to be embedded in the flavor response. This will lead to a confounding of the response and potential misinterpretation of the results. A product's appearance will impact an individual's response to that product's taste, etc. Regardless of what one may like to believe or has been told, responses to a product are the result of interactions of various sensory messages, independent of the source. To avoid obtaining incomplete product information, it is important to design studies that take this knowledge into account. The familiar request to "field a test but tell the subjects to ignore the color as that will be corrected later" is a sure sign of future problems. This issue will be discussed in a subsequent chapter but is mentioned here to emphasize its importance in the overall evaluation strategy and the seeming lack of appreciation of its consequences. When the sensory professional reports results to managers who will not ignore the color, what kind of response will be offered? The definition also reminds everyone that sensory evaluation is derived from several different disciplines, but emphasizes the behavioral basis of perception. This involvement of different disciplines may help to explain the difficulty entailed in delineating the functions of sensory resources within a business environment. These disciplines include experimental, social, behavioral, and physiological psychology, statistics, home economics, marketing research, and a working knowledge of science product's technology.

As the definition implies, sensory evaluation involves the measurement and evaluation of the sensory properties of foods and other materials. It also involves the analysis and the interpretation of the responses by the sensory professional—that is, that individual who provides the connection between the internal world of technology and product development and the external world of the marketplace, within the constraints of a product marketing brief. These connections are essential such that the processing and development specialists can anticipate the impact of product changes in the marketplace. Similarly, the marketing and brand specialists must be confident that the sensory properties are consistent with the intended target and with the communication delivered to the consumer through advertising. They also must be confident that there are no sensory deficiencies that lead to a market failure. Linking of sensory testing with other business functions is essential, just as it is essential for the sensory professional to understand a product's market strategy. In recent years, other business units have expressed interest in using sensory information. For example, quality control/quality assurance professionals have initiated efforts to include sensory information into the quality equation. Here too, it has been found that sensory information is cost-effective. However, it has its own set of challenges that will be discussed in more detail later in this book. Thus, sensory evaluation should be viewed in much broader terms. Its contributions far exceed the questions of which flavor is best or whether ingredient A can be replaced with ingredient B. In this discussion and in those that follow, we assign specific responsibilities to sensory scientists, above the collecting of responses from selected individuals in a test situation.

This concept is especially important in balancing the impact of consumer information obtained by marketing research/consumer insights. In recent years, there has been a growing interest in and much greater use of sophisticated physiological and psychological approaches in measuring consumer behavior. Although the initial interest centered on advertising, it was only natural that it should also be applied to consumers' reactions to the products. Such approaches have had a positive impact on product success in the marketplace; it would be naive to assume that such information can replace product sensory information. There are those who advocate bypassing sensory testing and relying instead on larger scale consumer testing into which a sensory component is incorporated. Such an approach has much appeal because it could save time (but not cost) and works directly with "real" consumers; however, it has its own risk. In any population there exists a wide range of sensitivities to product differences and an equally diverse range of preferences. Approximately 30% of any consumer population cannot discriminate at better than chance among products they regularly consume. As we discuss later in this book, it takes approximately 7–10 hours to teach consumers how to use their senses and identify which consumers can discriminate differences at better than chance. It requires another 7–10 hours for these subjects to learn how to verbalize their perceptions. So bypassing the sensory test by asking consumers to respond to sensory questions or having them verbalize their perceptions has significant risk. It dramatically increases the likelihood of concluding there is no difference when, in fact, there is. The universal scale is another idea that adds to the potential for confusion. As already noted, it would be extremely useful to have one scale to obtain all needed information; it is unrealistic to assume such a solution to account for the complexities of consumer behavior and the very wide range of differences in sensitivity.

There is no question that there are important connections, and in some situations overlap, between sensory and market information. However, one cannot substitute for the other. There are situations in which the two kinds of information are obtained using an appropriate design; however, in many instances, they are clearly separated. This topic is discussed later in the chapters on multivariate designs. Briefly, sensory tests focus on product formulation differences, the impact of technology, packaging, and shelf life, etc. Sensory also can connect these variables with consumer attitudes, purchase intent, benefits and uses, etc. Market research/consumer insights will focus on the basis for a product preference, purchase intent, imagery, and so forth. Product attribute information also might be obtained; however, these latter responses are not a substitute for results from a trained descriptive panel. This and related issues are discussed in more detail in Chapter 7. Similarly, the results of a sensory acceptance test are not a substitute for a marketing research test with an appropriate population of qualified consumers. A failure to appreciate these differences has been and will continue to be a source of conflict and impact the growth and development of sensory resources in a company and also the ability of a company to develop and market successful products. These issues are discussed again in subsequent chapters of this book.

1.5 **A physiological and psychological perspective**

Sensory evaluation principles have their origin in physiology, psychology, and psychophysics. Information derived from experiments with the senses has provided a greater appreciation for their properties, and this greater appreciation, in turn, has had a major influence on test procedures and on the measurement of human responses to stimuli. Although sources of information on sensory evaluation have improved in recent years, much information on the physiology of the senses and the behavioral aspects of the perceptual process has been available for considerably longer (Geldard, 1972; Granit, 1955; Guilford, 1954; Harper, 1972; Laming, 1986; McBride and MacFie, 1990; Morgan and Stellar, 1950; Poynder, 1974; Tamar, 1972). In this discussion, we identify our meaning of the word "senses". As Geldard pointed out, classically the "five special senses" are vision, audition, taste, smell, and touch. The latter designation includes the senses of temperature, pain, pressure, and so forth. Numerous efforts have been made to reclassify the senses beyond the original five, but they are the ones we have chosen to consider here. In addition, the inclusion of "umami" as another of the basic tastes, as well a fat receptor, deserves some consideration (see Chapter 6).

From study of the physiology and anatomy of the systems, we know that each sense modality has its own unique receptors and neural pathways to higher and more complex structures in the brain (Granit, 1955; Morgan and Stellar, 1950; Pfaffman *et al.*, 1954; Tamar, 1972). At the periphery, receptors for a specific sense (e.g. visual and gustatory) respond to a specific type of stimulation that is unique to that system. That is, a gustatory stimulus does not stimulate visual receptors. However, when the information is transmitted to higher centers in the brain, considerable integration occurs. Comprehension of how sensory information is processed and integrated is important in understanding the evaluation process (McBride and MacFie, 1990; Stone and Pangborn, 1968). What this means when translated into the practical business of product evaluation is that products are a complex source of stimulation and that stimulation will not be exclusive to a single sense, such as vision or taste. As already mentioned, failure to appreciate the consequences of this very fundamental component of sensory evaluation continues to have serious consequences (a perusal of the current sensory literature provides ample evidence of these practices). Consider an evaluation of a strawberry jam that has visual, aroma, taste, and textural properties. Requiring subjects to respond only to textural attributes (and ignore all other stimuli) will lead to partial or misinformation about the products, at best. Assuming that subjects (or anyone for that matter) are capable of mentally blocking stimuli or can be trained to respond in this way is wishful thinking. Response to all other perceptions will be embedded in the textural responses, which in turn lead to increased variability and decreased sensitivity. This approach ignores basic sensory processes and the manner in which the brain integrates information and, combined with memory, produces a response. Probably more harm is done to the science of sensory evaluation and its credibility when procedures and practices are modified based on faulty assumptions about human behavior in an effort to eliminate a problem that usually has nothing to do with behavior. Use of blindfolds, rooms with colored lights, and dark glasses are examples that quickly come to mind. These practices reflect a poor understanding of the perceptual process and the mistaken belief that by not measuring some product sensory characteristics, they can be ignored. These practices are not recommended because they create problems leading to increased risk in decision making. In subsequent chapters, solutions to these problems are proposed along with some practical examples.

The psychophysical roots for sensory evaluation can be traced to the work of Weber (cited in Boring, 1950), which was initiated during the middle of the 19th century. However, it could be argued that it was Fechner (Boring, 1950), building on the experimental observations of Weber, who believed he saw in these observations a means of linking the physical and psychological worlds that gave rise to the field of psychophysics. Much has been written about this topic as well as the early history of psychology in general. The interested reader will find the book by Boring especially useful in describing this early history. For purposes of this discussion, we call attention to just a few of these developments because of their relevance to sensory evaluation. This is not intended to be a review, because such reviews are readily available in the psychophysical literature, but rather to remind the reader, in an abbreviated manner, of the strong ties between sensory evaluation and experimental psychology. For a more contemporary discussion about the perceptual process, the reader is referred to Laming (1986), McBride and MacFie (1990), and Lawless and Heymann (2010).

Fechner was most interested in the philosophical issues associated with the measurement of sensation and its relation to a stimulus. He proposed that because sensation could not be measured directly, it was necessary to measure sensitivity by means of differential changes. This conclusion was based on the experimental observations of Weber. By determining the detectable amount of difference between two stimuli (the just-noticeable-difference or JND), Fechner sought to establish a unit measure of sensation. He proposed that each JND would be equivalent to a unit of sensation and that the JNDs would be equal. From this point, an equation was formulated relating response to stimulus:

$$S = k \log R$$

where S is the magnitude of sensation, k is a constant, and R is the magnitude of the stimulus.

As Boring emphasized, Fechner referred to this as Weber's law, now known as the Weber–Fechner law or the psychophysical law. This initiated not only the field of psychophysics but also a long series of arguments as to the true relationship between stimulus and response and the development of a unified theory of perception. For many years, it was argued that one could not measure sensory magnitudes and therefore such a psychophysical law was meaningless. However, the most concerted attacks on the Fechnerian approach were made by Stevens (1951) and his co-workers (for more detailed discussions on this topic, see Cain and Marks, 1971; Lawless and Heymann, 2010), who advocated a somewhat different explanation for the relationships of stimulus and response. Stevens proposed that equal stimulus ratios result in equal sensation ratios rather than equal sensation differences as proposed by Fechner. Mathematically, as proposed by Stevens, the psychophysical power law was as follows:

$$R = kS^n$$

and

$$\log R = n \log S + \log k$$

where R is the response, k is a constant, S is the stimulus concentration, and n is the modality-dependent exponent.

The formulation of this law had a renewed and stimulating effect on the field, as manifested by hundreds of publications describing responses to a variety of stimuli including commercial products and setting off numerous debates as to the numerical values of the power functions for the various sensory modalities. Certainly the development of signal detection theory has had a major impact on our knowledge of the perceptual process in general and on sensory evaluation in particular. It too has its opponents who seek a single unified theory of perception. However, as Stevens (1962) observed, such laws generally attract criticism as well as support, and the power law was no exception (Anderson, 1970). In fact, the extent of criticism directed at any one theory of perception is no greater or less than that directed at any other theory, and to date, there has been no satisfactory resolution. The interested reader should study the previously cited work of Anderson and Laming, and for more on this issue from a sensory evaluation perspective, see Giovanni and Pangborn (1983). The observations of Weber, however, warrant further comment because of their importance to product evaluation. Basically, Weber noted that the perception of the difference between two products was a constant, related to the ratio of the difference, and expressed mathematically as

$$K = \frac{\Delta R}{R}$$

where R is the magnitude of the stimulus, and K is the constant for the JND. Experiments on a variety of stimuli and particularly those involving food and food ingredients have shown generally good agreement with Weber's original observations (Cain, 1977; Luce and Edwards, 1958; Schutz and Pilgrim, 1957; Stone, 1963; Wenzel, 1949). As with other developments, there are exceptions, and not all experimental results are in complete agreement with this mathematical expression. Nonetheless, the JND has found widespread application in the sensory evaluation of products, as described in Chapter 5.

Among the contributions to sensory evaluation by psychologists, the work of Thurstone (1959) is especially noteworthy. Many of his experiments involved foods (rather than less complex one-dimensional stimuli) or concerned the measurement of attitudes—both topics that are of particular interest to the manufacturer of consumer goods. Thurstone formulated the law of comparative judgment enabling use of multiple paired comparisons to yield numerical estimates of preferences for different products. From this and other pioneering research evolved many of the procedures and practices used today by sensory professionals. Today, renewed interest in Thurstonian psychophysics is welcomed; however, these authors have always considered it an integral part of the perceptual process.

Psychology has contributed substantially to our understanding of the product evaluation process; however, it would be incorrect to characterize sensory science solely as a branch of psychology.

As the adherents to the various schools of psychology or particular approaches espouse their causes, it is easy to confuse research on scaling or perception with assessment of product success without appreciating the differences and the risks. It is easy to mistake developments in psychology as having immediate and direct consequences for sensory evaluation and to blur the lines between the two. This has been particularly evident by those who implied use of magnitude estimation, a type of ratio scale, as the best procedure for obtaining meaningful numerical estimates of the perceived intensities of various stimuli (Stevens, 1962). It was claimed that use of this scaling method enabled one to obtain unbiased judgments, allowed for use of higher order statistics, and allowed for a much greater level of sensitivity than achievable by other scales. As discussed in Chapter 3, attempts to demonstrate this scale superiority in sensory evaluation applications have been unsuccessful (Moskowitz and Sidel, 1971). What may be demonstrated with a single stimulus in a controlled experiment does not necessarily translate to the evaluation of more complex stimuli such as foods and other consumer products. That such superiority has not yet been demonstrated with any scale should not be surprising, particularly from an applications perspective. As Guilford (1954), Nunnally (1978), and others have noted, having a less than perfect scale does not invalidate results, and any risks that such use entailed are not as serious as one might expect. As it happens, the risks are very small, particularly in regard to the product evaluation process. In Chapter 3, this topic is discussed in more detail. Nonetheless, it should be clear that sensory evaluation makes use of psychology, just as it makes use of physiology, mathematics, and statistics in order to achieve specific test objectives. The discipline operates without the restrictions of different theories of perception competing to explain the mechanisms of perception, a task of psychophysicists and physiological psychologists. The sensory professional must be sufficiently well-informed in these disciplines to identify their relevance to the organization, fielding, and analyses of a sensory test, without losing sight of sensory goals and objectives.

The remaining chapters of this book emphasize the organization and operation of a sensory evaluation program, taking into account both the short-term and long-term issues that directly affect the success of a program.

The Organization and Operation of a Sensory Evaluation Program

CHAPTER OUTLINE

2.1 Introduction

An organized sensory testing program with full responsibility for its actions is essential for its success within a company and for the success of a company's products, considering both short- and long-term objectives (Sidel and Stone, 2006;

H. Stone, R. Bleibaum, H. A. Thomas: Sensory Evaluation Practices, fourth edition.
DOI: http://dx.doi.org/10.1016/B978-0-12-382086-0.00002-9

Sidel *et al.*, 1975; Stone and Sidel, 1995). Operating within an organized structure (with management approval) formally recognizes the function and facilitates staff awareness of product plans and strategies. It also provides a means for enhancing the potential of sensory evaluation. With formal recognition and defined areas of responsibility, there is increased likelihood of involvement in new projects, more opportunity to provide a sensory perspective on projects, more use of sensory resources, and generally more opportunities to partner in the overall product evaluation process. There is ample evidence that sensory evaluation contributes directly and indirectly to company profitability—that is, providing actionable information that can help grow brands. The ability to field a test in a few hours and directly advise a brand manager that a particular ingredient change can be made with little or no impact on the sensory characteristics of the product is one example. Enabling a technologist to observe how consumers react to formulation changes in a real-time basis reduces the time to reformulate, and that alone has considerable market value. However, such a sensory capability does not occur by chance; rather, it comes about through an organized effort and a reputation for being responsive to business priorities. The ability to anticipate needs and quickly develop the needed resources and assess new developments can only occur if one plans and builds toward those objectives. In effect, one has to be proactive versus reactive. Companies appreciate the importance of having an organized program for product evaluation that is aligned with a company's business plans and specific project objectives (Eggert, 1989; Sidel and Stone, 2006; Stone and Sidel, 1995). Successful sensory research programs that service brand managers and product technologists demonstrate skillful planning, strategic thinking, and an ability to provide information that is understood and actionable, at the right time. As noted in Chapter 1, as more companies experience market success based on use of sensory information, the more likely there will be support for a sensory program and equally support for sensory professionals.

Although the concepts (and the practices) of planning and developing an organizational structure are essential for the survival of a sensory program, it is clear that many sensory professionals do not appreciate their importance to the business or use their resources to best advantage. This is particularly important in the current business environment, in which mergers have created very large global organizations with large and diverse product lines. In many of these organizations, product development and testing is done on a global basis in ways that were unanticipated a few years ago, particularly with regard to responsiveness and reporting actionable information. Restructuring of large business units based on specific product categories has also resulted in separation of resources or their unavailability because they are now part of another business unit. As noted previously, it is not unusual for product development to be conducted in a location far removed from where the sensory staff is located. This presents challenges for sensory professionals, especially if there is no formal organized program to promote consistent use of the resources or serve as a role model for other business units. As Blake (1978), Tushman and Anderson (1997), and others have emphasized, companies must be sufficiently flexible so as to respond quickly and able to access companywide resources. Although it is not our intention to provide an analysis or an evaluation of various business management practices (nor is it appropriate), there is no question that sensory professionals must be aware of their companies' business focus and the strategies for their various brands. A key to success is flexibility—to be able to adjust to any changes that occur. For more background on these issues, the reader is referred to Fuller (1994), Smith (2007), Smith and Reinertsen (1998), and Tushman and Anderson (1997). The books by Smith, and Smith and Reinertsen are particularly relevant because they address the changing nature of the development process in a time-constrained environment. For discussion about critical thinking associated with brand management, the reader is directed to Aaker (2010).

It should come as no surprise that there are different views of what sensory evaluation is, who it serves, etc. Each of these views may require a somewhat different stance (organizationally and operationally) in terms of testing, reporting, and so on. Sensory resources can be used in different product applications within a company, and the interaction and dialogue with each group will vary, depending on the problem, how the information will be used, the extent to which sensory will participate in the project planning and development process, and in the utilization of the results. For example, tests requested by product development will be handled differently from those requested by quality control or marketing. In fact, the entire process (from request to issuance of a recommendation) should be part of a sensory program's operational plan. Such a plan should call attention to such issues as (1) whether a problem warrants formal sensory analysis (as distinct from bench screening or small-scale tests with no replication), (2) what priority will be assigned, (3) what is the product source (obtained from the marketplace, production but before distribution, pilot plant, bench top), (4) what is the stage of the project (early or final), (5) what are the economic implications, (6) where will the testing be done, (7) how will the results be used, (8) the impact of other product or product-related information, and (9) how the information will be communicated and to whom. All these issues must be considered before a decision is made as to how best to satisfy a specific request. Of course, this assumes the sensory staff has the authority to accept, modify, or, in the extreme, reject a request. Not surprisingly, the decisions that are made will impact how the information will be received and, ultimately, how the sensory program is viewed by others—that is, its responsiveness, cooperation, and credibility.

This need for strategic thinking is not limited to companies with a large sensory staff; a staff of one or two individuals may not have as many test requests, but the same principles apply. The key to a successful program is to identify the strengths and weaknesses of one's program (in the context of that company's business strategy) and to develop capabilities based on the available resources (people and support services). If capabilities are limited, then some trade-offs will be necessary until such time as the needed resources have been acquired through education, staff additions, or using an outside service. Once organized, however, there is a greater likelihood that the program will be considered as more credible. Of course, providing reliable, valid information that is actionable will have an even more significant impact. Each requires the other to be effective. Finally, managers will better understand sensory evaluation information when presented in the context of that specific project.

In describing the organization of sensory resources, our focus is on the sensory professional developing an organizational plan and an operational strategy, including responsibility for its implementation. This approach is in contrast with those of companies in which an individual not familiar with sensory evaluation decides on the organizational plan without any appreciation for its requirements let alone its unique role between technology and the consumer. Successful sensory programs use a project management approach based on assigning an individual with the requisite experience rather than basing the assignment on knowledge of a specific test method. This approach is by no means unique or innovative, but it has proven to be successful because it enables all sensory professionals to be familiar with all stages of a project, and perhaps most important, it minimizes sensory staff "fitting" a problem to a method. These issues are discussed in detail in subsequent sections; they are mentioned here to emphasize their importance to the overall success of a sensory program.

The success of this approach comes about, first, through a management that is aware of the value of sensory information and, second, because of a business approach taken by the sensory professional. Time and cost are critical to success for sensory staff and resources. It is especially important to recognize that this approach, when put in practice, will more than compensate a company for its investment, most likely within a few months of initiation. Companies have found that every dollar invested in sensory resources returns 10 dollars in savings. Conversely, highly skilled sensory professionals become frustrated and ineffective in environments that neither formally recognize the value of information nor provide for an appropriate organizational and reporting structure.

2.2 Organizing a sensory evaluation program

Organizing sensory activities begins with a plan to make management more aware of the value of sensory information. Generally, all managers with technical responsibilities have some awareness of sensory evaluation. This may have come through their own practical experience, reading the trade or technical literature, discussions with other managers, or from meetings at which the sensory contribution resulted in a success. This information is useful as a starting point, but sensory professionals should consider this only as background and make no assumptions regarding an individual's working knowledge of sensory evaluation. In some situations, that information could be negative; that is, a sensory test is remembered because it was unsuccessful, it did not solve the problem, etc. Sensory professionals should expect to be challenged on some sensory issues whenever results do not meet expectations.

Table 2.1 Sensory Evaluation Activities within a Company

Benchtop screening
Product improvement
Monitoring competition
Product optimization
Product development
Product reformulation/cost reduction
Product sensory specification
Quality control
Raw materials specifications
Storage stability
Process/ingredient/analytical/sensory relationships
Plant audits
Postmarketing audits
Extended use testing
Legal issues and advertising claims substantiation

Therefore, the initial step in this process is to develop a plan, taking into account what has been done, what is being done now, and what could be done in the future to enhance existing product knowledge. Table 2.1 lists business activities to which sensory can contribute, directly or indirectly. All of these are important to a company. For some companies, the emphasis may be on the marketplace—that is, on new products, cost reduction and reformulation, line extensions, etc. For other companies, quality control is a primary focus. The list is intended to be general, with no designation as to where tests will take place, who will serve as subjects, or which methods will be used. The point is that managers first must be made aware of the different ways in which sensory information can be obtained and how it contributes to product knowledge. This plan also should include, where available, some technical details, such as the product focus and the link with product technology; proposed budgets, depending on how a company accounts for personnel participating as subjects if used for that purpose; timing to achieve specific objectives; and some indication of the financial benefits. Financial benefits and timing are particularly important issues for sensory evaluation.

Most individuals consider sensory evaluation as a resource that costs money to develop and operate, but in fact it can be shown to more than pay for itself. This is demonstrated through its ability to screen large numbers of formulations quickly, thereby reducing the number of products requiring larger scale and more costly external testing, and in ingredient cost reduction programs through the use of the discrimination model. This latter topic is discussed in more detail in the next section.

Before continuing with the organizational plan, it is useful to briefly comment about the most appropriate location of the sensory function within a company.

To some, sensory evaluation is another group competing for available funds and space; to others, it is one more potential group to manage and thereby to have a larger staff. Although such issues may seem trivial, they form part of the background from which recommendations are prepared regarding the organization and location of a sensory function.

In most companies, sensory resources are located within a technical function; however, they are sometimes part of a marketing research/consumer insights group. There is no rule stipulating that sensory resources must be located in a particular area, but because most sensory professionals have scientific/technical training and most often are providing services to technical groups, sensory resources are frequently placed within the technical function. Historically, the company flavor expert was usually a part of the production and/or technical areas, thus providing additional justification for positioning of sensory resources in that part of the company. In some organizations, it was assumed that the two activities were essentially the same—a faulty perception that has led to all sorts of problems that we discuss in more detail in another chapter.

Most requests for sensory services come from product developers/technologists; however, the source of the request is more likely a brand manager or his or her support (marketing research/consumer insights). Sensory resources have to be positioned high enough to be able to participate in the various meetings at which product and project decisions are made. Sensory resources positioned too low will likely be less effective, less likely to be included in meetings, and not familiar with schedules and priorities.

As previously noted, sensory results can have an impact well beyond a specific test objective. For example, it was noticed that different acceptance ratings were obtained for the same product produced at different manufacturing locations. Assuming this was not the purpose for the test, it would be surprising if it did not raise questions; for example, were the comparisons valid? How was the testing managed? Plant managers can be defensive if they think the results could be used as a basis for making changes at a plant. Sensory staff needs to be able to discuss these kinds of results and expect management support. A program's credibility depends on it.

Although one would like to believe that these issues should not be of concern in the assessment of results, the realities of the situation suggest otherwise. Obviously, the experience and skill of a sensory professional will help to dispel some of these challenges. Nevertheless, positioning is an important issue, as already noted, it may be a barrier to information crucial to the design of the test or to the true reasons for a particular request. Thus, the reporting structure of sensory resources within the organization is important. Sensory staff must present themselves as providing unique product information, and this is best achieved when it is independent of the groups that it serves. As already indicated, most company sensory resources are positioned within the technical function. This is reasonable provided project reporting goes directly to requesters instead of through an administrative pathway. In other companies, sensory resources are positioned as part of consumer insights or related product marketing services. Here, too, reporting lines of communication must be direct to minimize political issues as much as possible (e.g. protecting one's turf).

In situations in which managers are undecided about the placement of sensory resources, its strategy and the professional's business skills will be tested. Managers must be comfortable knowing that sensory capabilities can contribute to the product development effort, can help marketing use sensory information effectively, etc. The final decision on sensory program placement within the organization may be postponed temporarily until sufficient experience has been obtained relative to the business needs. This particular approach is most desirable for those individuals starting their sensory programs for the first time and who have not yet had the opportunity to demonstrate value of their information. Of course, temporary situations can become permanent unless the sensory professional maintains an active presence, keeping management informed as to the program's successes and its contributions to the company.

In addition to establishing an organizational location for sensory evaluation within a company, resources must be capable of responding to current as well as anticipated needs. The following elements form the foundation for an effective sensory evaluation program:

1. Stated and approved goals and objectives
2. Defined program strategy and business plan
3. Professional staff
4. Test facilities
5. Ability to use all test methods
6. Pool of qualified subjects
7. Established subject screening procedures
8. Established subject performance monitoring procedures
9. Standard test request and reporting procedures where appropriate
10. Online data processing capabilities
11. Formal operations manual
12. Planning and research program

For some readers, this list may simply be a reaffirmation of their own resources; however, in all probability, only a small minority will have all 12 in place and fully operational. Throughout the years, we have seen more companies adopt these requirements, reflecting a gradual increase in test activities and a need to be better organized to respond. The crucial point is that a successful sensory program must include all of these resources if it is to meet and exceed expectations within a company. For some individuals and companies, meeting the requirements is relatively easy, whereas for others, it will take time—as much as a year or more. This does not mean that an operational program will be a year or more before actionable information will be available. Rather, it means that the full and effective use of resources often takes that long because users (of the service) do not always understand the information or how to use sensory resources to best advantage.

In addition, it usually takes time for sensory professionals to establish how best to communicate results to different requestors. With the availability of advanced statistical software and use of multivariate analyses, the use of complex graphical displays, mapping, and so forth adds to this communications challenge. Not everyone understands the displays, and without consideration given to how best to communicate their meaning, the impact of the information will be lost. That is a sensory responsibility, and without a clear, defined plan, program development will be limited. Sensory professionals must regularly educate their customers to ensure that there is a good understanding of sensory information, beyond a simple statement such as "Product A was significantly preferred to product B." In addition to developing and maintaining lines of communication, business priorities will change from year to year, sometimes in unpredictable ways, necessitating a shift in testing emphasis. Thus, the time span should not be considered as excessively long or unusual. The point here is that test methods take relatively little time (a few months) to organize and provide actionable results. The proper positioning of the capability and its optimal effectiveness as a contributor to product success takes considerably longer and is more challenging. One goal of this and succeeding sections is to provide readers with some background information about all these interrelated issues and provide ideas that are applicable to their companies.

2.2.1 Goals and objectives

Generally, sensory professionals and their research counterparts do not fully appreciate the importance of having defined and management-approved goals and objectives for a sensory program. Goals and objectives may be likened to a license to operate, a charter that defines where and how sensory will operate. Perhaps the most fascinating aspect of the process of defining goals and objectives is the diversity of opinions and definitions expounded by the individuals who participate in the process. To the extent possible, sensory professionals should formulate the goals and objectives as an initial step and be able to provide supporting materials should they be required by those managers who have an input but do not necessarily understand what it is that sensory can and cannot do for them. There is a need to formulate mutually agreeable goals and objectives that reflect current company business plans as well as the overall sensory needs from a technical perspective. This process should not be carried out in isolation if the result is to be workable and the impact is to be lasting. Although goals and objectives will be modified if they become unworkable, such changes, if not carefully positioned, will have an unsettling effect on managers (whether the sensory evaluation professionals know what they are doing) and an unsettling effect on those groups that work with sensory evaluation and may cause them to question the true purposes of the changes.

Therefore, the challenge is to develop goals and objectives that, on the one hand, provide sensory evaluation with sufficient freedom of movement to operate effectively and, on the other hand, do not appear to threaten and/or replace other product information sources. For example, a purpose for sensory test results is not to tell the technical staff how to formulate products (a common misconception), nor is it to replace large-scale consumer tests or to determine which new product should be marketed. However, it would also be inappropriate to restrict sensory goals and objectives to specific test methods or solely to measuring employee reactions to products—that is, to provide barriers based on test type or location of data source. Such barriers reflect a lack of understanding of the true purpose of sensory evaluation and an inefficient use of resources. It is far more useful if these goals and objectives are defined in terms of what sensory evaluation does and what it is capable of doing, as listed here:

1. Provide quantitative information about the sensory properties of company and competitive products of interest
2. Provide timely information about the effects of ingredient and processing changes on product sensory properties
3. Maintain a pool of qualified subjects
4. Develop methods for products with atypical properties
5. Develop methods and procedures for relating sensory, physical, and chemical data with product differences
6. Maintain awareness of new developments in product evaluation and their application to the company
7. Provide assistance to other groups in the company, on request
8. Ensure that no product of the company fails because of a sensory deficiency

Based on this listing, the sensory staff should be able to develop goals and objectives that are reasonable and achievable within a specified time period. It is important to acknowledge that regardless of how these goals and objectives are stated, they do not specify a particular test method, test location, or type of individual to be included or excluded as a subject. For example, specifying that only technical or professional staff members can participate as subjects or that an individual must have worked for the company for a certain period of time is self-defeating. Such restrictions will clearly work against the development of a viable sensory testing function.

Developing goals and objectives is but one step in the overall approach, a step that may prove to be more difficult to achieve than one realizes. Managers expect sensory staff to provide a reasonable explanation for the stated goals, a timetable, and some indication of the financial commitment. Considerable preparation will be needed, including written justification and supporting documents. The continuation of test activities during this preparation phase will help to reinforce the usefulness of sensory resources. The key to success in this endeavor resides primarily with the skill of the sensory professional in identifying the necessary information and in communicating directly with managers about the goals and objectives and their benefits. When communicating, it is naive to assume that the listener, whether or not technically trained, is more familiar with sensory evaluation than any other employee of a company. In fact, many individuals have considerable misinformation about sensory evaluation that can and will create problems in establishing an adequate sensory testing capability. Inasmuch as we are all equipped with sensory receptors there exists in all of us an expert element or, at the least, an attitude expressed as *de gustibus et de coloris non est disputandum*. This is further complicated by the position/title that an individual holds in a company; the higher the position, the more correct are one's sensory skills. Communications are also an extremely important issue, one that appears to cause considerable difficulty for many sensory professionals. Most managers have little interest in knowing test details; rather, they are interested in what actions are recommended. Ultimately, of course, both managers and sensory professionals must agree on the goals and objectives for a program. A failure to do so will reduce the effectiveness of sensory evaluation and will undermine its independence within a company. On the other hand, establishing goals and objectives does not guarantee a successful program.

2.2.2 Program strategy

Every sensory program has to develop an overall strategy that is consistent with a company's business plans and with specific brand plans. Whereas business plans tend to be expressed in general terms, brand managers focus on very specific targets such as product benefits and increased sales/profits. As a further complication, product market strategies can and will change as a result of new technologies, competition, an acquisition, and so forth. Sensory professionals are made aware of these developments often as a result of attendance at team meetings, through dialogue with technologists, or some other means. This information provides a basis for reviewing testing schedules, availability of resources, and the amount of time and effort that will be needed to develop resources not currently available. Without this information, sensory professionals are at a disadvantage. Correcting the information gap means reaching out to product managers and others who have the relevant information. It is also a reminder of the role that sensory plays in the business of product development, formulation, production, and marketing. Everyday sensory strategy refers to managing test requests, determining that products warrant testing, how priorities are established and by whom, policies and procedures for communicating results, daily operating procedures, and so forth. In the matter of communicating results, the sensory staff needs to know the kinds of information that the various groups want to see at a meeting. For example, some groups are interested in what the results mean versus a detailed description of the analysis and the level of statistical significance. Such an approach makes for more focused discussions and, in the long term, better satisfied managers.

Having an operational strategy also helps to minimize the impact of external pressures to modify established procedures or to involve sensory in a project for which it is not adequately staffed or qualified. A typical example is the request to use a specific method and/or subjects, or for the requestor to do his or her own analysis and/or interpretation of results. Other practices include the insistence that a test be conducted regardless of merit or that a test be repeated because the result was inconsistent with a requestor's expectations.

An operational strategy also helps define each step in the testing process, beginning with a review of the request and concluding that a test is warranted. Practically speaking, it should begin in the project planning process, even before a specific request is under consideration. Participating in the planning enables the sensory professional to put forward ideas about testing, to suggest alternative designs, to have an impact on what additional products could be included in a test, to have some idea as to the timing, and so forth. As depicted schematically in Figure 2.1, the sequence of events begins with receipt of a request and culminates in issuing a report with specific action steps and should include a presentation to the project team.

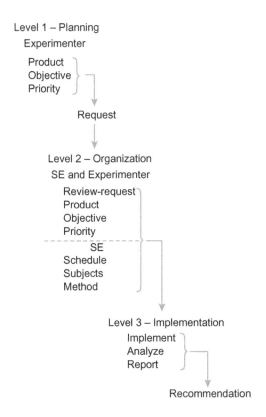

FIGURE 2.1

A representation of the stages involved in the testing process consistent with a defined strategy. Level 1 is the responsibility of the experimenters requesting assistance. Level 2 involves the sensory evaluation professional reviewing the request with the experimenter. On completion of this task, the test is scheduled. Level 3 involves data collection, analysis, and a written report leading to actionable recommendations.

The first level in Figure 2.1 is the planning process. As noted previously, it is the responsibility of the sensory professionals working with requestors to ensure that products and the test objective(s) are consistent and that each request has been assigned a priority. This is usually accomplished in a face-to-face meeting followed by an electronic confirmation to provide another check on the plan. Following a standardized protocol minimizes the likelihood of testing the wrong products or the wrong population, and it increases the likelihood of a successful outcome to the request. The basis for a request must be taken into account in its entirety. Specifically, what question or questions will be answered, how will results be used, and is the request one of a series for this product category or is it a single request? Other important information includes where the products are located, how they were prepared (at the bench, pilot plant, or manufacturing location if more than one), when products will be provided for the test, and who is responsible for their delivery. Although this discussion seems at first to be tedious and intrusive, the success of a program is built on providing the correct information, and unless a thorough effort is applied at the beginning, the likelihood for success is lessened. Obviously, with time, most everyone involved in the process learns what is needed and requests proceed more quickly. Long-term success of a program is based on a well-defined strategy that enables requests to be processed quickly and facilitates the working relationships between sensory and requesting groups.

2.2.3 Professional staff

Developing a capable staff is a challenge for a company particularly when managers are not familiar with sensory evaluation or the current sensory staff lacks the depth and breadth of experience necessary to develop needed professional skills. As noted previously, sensory evaluation is the application of knowledge and skills derived from several different scientific and technical disciplines, such as physiology, chemistry, mathematics and statistics, experimental psychology, human behavior, and knowledge about product formulation and processing practices. In recent years, the increased availability of academic courses, advanced degrees, and distance learning certificate programs has made finding qualified professionals easier.

Sensory professionals must be sufficiently knowledgeable in the topics previously mentioned. They must also be able to train a staff to the extent that they can perform their work and fully understand the rationale for sensory principles and practices and to know when to seek assistance.

How does one find such individuals, and particularly sensory professionals? For a company with limited or no recent experience with sensory evaluation, the first action is to contact a local university or one that is known to have a sensory program or to recruit an individual from another company. The ideal candidate should have a science background and course work in experimental psychology, physiology, and statistics. If such individuals are not available, one might consider an individual with product development experience with the understanding that additional training will be needed (e.g. distance learning classes).

The options are numerous; however, more practical considerations may dictate that before deciding on the type of professional, some consideration should be given to having an external entity review existing and future activities before deciding on the type of skill required. This makes it easier for human resources to focus their efforts on finding the best person for the job. This process also helps to estimate the number of support staff and their skills. Of course, hiring an academically trained sensory professional will expedite the process.

The sensory professional–manager has numerous responsibilities, all of which are equally important to the success of a testing capability. These include the following:

1. Organize and administrate all departmental/group activities (staff, test requests, timing, priorities, and budgets)
2. Plan and develop resources not already available (tests, subjects, and database)
3. Advise on feasibility of test requests
4. Select test procedures, experimental designs, and analyses
5. Supervise/delegate testing
6. Determine need for special panels
7. Ensure smooth operation of test program
8. Train subordinates to handle routine assignments
9. Responsible for reporting results
10. Maintenance of all necessary supplies and support services
11. Provide progress reports
12. Develop and update all program plans
13. Research new methods
14. Maintain regular communications with other departments that use sensory information
15. Partnerships with suppliers

This list clearly emphasizes the relative importance of management skills in addition to knowledge about sensory evaluation. Because results from sensory tests are used in business decisions, managing the information is especially important. The information may be used in a variety of situations. For example, ingredient replacement is a typical situation in which sensory information is needed. Although the question answered by the test may seem simple—that is, product "A" was perceived as different from product "B"—the result often has more far-reaching implications: Does the difference change preference and purchase intent? What are the potential ingredient cost savings? Is shelf life changed? Was product made in the pilot plant? With this information, the sensory staff can prepare a report that is actionable relative to the initial discussions with the requestors. The situations are almost limitless, but all require decisions, usually on the spot, and all are potential sources for confrontation or missed opportunities if not handled with skill.

In general, it is the exception rather than the rule to find a sensory professional with good management skills. Most sensory professionals are capable of organizing and fielding a test in a technical sense but experience considerable more difficulty functioning effectively in project meetings with marketing, consumer insights, production, and others, in which strategy and communication skills can be more important than results from a test. It is all about thinking strategically. This does not mean that a sensory professional is incapable of being, or cannot be trained to be, an effective manager. Because the activity is relatively new, there are few examples that can be cited and still fewer professionals with the experience and needed management skills. There has been improvement in this situation as professionals gain experience functioning in a business environment. Sensory professionals need to develop communication skills through their own efforts and particularly by attendance at company-sponsored training programs. These issues are independent of company size; that is, selecting a test method, communicating results, organizing resource, etc. are no different whether one is working in a small or a large company. Some companies may not want to wait for a sensory professional to learn how to be an effective manager and instead assign an individual from another area—someone with proven management skills. Although such a move may satisfy management, it will not necessarily satisfy the development of sensory resources or of the sensory staff, at least not in the long term. In practice, it has been observed that a manager with little or no familiarity with sensory evaluation is more vulnerable to a variety of pressures, to agree to test requests that are not consistent with accepted sensory practices, to acquire a methodology without appreciating the risks, and so forth. When such a system is structured on an interim basis, to allow the sensory professional to gain the necessary skills, it may be a reasonable option. In the long term, and in the best interests of a sensory program, a professional with good sensory skills should manage the program. As already noted, companies will provide management training to those exhibiting potential, including attending special training programs offered at various business schools.

The sensory professional carries out the duties assigned by the manager, including the following: establish test schedules; assign responsibility on a test-by-test basis; supervise all phases of product preparation and testing; report preparation; maintenance of records, supplies, and support services; maintain the department library; coordinate new subject orientation, screening, and motivation; plan and supervise research activities; meet with requestors to review future tests as well as to discuss results; and perform all other duties delegated by the manager.

In addition to a manager/sensory professional, a minimal department staff will include two analysts or technologists. Although some companies might not prefer the designation of technologist, we believe that the responsibilities justify this type of professional designation. Responsibilities include preparing test plans and scorecards, scheduling subjects, supervising and/or serving, collating responses, routine analyses, maintaining records, and maintenance of facilities. Sensory technologists should have a science background, preferably as an undergraduate degree; knowledge of sensory principles and practices is desirable, as is training in statistics and experimental psychology. For evaluation of foods, knowledge of recipe development and product preparation will also be helpful. It is becoming more common for individuals to have these qualifications, making the hiring process easier.

In some situations, the sensory technologist position is considered interim, or entry level. Demonstrated performance leads to promotion and is considered permanent (i.e. not necessarily interim). The intention is to seek an individual who will continue in that position. We believe that the sensory technologist position should be permanent and should have a professional designation. If an individual has demonstrated the necessary capabilities, then promotion is justified and an opening for a new sensory person is created. The importance of the sensory technologist cannot be underestimated, inasmuch as this individual is in direct contact with subjects on a daily basis, observing their reactions before and after each test. In addition, the sensory technologist prepares and serves products and generally functions as an "early warning system" if there are unanticipated problems during a test. Currently, distance learning programs offered by a few universities enable the technologist to gain further knowledge about sensory evaluation along with a certificate of achievement. This helps to minimize employee turnover and adds to the skills of the sensory function as a whole. In addition to a professional staff, a sensory program needs an individual familiar with data capture systems because of the move away from paper ballots. Substantial numbers of reports with text, tables and graphics, and large volumes of records are produced, requiring staff support. Although these tasks can be performed by the professional staff, it is an inefficient use of their time.

A typical staff should consist of a core group of five persons: a sensory professional–manager, a sensory professional, two technologists, and one or two support staff. Companies familiar with the use of sensory testing procedures and with an extensive product line may have a sensory evaluation staff consisting of five or more professionals and an appropriate number of technologists. Companies new to sensory testing may have only two individuals—a sensory professional–manager and a technologist. This is minimal, and its overall impact will, of necessity, be limited. Either fewer tests are done each week or the amount of dialogue with requestors (before and after a test) is limited. Obviously, no single formula or rule can apply to all situations; thus, one cannot state the optimal number and type of staff necessary to operate a sensory program. Some companies may choose to contract with external services for most testing needs, in which case the staff needs are adjusted accordingly; however, there can be no compromise on their sensory knowledge.

It is possible to estimate the number of professional and technical personnel needed in a program by indirect means. With an entirely new program, there are few criteria to use in estimating service level; nonetheless, it is surprising how much information can still be obtained through discussions with those that have requested tests in the past or have product information needs. From this information, one can develop three levels of test volume, including an estimate of staff and time requirements. Considered in another way, if a program starts out with a limited service, such as 4 or 5 tests per week, and then expands to 8–10 tests per week, what are the limiting factors? One can make some reasonable estimates assuming there are no subject limits, data capture and analyses are completed electronically, and the facility has an adequate number of booths. Other estimates can be developed based on limiting the aforementioned resources. All these estimates require a conversation between the sensory professional and requestors—past, present, and future. Successful programs are proactive and not reactive.

A program capable of providing approximately 200 tests per year will need a full-time staff of a sensory professional, a technologist, and an assistant at 50% time, as a minimum. The professional's time is spent meeting with requesters, planning tests with the technologist, performing data interpretation, preparing reports, and communicating results. The technologist's time is spent organizing and fielding tests, contacting and qualifying subjects, scheduling subjects, reviewing results, and maintaining records. The assistant's time is spent preparing products for a test, serving, and performing general cleanup and related activities. A volume of 200 tests per year assumes an average distribution in terms of the types of tests—discrimination, descriptive, and acceptance—as follows:

Discrimination	10%
Descriptive	50%
Acceptance	40%

This list of test type by volume is an estimate, and each test type requires different effort, more or less subjects, different amounts of time to complete a test, and different amounts of effort to review and report results. Of these, staff time for interpretation of descriptive test results will take the most time compared with results from a difference test. Increasing volume to approximately 300 tests per year will create considerable pressure on a staff of 2.5 by (1) increasing the time between request and report and (2) decreasing the amount of time given to each request and to discussing results. From a management perspective, the issue of staff growth must now be addressed if the services provided are to continue to grow (assuming the demand is there). In general, one sensory professional should be able to supervise the work of two or three technologists, so the addition of a second or third technologist would be manageable. Note that an increase in test volume is expected at this stage because the usefulness of the information and its contributions to overall company profitability is recognized and the demand is there. Increasing volume beyond this (e.g. to 400 tests or more) will require additional professional staff and another assistant (or switching the assistant from half- to full-time). Having two allows for all-day coverage of testing and is an advantage versus having one on a full-time basis. The increased volume of tests demands more professional attention; hence the staff need for the professional rather than more technologists. As test volume increases, more attention must also be given to the nature of the tests (e.g. descriptive and discriminative), the amount of time that the sensory professional needs for project planning, discussing results, and the extent of involvement in related test activities such as those that incorporate consumer attitudinal information, and so forth. All this activity is incorporated into the department's growth plan as originally developed (and revised), and it is used as a guide for further staff expansion.

As companies expand their markets, and the demand for product sensory information increases, the volume of testing will grow but not necessarily in a continuously upward manner, and the testing locations will also grow beyond a domestic market focus. All this creates further pressure on sensory staff to have plans to accommodate this change. As noted previously, this growth in test volume is finite, in part because the number of variables that are tested is finite. In addition, product data files should provide sufficient information that identifies variables that have little or no effect or their effects are known to negatively affect preferences. Often, one encounters plans for testing that were considered and rejected 3 or 4 years ago but have re-emerged with a new team (except for sensory).

Basically, the growth in demand for sensory information reflects the success experienced by users of the information. With a sensory professional supervising two technologists and two assistants, this group should be able to manage approximately 300 tests per year. Note that the volume of tests may actually be less than the number quoted, but their complexity will require much more time to field, interpret, and report. In situations in which shifts occur in the number and types of tests, the sensory professional needs to be able to allocate resources to reflect priorities and to keep management informed. Having a designated reporting structure as early as possible is important because this person serves as the focal point for the program and "sets the stage" for the future, including the process of facilitating staff additions when necessary. Of course, the type of testing will require staff additions that favor one type or another skill set to best satisfy a particular need. For example, increased use of experimental design studies and larger numbers of products per test with a concomitant increase in complexity of the analyses and interpretation will require more professional-level involvement. This will lead to an obvious growth in the sensory professional staff rather than increasing the number of analysts.

A topic that is less common among sensory professionals is the strategic use of vendors or vendor/partners to supplement their research needs. First, we might ask, "Why use vendors at all if we have an informed and capable sensory staff?" Often, organizations are not wholly self-contained and need outside assistance beyond their internal corporate resources. The use of outside resources enables an organization to "hire out" sensory and consumer research to selected professionals who provide additional resources and/or new expertise. Outside vendors also make it possible to hire, on a part-time basis, experts otherwise unavailable within the company. We further discuss this topic later in this chapter.

2.2.4 Facilities

One of the fundamental requirements of a sensory program is the availability of a facility dedicated to testing. A specific area should be set aside for this purpose, just as space is allocated for other company activities such as a chemistry laboratory. Most companies recognize the need for a test facility for sensory evaluation and have provided space for this purpose. In the past, many design considerations created facilities that were inconvenient to use, thus slowing the testing process. Recently, sensory professionals have been asked to actively participate with those involved in designing test facilities, which is a reflection of the advances made in the field. The literature was of limited value until the mid-1980s, when the first edition of this book was published containing detailed schematic diagrams. In 1986, the American Society for Testing and Materials (ASTM), Committee E18 on Sensory Evaluation, published a monograph (Eggert and Zook, 1986) on the design of a sensory facility, including guidelines for size of a facility, suggested booth dimensions, and related information. Several schematic diagrams are included as a supplement to assist in the planning of a facility. This latter document was updated with a second edition (Kuesten and Kruse, 2008) taking into account the many changes in the field and especially the impact of direct data entry on booth design. As with any major capital expenditure, it is very important to emphasize the importance of following established guidelines. Obtaining funds to correct deficiencies is very difficult after a facility is in operation.

We believe that the sensory facility is as important as any instrument used by a scientist and therefore warrants serious consideration. Of particular importance are ventilation, lighting, traffic patterns and locations, product preparation, subject communications, and experimenter comfort. Deficiencies in most facility designs occur for several reasons. In most cases, a facility is ideally described as a quiet area free from distraction, with controlled lighting, partitions between subjects to minimize visual contact, neutral colors for the walls, and odor-free surfaces wherever possible.

It is not unusual to observe two companies having comparable numbers of staff but with major differences in test capacity or the converse—that is, similar test volumes but different numbers of staff. Facility design is usually the primary difference. In some food companies, inadequate ventilation necessitates a long delay between tests to allow for removal of cooking odors, which reduces test capacity. Alternatively, one facility has its own entrance, reception area, and is easily accessible from the outside, whereas the other is difficult to reach, requiring longer walks for the subjects. During approximately the past 40 years, we have observed that architects and building contractors appreciate receiving as much assistance as possible in the design and construction of a facility. The objective is to have a facility that is flexible enough to handle current and future testing activities as well as to provide a workable environment for the staff.

Before making a commitment to actual drawings, it is important to develop an information base about testing in your company. This should include information about the amount and location of available space, current and anticipated volume of testing on a daily and monthly basis, current and anticipated types of products, current and future availability of subjects, and types of tests. With this information, one can identify some of the other details and prepare drawings that will be useful to an architect. Table 2.2 is a guide relating space, number of booths, number of tests, staff, and number of subjects.

Table 2.2 A Guide for Allocating Space for Sensory Testing[a]

Area (ft²)	No. of Booths	No. of Staff	Annual Volume of Testing	No. of Subjects
400	5–6	1–2	200–300	100–200
600	6	2–3	300–400	200
800	6–8	4	400–600	300–400
1000	8	5–6	700–800	400–500
1500–2000	12	6–7	>1000	500–800
3000–4000	24	7–8	>1500	>800

[a]The entries are estimates of the amount of space, booths, and staff that are capable of doing a specified number of tests. Additional information about the use of the table can be found in the text.

Basically, a certain amount of space is necessary for a sensory evaluation facility; the space requirements range from a minimum of 400 to approximately 4000 ft² (or more, in some instances). However, the number of booths and number of staff do not increase in relation to the space allocation. Obviously, efficiencies are to be gained through size, but the greatest advantages are realized in terms of the types of tests that are possible within a larger facility. In fact, it is often more efficient if sets of booths are located in a facility rather than in a single bank of 20 or more. Of course, having booths in two locations does not mean two physically different parts of a building, which is quite a different arrangement (referred to as a satellite facility with its own reception and product preparation areas). Although an arrangement of 12 booths in a single bank may have initial appeal, it could prove to be inefficient, particularly if there are an insufficient number of servers. Serving requires more than one staff person unless one uses rollerblades! If products are heat sensitive, temperature control will be difficult; keeping track of when subjects arrive for a test is not easy, and generally the preparation area must be oversized (this can be inefficient and a costly use of space). Finally, because most in-company testing uses small panels, rarely exceeding 40 subjects per test, there may be little need for more than 10–12 booths in any one location (serving as many as 24–32 subjects in 60 minutes, assuming 15 minutes per subject).

With the increased reliance on electronic data capture and the ability to monitor a subject's progress during a test, serving efficiencies are achievable, thus enabling a higher test frequency. However, testing efficiencies are also controlled by the availability of subjects.

Most companies have reduced the number of employees available in their technology centers, and this makes use of nonemployees an especially attractive alternative. With fewer employees available, one could also make use of satellite facilities where nonemployees are located. In recent years, companies have switched to reliance on local residents for their subjects. If this is a possible option, then access to the testing site becomes important. It should also be emphasized that the guidelines make no differentiation between types of products or types of tests (both of which are discussed later). Each of these considerations will alter the input–output listings in the table. Nevertheless, one can proceed to develop the basic information according to the five primary criteria.

Space may be a problem only if the area is already designated and if total space is less than that estimated to be needed for the current testing volume, or if it is located at some distance from subjects. The emphasis on locating the facility in a quiet area is secondary to accessibility. A location on the fifth floor in one corner away from most people is undesirable. No one wants to walk for 15 minutes to take a test that can be completed in 5 minutes. One should begin with a map (or blueprint) of the building and make notations on the number of people in each location (this may require a little extra effort, but it is worthwhile). The primary purpose of this exercise is to locate the facility as close as possible to the largest concentration of potential subjects. It also may be necessary to trade some space for accessibility. Management may offer twice the area in a location adjacent to a pilot plant or some other out-of-the-way place. Although the offer may be tempting, sensory staff should review its proposal with management to make clear the importance of accessibility.

When considering space, it is useful to be able to characterize it in some detail. When plans are in preparation, the ability to provide detailed requirements, including dimensions for booths and counter height, will demonstrate sensory responsiveness and help to minimize faulty designs. The typical facility can be separated into six distinct areas, as shown in Table 2.3. Also included are approximate amounts of space for each of these areas. The space allocations in square feet are realistic for most food and beverage testing activities. With limited cooking, the preparation area can be reduced and some other area expanded. If extensive product preparation or storage is necessary, then this area would be expanded. For product requiring no or only limited preparation, the area could be reduced and the space used elsewhere in the facility. A panel discussion area may also not be available, and a nearby conference room may be proposed for this purpose. This is feasible provided it is accessible and has adequate ventilation, lighting, and so forth.

Another consideration is the pattern of subject movement to and from a facility. Subjects should pass through the reception area on their way to a test, but they should not pass through or see into the preparation area. Similarly, visitors to a facility should have separate access to the offices and preparation area of the facility. Thus, the issue of space has many ramifications beyond the specific question of total area.

Space requirements for data processing have changed significantly as a result of direct data entry. Most companies have moved to use of these systems, thus obviating the need for the traditional paper scorecard. The relative merits of such systems are discussed later in this chapter and are mentioned here only as they relate to the space and wiring requirements. With the development of wireless technology, subjects can more easily test without major facilities changes or, for that matter, they can evaluate products without the need for a facility.

Table 2.3 Individual Areas in the Test Facility and the Approximate Square Footage[a]

	Approximate Space Allocation (ft²)	
Area	6 Booths	12 Booths
Subject reception	50	100
Booths	100	250
Panel discussion room	400	400
Preparation, holding, and storage	350	400–600
Sensory technicians, data processing, administration	75	150
Sensory scientists desk/office	150	200–300

[a]The allocation of space is relative and will change depending on the types of products and the daily volume of testing. See text for further explanation.

Because most tests involve limited numbers of subjects, usually 25 or fewer per test, one could field as many as four or more tests and use approximately 80–100 subjects. Because it is possible for an individual to participate once a day, 5 days a week, a potential maximum volume can quickly be estimated at 20 tests per week × 50 weeks = 1000 per year.

Of course, a volume of 20 tests per week will require a large panel pool (more than 500) to minimize subject overuse and management complaints about an individual's time spent away from his or her regular work routine. Ideally, an individual should expect to participate approximately four times per week, with holidays of approximately 2 weeks every 6 weeks. In practice, most individuals participate two or three times per week, with occasional periods of several weeks of no participation and other times when a test design requires daily participation for 2 or more consecutive weeks. The issues of how large the pool should be and frequency of use are explored in more detail in the section on subjects. The numbers used here are for illustrative purposes, relative to facilities design. Of course, this assumes that there is a demand of that magnitude.

Assume the following:

Ten tests per week use 25 subjects per test.
Ten tests per week use 50 subjects per test.
Ten minutes residence time per booth per subject (6 booths).

Thus, the total booth time is determined to be

$$\frac{(10 \text{ tests} \times 25 \text{ subjects} + 10 \text{ tests} \times 50 \text{ subjects}) \times 10 \text{ minutes/test}}{6 \text{ booths}} = 1250 \text{ minutes/booth/week}$$

If we assume 5-hour test time per day × 60 minutes = 300 minutes/day test time, then 4.2 days are required, which is within the typical workweek.

Therefore, we conclude that 6 booths would be adequate for up to 1000 tests per year. As noted in Table 2.2, however, 8 (or more) booths are recommended once volume reaches approximately 800 tests per year, and two banks of 6 booths each are recommended when the volume reaches 1000 tests per year. This increase in the number of booths reflects the practicalities of depending on large numbers of volunteers. Perhaps most important, however, the requirement for replicate responses in many tests will increase the number of visits to the booth by each subject. To minimize potential scheduling problems and to increase flexibility, the additional booths are recommended. It is best to start out planning for 12 booths based on availability of space and an adequate budget.

The types of products that will be evaluated are a related issue. The more different types of products that are to be tested, especially those that require preparation, the more space must be allocated for this purpose. For example, frozen prepared foods will require freezers as well as heating and related holding equipment. This equipment will reflect the primary methods by which the products will be prepared (by the consumer, the preparation staff, or others). Sensory evaluation should prepare a list of needed equipment with space and power requirements so that it can be reconciled with the space and power allocations. Installation of such equipment at a distant location is not acceptable because it will mean transporting products from one area to another. In addition to product preparation requirements, one must also consider the available areas for test preparation; that is, sufficient storage space must be allocated for serving containers, utensils, and so forth.

The types of tests will also have an effect on the allocation of space as well as on subject requirements. If there is an equal distribution between small and large panel tests, then the previously mentioned numbers may be used as a guide. However, in situations in which product acceptance tests are rarely done, an increase in test volume with no increase in number of subjects could be expected. Discrimination tests require approximately 20–25 subjects per test; thus, three tests per day could be done in approximately 4–6 hours. The sensory professional must consider past history relative to the test pattern; anticipated requirements must be reviewed and then a plan prepared.

Using this information, sensory staff should develop a comprehensive plan, describing basic information about the test program, desirable space requirements, a listing of equipment, and appropriate schematic drawings. Once management has given approval to this preliminary plan, the next step is the development of working plans that can be incorporated into the architect's drawings that serve as the basis for the eventual construction.

Working with an architect requires considerable patience and a tremendous amount of preparation. It is not a matter of doing the architect's work so much as delineating the specialized nature of a sensory evaluation facility. This preliminary effort will facilitate the development of plans, the bidding, and the actual construction. If the facility is to be located in an existing structure, then it must be visited and regularly inspected. Of course, if the facility is to be located in a new structure, one works from blueprints, but attention to detail is just as important and an effort should be made to visit the site during construction. Once dimensions are known, one develops schematic diagrams including dimensions for each piece of equipment, for all cabinets, and so forth.

2.2.4.1 Booth area

Most product tests use individual booths for data collection. However, there will be situations and products for which a booth is not possible (e.g. personal care products and quick service restaurant foods). In these situations, the subjects could use a tablet or other handheld electronic device to record responses. The responses can be directly linked to the web or, when finished, the data can be electronically transmitted.

A booth consists of a countertop with walls that extend from ceiling to floor on three sides and approximately 18 in. beyond the serving counter surface (so subjects cannot view their neighbors), with a small door at the counter surface. The subject seated facing this door signals the experimenter to serve the product. Figures 2.2 and 2.3 are schematic diagrams of a booth and booth area, including recommended dimensions.

We recommend the dimensions shown in Figure 2.3 and find that they are quite adequate for all products. The booth depth of 18 in. is adequate except when "bread box" openings are used, which will require an additional 6–8 in. The booth width of 27 in. is sufficient without producing a claustrophobic effect; increasing this width beyond 30 in. may be a waste of space. Note that the counter is between 32 and 36 in. from the floor. This height is desirable because it makes serving easier and reduces server problems with lower back pain. Adjustable chairs enable the subject to select a comfortable height for a test. The length of the booth room will depend on the number of booths, but if possible additional space (2 or 3 ft) should be allocated as a buffer from the reception area. The partitions extend from floor to ceiling and 18 in. beyond the booth counter. Approximately 4 ft above these counters is a soffit that extends the length of the booths and out approximately 18 in. (even with the counters).

This soffit accommodates booth lighting fixtures and ventilation ducts. The booth room should be not less than 6 ft wide, with 3 ft allocated for partitions and counters and the remaining space allocated for passage by the subjects. Additional space is unnecessary. Having a small sink for oral rinsing in each booth is not recommended. Sinks are a potential source of odors and noise, require regular maintenance, and will increase construction costs. Where possible, the use of covered, disposable containers for expectoration is a more acceptable alternative. However, if one is evaluating personal care products, this situation may necessitate availability of sinks.

Lighting in the booth area is fluorescent, except in the booths themselves, where incandescent lighting is recommended. This lighting should be sufficient to provide 100–110 ft-candles (or their equivalent) of shadow-free light at the counter surface. These incandescent lights are located slightly to the front of the soffit and are tilted toward the seated subject to minimize shadows on the products. Opalite diffusion

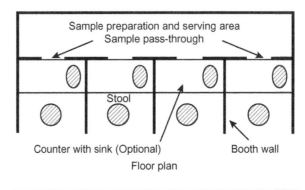

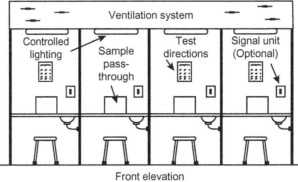

FIGURE 2.2

Schematic diagrams of the booth area in a sensory evaluation facility (not drawn to scale).

glass can be placed beneath the bulbs to eliminate any spotlight effect. Off–on and dimmer switches for the booths are located in the experimenter area.

In addition to the aforementioned lighting, it is not unusual to be asked about the use of other colors, such as red lights. There are several reasons why they should not be used or even installed. Its use is often justified as a way of masking product visual differences. Technologists often state that the color difference will be fixed later. Such claims rarely, if ever, are realized. Where sensory testing is concerned, dependency is the rule and independence is the exception. That is, a product's appearance influences the consumer's expectation as to that product's aroma, taste, etc. Failing to capture all the information makes it likely that useful product information will be lost and recommendations may not be correct.

It has the potential for increasing variability because it is different from the typical test situation so one has to wonder why creating a more artificial situation has merit. A practical problem is the situation in which brand managers and consumer insights staff discuss test results and sample the products in a conference room under typical building illumination. Is the meeting room lighting changed to red? Sensory staff must be careful about reporting results and how they were obtained

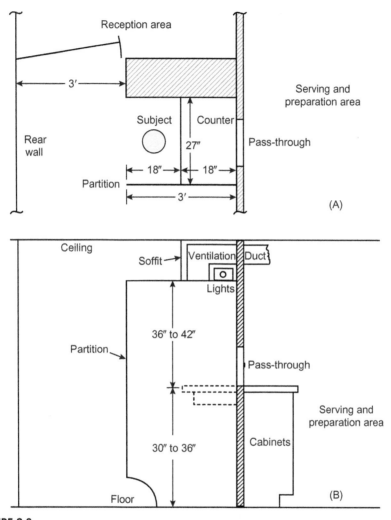

FIGURE 2.3

Detailed description of the booth area including recommended dimensions. (A) The booths from an overhead view; (B) side view of the same area.

(using red illumination or visual masking). Despite reminders that the test was done under red illumination to mask this difference, there will be doubt expressed about the credibility of such a result, and the long-term credibility of sensory evaluation could be questioned. Although there are numerous procedures and practices that are unique to sensory evaluation (e.g. three- or four-digit coding and balanced designs), the idea of masking a difference represents a very difficult concept for managers to accept (or at least it should do). As already observed, product technologists' claims that this difference will be corrected are rarely achieved, creating other problems in the future. We have observed situations in which red illumination was used in the booths, but white illumination in the preparation area enabled the subject to see the difference each time a pass-through door was opened. In some situations, colored containers have been used as a means of neutralizing visual differences; however, one must be very confident that the difference can be masked. In Chapter 5, we describe use of a paired procedure in which a monadic serving procedure will minimize the color difference. This permits use of the typical white illumination in the booth, and as we will describe, it is applicable to many product situations. In the long term, we believe that taking advantage of appropriate test design and measuring visual differences is the preferred course of action. It is the responsibility of the sensory professional to design a test with obvious product color differences without resorting to procedures that confound the results.

In some instances, the product preparation area will be used as a primary location for bench screening prior to testing. In this case, the lighting should be based on the U.S. Department of Agriculture guidelines for approved illumination (see File code 131-A-31, January, 1994, or any newer version).

The ventilation for this room and especially for the booth area is quite critical and is probably the most costly (and most important) investment that is made besides wiring the facility for high-speed or broadband Internet. If one aspect of a facility needs special attention, it is the ventilation. Most architects and engineering personnel cannot be expected to appreciate this need and will generally state that the building ventilation is more than adequate—Do not believe it! For example, the ventilation requirements must take into account all booths being occupied, the number of people in the preparation areas, the likelihood that all heating equipment (ovens, steam tables, etc.) and computers will be in use throughout an 8-hour time period, etc. The booth area must have a slight positive pressure relative to other areas; in addition, individual exhaust ventilation ducts should be located in each booth (in the soffit, if part of the design). Air turnover in the booths should occur every 30 seconds as a minimum; however, this timing can vary depending on the product category. For example, a more rapid turnover of every 20 seconds will be needed for tobacco and for flavors and fragrances, with a special exhaust in case of spills or excess smoke buildup. By confining the improved ventilation to the booths and booth area, the capital expenditure should be reasonable. Obviously, for products that do not contribute odor or fragrances, no additional ventilation may be necessary; however, the facility should still have good ventilation. Because ventilation systems also have filters, every effort should be made to ensure that they are accessible for frequent changing. We recommend changing them at least once per month unless they are very dirty, in which case weekly changes will be necessary.

Finally, when in doubt about the suggested ventilation rates listed here, the ASTM MNL 60 2008 publication should be referenced.

Facing each subject seated at a booth will be a small door, referred to as a "sample pass-through door." The door opens either up or to one side, allowing samples to be passed to the subject (Figure 2.4). A variety of designs have been developed,

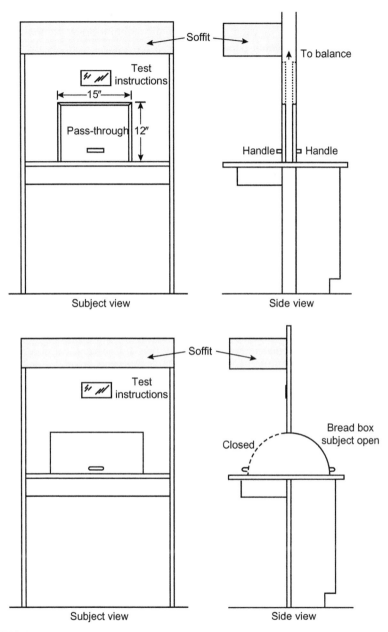

FIGURE 2.4

Schematic diagrams of the two most common types of product pass-through doors.

and although most look attractive or appear simple to operate, they have the potential to create problems. For example, horizontally sliding doors or those hinged at the top usually stick, catch papers, or have a tendency to fall on the samples. Another type is based on a lazy Susan design but with a center barrier so that when it swings around there is space for the product. Unfortunately, these units loosen and after a while they move like a merry-go-round and can spin product out, causing spills. The most commonly used door types are the "bread box" and the balanced guillotine. The bread box is so constructed that when one side is open, the other is closed. This minimizes the likelihood of the subject having any view of the preparation area.

Major disadvantages of the bread box include the increased amount of space required on the subject's side when the door is down, the height of the box relative to serving containers, and the inability to easily communicate with subjects. Because these containers are made of stainless steel for durability and ease of cleaning, they are costly. The use of the bread box delivery will require a deeper counter (from 18 to 22–24 in.). Before fabricating, however, it will be necessary to check the height of all serving containers to make certain they will fit. Because of these disadvantages, we do not recommend use of the bread box door and find that bread boxes are losing favor.

The balanced guillotine-type door is newer; earlier versions were held open with blocks or pegs. Unfortunately, there was a tendency for them to crash on one's fingers or on the products (hence the name "guillotine"). Recently, we observed considerable success with a guillotine door that is counterbalanced and suspended with high-quality wire (airplane) cable to minimize breakdowns. The use of food-grade, white polymeric belting material for the door, approximately ⅜ to ½ in. in thickness, makes them easy to clean. For ease of removal for washing and maintenance, the experimenter-side trim is bolted. Alternatively, one can create a door without the wires that is not counterbalanced and uses a small ledge on which to rest the door when open. Such a door is quite inexpensive to make and will last for decades. These comments regarding door opening and closing design could be impacted in situations in which there are laws regarding safety. We recently encountered such a problem designing a facility in the European Union, necessitating a change in how the doors are to be operated. In these latter situations, the sensory professional should be able to accommodate such restrictions by referring to the function, the advantages of a particular design, and referencing the ASTM document or a similar guide published by the International Organization for Standardization.

The booth arrangement is primarily intended for products that lend themselves to this type of evaluation. Some products, such as home care, personal care, or pet care products, may require a different test environment; however, there should be no visual contact between subjects, minimal contact with the experimenter, ventilation to remove odors and fragrances, and constant lighting. For example, the evaluation of a room air freshener may require construction of a series of containers of known volume (e.g. $27\,\text{ft}^3$) with small doors to enable the subject to open and sniff the contents without disturbing subjects at adjacent test stations. In addition, subjects can move from station to station, thus making more efficient use of the facility. The key to a successful facility design is to develop an understanding of the reasons for the various parts and to then incorporate them into the particular booth design. Here, too, however, the use of booths may not be necessary if subjects evaluate products in their own home and transmit results via the Internet.

The booth area should be adjacent to a subject reception area. This multi-purpose area provides a sound barrier to outside noises and enables the experimenter to monitor arrivals; it is a space for posting test results and for serving rewards for test participation; and it serves as a barrier to subjects wandering into the product preparation area or prematurely disrupting other subjects.

The traffic pattern is also important. Subjects must be able to get to the booths, check on their specific assignment, and determine in which booth they are to be seated. The experimenter (or some designated individual) should also be able to monitor arrivals. Once preliminary drawings have been completed, the sensory staff should review them relative to the flow of people and to the location of specific equipment, and so forth. This state of preparation may best be accomplished by a mock-up or computer-aided design to ensure that there is adequate space.

2.2.4.2 Preparation area

The preparation area is more difficult to define because it is highly dependent on the types of products and equipment and on the amount and type of preparation required. Figure 2.5 shows two schematic diagrams for an entire sensory facility with six areas,

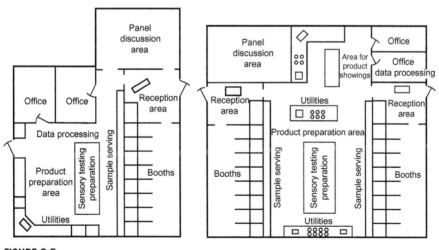

FIGURE 2.5

Two examples of sensory test facilities showing the various activities within the department. The Tragon website shows a new style of facility with a large floor plan (http://tragon.com/where/redwood-shores-lower.php).

designated as reception, product preparation, panel discussion, and so forth. Of the six, three are continuous (reception, booths, and preparation), whereas the others can be moved around to fit the available space. The design on the right has two product preparation areas to reflect needs of a large product portfolio. Approximately 300–400 ft² should be allocated to the area. Critical factors in determining the size of the preparation area include the expected amount of cooking and the size of the units necessary to achieve this and also storage requirements for these products.

As noted previously, this space can be modified for products such as tobacco, fabrics, chewing gum, and candy, where product preparation is minimal and limited to storage and test setup arrangements. However, the individual confronted with a limited amount of space must be even more resourceful. It is not so much a question of the space being too small but, rather, of organizing it to allow for the most efficient operation. For example, the space below the serving counter (on the experimenter side) can be used for storage of supplies as well as for refrigerator and freezer space. Similarly, the space above this counter can be used for storage, provided the cabinets are not so low as to constitute a hazard to the experimenter when serving. Counter space is especially important, and there can never be enough of this type of space. In areas with minimal space, it may be useful to have drop-down counters set in place for product setup or to prepare scorecards and products on stacking trays. For products that require heat or refrigeration, one must be sure there is adequate equipment and space.

Storage cabinets and counters should be made from any polymeric material that is easily cleaned and maintained. Although the booth area requires careful attention with regard to selection of color, one can be more relaxed with regard to the preparation area; counters continuous with the booths should be of the same color because a contrast in color could affect the appearance of a product as it is served to the subject.

Ventilation in the preparation area is especially important for those products that contain volatile chemicals. All heating areas should have individual exhaust units (approximately 350–400 CFM), and estimates of environmental and electrical load requirements should always be based on the "worst case" point of view (all heaters operating at once). Use of three-phase wiring for heating outlets is highly recommended, as is minimal use of shared wiring for these outlets. Lighting in the preparation area should be such that products can be examined with ease and, where appropriate, special lighting should be available—for example, northern, supermarket, showroom, and retail outlet.

2.2.4.3 Data processing

In the past two decades, significant changes have occurred in the use of electronic systems for scorecard preparation, test designs, response capture, data analysis and record keeping. These developments have been a long time evolving (Aust, 1984; Brady, 1984; Brady *et al.*, 1985; Gordin, 1987; Guinard *et al.*, 1985; Pecore, 1984; Russell, 1984; Savoca, 1984; Stone and Drexhage, 1985; Winn, 1988). Additional developments continue to occur such as the use of tablets and the Internet.

This removes the tether connecting a subject to a booth. This should not be construed to mean that booths will no longer be needed; however, there will be opportunities to perform field tests away from the laboratory environment. This also means that space dedicated to the manual preparation of scorecards and a data center will be eliminated, with these activities completed at a staff person's workstation. With the growth in cloud computing and the power of PCs, hardware and software requirements have changed in terms of speed and cost. However, these advantages have shifted the burden of test design and analyses to the individual staff persons. It also has meant that staff has to be more conversant with the software and have ready access to a person specialized in managing the system. In today's wired world, tests are fielded globally and data sent via a secure system allowing for analyses as live results or within a few hours. This collapses the time requirements related to obtaining output; however, it does not reduce the requirements for interpreting and reporting of results. As mentioned in Chapter 1 and discussed later in this book, it does not obviate the need for thinking about the meaning of the results.

If one chooses to use direct data entry, sensory staff need to do some investigation because there are several options, some of which offer much more than just capturing responses, including test design and analyses. The major challenge for sensory staff is to decide if the specific features are best suited to your needs. Consider the widespread use of Latin Square designs, which are a feature listed in almost all systems but are not the best choice for sensory tests. Balanced block designs are more appropriate to ensure that all products are served equally often in each position and before and after all other products in the test. Because order effects occur in all tests, it is most appropriate to balance these effects across all products. Some systems also provide inflexible serving designs. In sensory research, there are instances in which a product may not be served in the specified order for any number of reasons. The software should be flexible and easily allow for these changes during fielding. Many of the software programs provide an inflexible, overly complicated, and error-prone system to deal with these real-world issues. In Chapter 4, we discuss this in more detail. In addition to designs, another feature is the use of default systems for the statistical analysis. Almost all sensory analytical tests involve scoring of one type or another, and analysis of variance is a core procedure. Although there are many analysis of variance models, not all are suitable for sensory data, and of particular concern is the inappropriate use of fixed effects models. When reviewing the additional features offered with electronic data collection and analysis systems, sensory professionals must be especially thorough in their review of these features to be sure they meet their requirements; otherwise, one will be investing in features that are not appropriate. This issue is explored in more detail in Chapters 4 and 6.

2.2.4.4 Satellite facility

The basic elements that are essential to a sensory facility within a company can also be achieved at an external location or through a research supplier. Such a situation is not atypical; one's internal resources are often physically limited or constrained, usually by an insufficient number of subjects available for testing. The use of satellite or external facilities represents a degree of freedom that is often overlooked in the expansion of resources. Such facilities can be located in areas where there are large numbers of potential subjects, such as in corporate offices, or off-site near high-density population centers, suburbs, or a shopping mall. These sites may be designed for specific uses, such as primarily for acceptance tests that require larger numbers of subjects (100 or more) per test. Lighting, ventilation, booths, and the other structural requirements remain valid for this facility, although there may be some modifications to meet the particular testing needs. The key is to provide an environment that is constant; that is, the lighting and ventilation are the same throughout the test.

A satellite or off-site facility may be nothing more than a large room with a series of tables with portable partitions to serve as booths. Such a facility will not be ideal; however, it may be suitable as long as lighting, temperature, and odors are controlled and preparation facilities are adequate. Each situation must be considered within the context of the overall plan for sensory testing services. Are the reductions in controls and the potential reduction in sensitivity offset by the need for information? Are the expected product differences of sufficient magnitude to be perceived in this test environment? Will management consider the satellite or off-site as a more permanent solution? The sensory professional must give consideration to all factors that will influence the current facility before deciding on the requirements of a satellite facility.

Another factor that is often overlooked is the extent to which a sensory testing facility is used for public information/public relations activities (e.g. tours). Sensory evaluation is one of the few technical activities in a company that most visitors and consumers can "understand." With this knowledge in mind, the facility, whether it is a single unit or has several satellites, should be designed accordingly. It may be helpful to discuss this with public relations and personnel representatives.

2.2.5 Test methods

In addition to an appropriate facility, sensory evaluation must have tools with which to work, including the methods used to evaluate the products (e.g. difference tests and acceptance tests). There are a substantial number of test methods (Amerine et al., 1965; ASTM, 1996; Sidel and Stone, 2006; Sidel et al., 1975; Stone and Sidel, 1995). In recent years, the focus has been on modifications to existing methods to improve sensitivity or adapting them to specific problems. Work by Schutz and others (Cardello and Schutz, 1996; Schutz and Cardello, 2001) on the development of enhancements to the acceptance/preference model shows substantial promise. This latter topic is discussed extensively in Chapter 7.

As shown in Table 2.4, methods were originally categorized as belonging to three groupings. This classification, just one of many that have been proposed, is used here for illustrative purposes and to make the reader aware of the different kinds of test methods that are available. In more recent years, we have chosen

Table 2.4 Categories of Tests and Examples of Methods Used in Sensory Evaluation

Category	Test Type
Discrimination	Paired comparison, duo–trio, triangle, etc.
Descriptive	Descriptive analysis (flavor profile, quantitative descriptive analysis, etc.)
Affective	Acceptance, preference, 9-point hedonic, LAM scale

to reduce the classification by one and talk about two categories—analytical and affective—and consider the discrimination and descriptive as two types of analytical methods. In subsequent chapters, each test type is described in detail. Some newer methods, such as napping, emotional descriptive analysis, holistic testing, and laddering, do not easily fit within the aforementioned categorization. They will be discussed in Chapters 6–8. The sensory professional must be familiar with all the methods in order to use them effectively, if at all. A thorough understanding of all methods will also reduce unwarranted reliance on a single method to solve all problems. Implicit in this categorization is the need to keep the methods separate; that is, each provides a specific kind of information and combining methods will only confuse the situation and is not recommended. There are different methods because they provide different kinds of information. We discuss this issue later in the book.

2.2.6 **Selection of subjects**

Besides test methods, the sensory professional needs people to serve as subjects. Historically, companies used employees as subjects and rarely recruited individuals from the local community. This practice changed as the value of sensory information was recognized and testing frequency and the demand for subjects increased. Companies realized that the need for sensory information could be met more quickly and without reliance on employees through the use of local residents. This approach enables testing to be completed sooner and with better budgetary control. Using local residents is becoming more common, but there are companies and test situations in which employees are more likely to be used. Smaller companies may use employees for cost and/or security reasons.

Regardless of the source, all subjects participating in sensory tests must be qualified to participate. One of the major impediments to progress for sensory capabilities is the use of subjects who are not qualified to participate in a test. Just as consumer insights have precise requirements for recruiting consumers for a test, including but not limited to criteria such as age range, gender, family size, household income, primary shopper, brand purchase, and frequency of use, so too there are precise requirements for subjects to participate in sensory tests. It is surprising how often this is overlooked, yet its impact on decision making is serious. Failure to use qualified subjects significantly increases the risk of a decision error, of which the most serious is Type 2 error, concluding there is no difference when, in fact, there is. It is discouraging how often one is asked whether it is necessary to use subjects qualified based on sensory skill. The response to the question about subject qualifications typically begins with the comment that the subjects are experienced, they have participated in tests for several years, and judgments are always obtained. This seems to be the basis for not bothering to qualify the subjects. Recently, results from some types of factor analysis of subjects are presented as evidence of the similarity of their responses. Unfortunately, examining responses from many of these tests reveals there were no repeated trials; i.e. each subject provided a single judgment. This latter point is discussed in the chapters on discrimination and descriptive testing. Suffice to note that sensory skill can and must be demonstrated directly without using the claim of qualification based on test experience. The other extreme is the subject trained to provide the same response every time a particular stimulus is presented. In other words, the subject is invariant. It is not known from where this claim derives, but it is inconsistent with knowledge of genetics, perception, and human behavior. Both views reflect a fundamental lack of understanding of human physiology and behavior, to wit, people are variable not only at the receptor level but also as a function of time of day, general attitude about the test, and the product will also be different.

More than three decades have passed since Pangborn (1980) published an important but largely overlooked review of the literature on the individual differences in sensory skills among people. Her research results were consistent with those reported by still earlier investigators that differences in sensitivity and in preferences were typical in any population of subjects. This raised several questions about the selection of subjects for sensory tests. Not surprisingly, it was also shown that preferences differed independent of the question of sensitivity. This latter topic is discussed in detail in Chapters 7 and 8. For sensory analytical testing—discrimination and descriptive—it is especially important that individual differences are taken into account. What is remarkable today is the lack of appreciation for the implications of not qualifying subjects and, by default, failing to include replication in the test design. It may happen out of a lack of awareness of the literature, assuming it can be accounted for by using large numbers of subjects, or basing selection on inappropriate criteria such as detection thresholds. The point is that if one expects to make an analytical decision about product differences, it is risky to base this on a single response from a person selected at random. After all, companies do not market products to a random population but, rather, to a specific population segment. Knowledge of who these people are is very important, whether it is their demographic profile, their attitudes about a brand, or some other aspect of the behavior. In addition, these people are generally more sensitive to differences among the products they regularly consume. This means that our choice of subjects should reflect the relevant profile as much as possible. In the following discussion, we explore this topic in detail, reflecting on the experience of many years of testing and decision making that has proven itself throughout the years.

Whether one is using employees or local residents, some basic principles are essential. When using employees, there are three requirements: minimizing use of the technical staff, not using project staff on tests of their products, and not providing financial payment for the testing beyond their annual compensation. Project staff tend to respond based on their knowledge of a project and not necessarily what they perceive. When using employees, a testing program should not limit selection to one part of the company. To attract volunteers and maintain their interest requires careful planning and a commitment on the part of one's company management. Employees should never be made to feel that they volunteer to participate as a requirement for employment or receive extra compensation. This latter approach typically creates problems when a subject is not meeting performance requirements and is excused for testing and then claims discrimination. This latter situation has occurred and likely will continue to occur as a result of inadequate planning at the outset. Subjects need to acknowledge that continued testing is based on meeting specified performance criteria regardless of what was thought at the outset. Individuals must be selected for a specific test based on their sensory skills rather than other criteria. There must be a formal program for subject selection so as to improve the general level of sensitivity, to match panel with problem, and to increase one's confidence in the conclusions derived from the results. Such programs are not created overnight and require regular monitoring to ensure that subjects are neither over- nor underused (easier stated than practiced). Subjects need regular contact with the sensory staff to ensure that their interest remains high. The following are guidelines on the subject selection and qualifying process:

1. Sensory skills vary from person to person and also within a person.
2. Most do not know what their ability is to smell, taste, or feel differences in a product. They need instructions in using their senses.
3. All need instructions on how to take a test.
4. Not all will qualify for all tests, nor should they be expected to.
5. They are rewarded for participation, not for correct scores.
6. Skills once acquired are forgotten if not used on a regular basis.
7. Skills can be overworked or fatigued.
8. Test performance can be influenced by numerous factors unrelated to the test or the product and include frequency of testing.
9. All information should be treated in a confidential manner.
10. Employees should not be paid to participate in a sensory test.
11. Test participation should always be on a volunteer basis.
12. Subject/product safety is of paramount importance and should precede all other considerations.
13. All subjects should be treated with respect.

These guidelines should be in writing and available to the sensory staff. Because there is a regular inflow and outflow of people (new employees and employee attrition), the value of written guidelines is obvious. Developing a qualified pool of subjects requires a special effort, regardless of past experiences or assumptions as to an individual's skill. In most companies in which testing has been in progress for many years, the problem can be especially difficult. From a manager's perspective, the need for more subjects may be viewed with disinterest or an inability to understand the need for any program to get more subjects. Ideally, an individual should not participate in more than two or three tests within 1 week, for not more than approximately 3 or 4 continuous weeks followed by 2 or more weeks of rest. Although there will be situations in which an individual could participate every day for 5 or 10 days for a special project, continued use on a daily basis can lead to problems in terms of both maintaining subject interest and the effect of the absences from that individual's primary job responsibility. These issues are discussed in Section 2.2.8; the current emphasis is on attracting employees to serve as subjects. For nonemployees (i.e. local residents), attracting people is best achieved using the Internet, local shopping guides, and by word of mouth. This topic is discussed later.

The initial goal of the sensory professional is to develop a plan for both managers and the sensory staff, describing the importance of recruiting and screening and how it will be achieved. Obviously, the documentation for management need not be detailed, but it should be sufficiently clear to be understood. The first task is a clear statement of the objective for seeking employee volunteers; for example, why all employees should be encouraged to volunteer, what benefits are accrued, and how this participation might disrupt their regular work activities. As many employees as possible should participate to minimize dependence on a few and to provide greater options in selection for tests. The larger the subject pool, the more tests can be done and the more rapid a response can be achieved. For example, a pool of 100 subjects will limit testing to approximately 300 tests per year, and it could be less because of work responsibilities, holidays, vacations, etc. Although it is possible to complete more tests with a pool of 100, there is a risk that their managers will resent the time away from their regular work or subjects will become test fatigued—that is, participating to such a degree that they become bored. With more subjects to draw from, such as 200, it will be possible to increase the number of tests, reduce the frequency of testing for an individual, as well as use less time from request to report or to field the same number of tests. The ability to respond quickly, to meet a deadline, and to provide rapid feedback will secure many benefits for a sensory program; hence the importance of having a large pool of qualified subjects.

Employee participation has some side benefits as it relates to the individual's regular work activities. This break is a welcome diversion from regular work activities, provided it is brief, and is often found to have a positive effect on motivation—contributing to company success. Although this benefit is more difficult to measure (compared with the increase in the total number of tests), it is a benefit and should be identified as such to managers.

All of these benefits and others that accrue to the company should be a part of any plan presented to managers. The goal is to have their approval such that one can approach individual area managers and, ultimately, potential subjects. Although manager participation (as subjects) is desirable, it is unrealistic to expect their regular participation. However, their willingness to permit solicitation of their staff is crucial to any permanent program.

The estimated time necessary to build a subject pool should also be taken into consideration. It usually takes at least 6 months to build and maintain a pool of approximately 200 subjects. This does not mean that it takes 6 months before one can start testing; rather, one can start testing when there are as few as 25 qualified subjects. If there already is a core group of approximately 25, it could take less time to convince former participants to return or that a new program will be worthwhile. The time needed to reach people through electronic mail or in face-to-face meetings will be minimal; however, the major time constraint is the screening/qualifying activities before a subject is considered qualified. In addition, all subjects who volunteer cannot be screened within the same time period, nor would it be reasonable or practical to attempt such an effort. Typically, one starts with approximately 20–25 individuals for analytical tests, discrimination and descriptive, so that there will be enough of them qualified to satisfy initial test requests. Once this group completes screening, a second group can begin the process. In this way, test requests can be satisfied without long delays, and the subjects have immediate opportunities to participate in a test, gain confidence in their abilities, and gain a sense of contributing to the program and the company's success. Too long a delay between the end of screening and the start of testing (e.g. 4 or more weeks) will likely have many subjects thinking they are no longer needed or they will forget the process. The concept is to structure activities so they can be managed to best advantage by the sensory professional and at the same time to not discourage the subjects.

Once individuals have been recruited, it is time to begin the actual screening process, keeping in mind that not everyone will meet the qualifying criteria primarily because they cannot demonstrate a minimum level of sensitivity to differences. It is interesting to note that approximately 30% of those who volunteer will not meet a minimum level of sensitivity and reliability, hence the need to start screening with more subjects than are needed. This observation has been shown regardless of where testing is done or who the subjects are in terms of age or gender. In general, females generally are more likely to qualify than males. Initially, the qualifying process will take time because there are no data to guide the process. Initial groups may be overscreened (i.e. they will participate in more tests than needed); however, once two or three groups have been screened, the process can be streamlined and the time requirements can be reduced to as little as 5 or 6 test hours. This is discussed in detail later in this section.

Once screening is in operation, it can be very useful to work with one's human resources department to include in a new employee's packet of information documents that describe the company's testing program. An e-mail contact indicating interest can also be included. In this way, there can be a flow of potential subjects available.

Previously, we mentioned that for some companies, sensory testing cannot be done using employees; there are too few employees, or work schedules or job performance criteria make it necessary to use local residents or partner with a testing agency. Recruiting local residents can be done using a field service or advertising in local newspapers, online listing services, shopping guides, or some combination of these. For more on this topic, the documents by ASTM (Anonymous, 1981a) and Resurreccion (1998) may be helpful, although these have yet to be updated to include social networking and web-based recruiting methods. In recent years, with the shifting of activities globally, most sensory facilities are located in areas where the number or availability of employees is limited. So now companies rely much more on local residents for their subjects. Regardless of the source, however, all volunteers must be qualified.

2.2.7 Subject screening procedures

Once individuals have indicated a willingness to be subjects, they go through a two-step process; the first step is to complete a product attitudinal survey, and second is to demonstrate they have the necessary sensory skills. The first, a survey of product likes and dislikes, as shown in Figure 2.6, includes some demographic information such as age groupings, general job classification, gender, and any food or skin allergies. The survey lists a wide variety of products (not necessarily by brand) of current or of future interest. Associated with each product is a 9-point hedonic scale (dislike extremely to like extremely) or the labeled affective magnitude scale as well as two additional categories—"never tried" and "won't try" (Peryam *et al.*, 1954; Schutz and Cardello, 2001).

The survey takes approximately 10 minutes to complete and return electronically. Completing it on paper will take longer, but it still needs to be returned as soon as is practical. The survey serves several purposes besides identifying potential subjects; it identifies individuals who are interested but travel for their work or have responsibilities that limit their availability, it identifies those with allergies specific to the company's products, and it identifies people whose responses are in the extreme. They either like or dislike every product at the extremes of the scale. Such people usually are poor discriminators. Another purpose is to identify those who wait for weeks before returning the document. Chances are very good that these are the same people who will not report for testing as scheduled or will in some way fail to follow instructions.

Once a sufficient number of surveys are returned (e.g. approximately 25–30), responses are collected, examined, and screening tests are scheduled relative to the expected product category test. For example, a planned test of orange juice will schedule those who are average or greater than average in consumption of this type of juice. The intent is to start the screening process with individuals who are representative of the typical consumer for that particular product category. This is especially important whether one is selecting subjects for a sensory analytical test or to measure product sensory acceptance.

Once this program is initiated, a database is established, keeping track of each person's frequency of participation and individual judgments by test method and product category. This record is kept current—that is, the most recent past testing history of approximately five or more tests. It enables the sensory staff to assess

SENSORY EVALUATION PRODUCT ATTITUDE SURVEY

To match your product preferences, usage, and sensory skills to the samples to be evaluated, please complete this questionnaire. All information will be maintained confidential.

PLEASE PRINT

Name _____ Department _____

Telephone Ext._____ Date _____

General Information

Female _____ Male _____

Under 34 yrs. 11 mos _____ 35 to 50 _____ Over 50 _____

Married _____ Single _____

Children 0 _____ 1 _____

2 _____ 3 _____

4 or more _____

1. Please indicate which, if any, of the following foods disagree with you. (allergy, discomfort, etc.)

Cheese (specify) _____ Poultry _____

Chocolate _____ Seafood _____

Eggs_____ Soy _____

Fruits (specify) _____ Spices (specify) _____

Meats (specify) _____ Vegetables (specify) _____

Milk _____

2. Please indicate if you are on a special diet.

Diabetic _____ Low Salt _____

High Calorie _____ No Special Diet _____

Low Calorie _____ Other (specify) _____

The following is a list of products of current, or perhaps of potential interest, arranged in categories. Each product has descriptive terms from *won't eat* or *never tried* to *like extremely* or *dislike extremely*. Using these descriptions as guidelines, please mark the box under each phrase that most closely describes your attitude about that particular food.

Categories	Won't Eat	Never Tried	Food Item	Like Extremely	Like Very Much	Like Moderately	Like Slightly	Neither Like nor Dislike	Dislike Slightly	Dislike Moderately	Dislike Very Much	Dislike Extremely
Baked Products & Desserts	☐☐☐	☐☐☐	Cakes / Cookies / Puddings	☐☐☐	☐☐☐	☐☐☐	☐☐☐	☐☐☐	☐☐☐	☐☐☐	☐☐☐	☐☐☐
Breakfast Foods	☐☐☐	☐☐☐	Pancakes / Toaster Pop-Ups / Donuts	☐☐☐	☐☐☐	☐☐☐	☐☐☐	☐☐☐	☐☐☐	☐☐☐	☐☐☐	☐☐☐
Beverages	☐☐☐	☐☐☐	Carbonated Soft Drinks / Coffee / Tea	☐☐☐	☐☐☐	☐☐☐	☐☐☐	☐☐☐	☐☐☐	☐☐☐	☐☐☐	☐☐☐
Juices	☐☐	☐☐	Citrus / Non-Citrus	☐☐	☐☐	☐☐	☐☐	☐☐	☐☐	☐☐	☐☐	☐☐
Canned Foods	☐☐☐	☐☐☐	Chili / Fruit / Spaghetti	☐☐☐	☐☐☐	☐☐☐	☐☐☐	☐☐☐	☐☐☐	☐☐☐	☐☐☐	☐☐☐

FIGURE 2.6

Example of the product attitude survey for use in screening prospective subjects. For data entry, code responses as follows: 1, dislike extremely; 9, like extremely; 10, never tried; 11, won't eat.

performance on an ongoing basis and address any concerns about performance and availability. Current software makes it easy to develop and maintain a file of results, to decide when retraining may be needed, to maintain regular contact with subjects, and to send reminder notices electronically.

In general, one should expect approximately 75–80% of the survey forms to be completed within a few days and screening tests scheduled within a week. This timing sequence is necessary because if it is not adhered to, some will forget that they submitted a form and comments to their associates will further lower interest in participation. This same effect occurs when using local residents. As already noted, one should plan to screen approximately 25–30 people at a time, assuming approximately 30% will fail, leaving approximately 20 available for testing. The strategy is to have a steady stream of people screened as well as others in screening such that within a week there will be as many as 35–40 available for testing. The screening process, as described here and used by the authors for many years, requires 4 or 5 hours. Usually, this time is divided equally into three 90-minute sessions on consecutive days. In many companies, having staff attend sessions of this duration can be prohibitive, so the sessions can be shorter and there will be more of them. In either situation, the goal is to be able to screen for potential subjects with minimal delay. For local residents, the scheduling is easier.

The second stage of screening is a very important step—the demonstration by the volunteers that they can actually discriminate differences in the products (or product category) that will be tested and are now qualified to participate in sensory tests.

Before discussing a recommended screening process, it is useful to briefly summarize the research behind the recommended screening and qualifying procedure. The early literature was instructive in identifying procedures that were and were not helpful for screening (Dawson *et al.*, 1963b; Giradot *et al.*, 1952; Mackey and Jones, 1954; Sawyer *et al.*, 1962). It was evident that many procedures had been studied; however, it was clear that selecting individuals for a test based on availability, knowledge about the products, years with a company, or a person's position in that company were not useful criteria. It was also clear that an individual's sensitivity to simple aqueous solutions of sweet, sour, salty, and bitter stimuli (i.e. thresholds) had no meaningful relationship to subsequent performance in the evaluation of typical and more complex stimuli such as foods and beverages. Individuals with the lowest thresholds were not predictive of performance in actual product evaluations. The correlation between threshold values and ability to differentiate among products was approximately 0.5. What is particularly interesting (and surprising) about this latter finding is the extent to which threshold tests continue to be used to this day as the basis for considering a subject qualified. One continues to read literature in which authors describe the basis for selection as threshold values. Unfortunately, much effort is expended on measuring thresholds to various stimuli for no practical purpose. The early research also suggested that predicting performance in a difference test based on results from a preliminary series of difference tests was encouraging, provided the products used were from the same category as those being tested.

Sequential analysis, a statistical approach, also was suggested as an approach to the selection of subjects. The procedure was described in a number of publications; however, the description in Amerine and Roessler (1983) is probably the easiest to follow. Another description of the method was presented by Bradley (1953). Basically, the method involves establishing statistical limits on a graph for accepting, rejecting, or continued testing of a subject. This method has been applied on a very limited basis possibly because there was no guidance relative to the level of difference or whether that difference could be changed as the testing progressed. However, the method was useful because it involved use of the discrimination model.

With this background in mind, it seemed clear that any screening process had to reflect certain basic parameters. First, all individuals interested in participating in analytical testing, discrimination and descriptive, had to demonstrate they could perceive differences in the products of interest at better than chance on a repeated trials basis. Second, learning to use one's senses required practice. Third, people had different levels of sensitivity and skill, and practice would not eliminate those differences. Fourth, screening had to identify individuals who were representative of the population of interest. Fifth, the process was planned to be a positive experience; individuals not qualifying would be assigned to the pool of subjects for preference testing.

If a person could not discriminate at better than chance among products that the person regularly consumed, he or she could not be expected to provide word descriptions or scores with any degree of sensitivity or reliability. The discrimination model is a special case of ranking, and individuals who could not rank products with any degree of consistency could not be relied on to score products with any degree of confidence. Their responses would be more variable and sensitivity decreased, leading to a greater likelihood of decision errors and especially Type 2 errors. Figure 2.7 is a graphical representation of the thinking on which the proposed qualifying procedure was developed.

This procedure acknowledges the wide range of sensory skills in any population and the differences in sensitivity based on amount of product consumed (with the high-frequency consumer being more sensitive than the low-frequency consumer), with the goal of identifying potential subjects who are representative of the discriminating segment of the population. All those identified would not be expected to be equally sensitive; rather, they would reflect that discriminating segment of the population—that segment that is of most interest to marketing and technology. Based on this approach, potential subjects were invited to participate in a series of discrimination tests starting with easy-to-detect differences and made increasingly more difficult as the testing progressed. This process included replication, and as the subjects learned to use their senses, they would be challenged as the differences were smaller. It was observed that after approximately 30 trials, one could identify those individuals who could discriminate differences at better than chance versus those whose skill was well below chance. Increasing the number of trials to 40 can be helpful depending on the nature of the product; that is, more trials were helpful

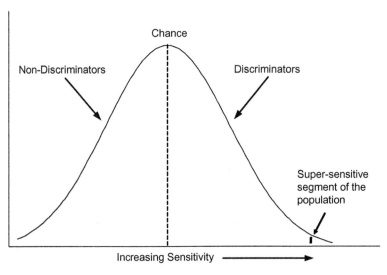

Chance

Non-Discriminators

Discriminators

Super-sensitive
segment of the
population

Increasing Sensitivity

FIGURE 2.7

Example of a normal distribution curve of the range of sensitivity in the population.

for the heterogeneous products versus 30 for the more homogeneous products such as liquids with no particulate matter. The decision regarding the number of screening tests is a responsibility of the sensory staff, taking into account the success to date for the volunteers and the extent of product variability.

This procedure for screening has several advantages. First, it is a progressive system; initial pairs are easy, enabling subjects to learn to use their senses and gain confidence, and as the testing progresses, the differences are smaller and more difficult to detect. It identifies subjects who cannot discriminate at better than chance or do not follow directions and require more time and attention. Although this approach does not guarantee that an individual, so qualified, will always detect differences, it does eliminate individuals who are insensitive, unreliable, and/or unable to follow directions. Empirically, we observed that decisions based on test results using this qualifying process have never resulted in a decision reversal based on subsequent testing or a marketplace decision.

Scoring has also been suggested for screening; however, no evidence has been published demonstrating its value other than providing subjects with experience using a scale. The process described here makes no *a priori* assumptions about an individual's level of skill and requires all individuals to participate regardless of prior test experience. As noted previously, subject interest is maintained by using a progressive system beginning with easy pairs. Determining which product comparisons are easy and which are difficult (to detect) is initially based on professional judgment. Sensory staff screen product pairs within the categories of interest and select those they consider to be easy versus those that are difficult; however, the true level of difficulty will be determined empirically. Once the first group has been screened, examination of the percentage of correct matches will make the ordering of sample pairs much easier for future screening. Initial selection of the product pairs is based on sensory staff judgment or a few informal tests. Selecting difficult pairs is easy—for example, two products from the same production line selected a few minutes apart. Note that the product pairs include all modalities relevant to the products (e.g. visual, aroma, texture, and handfeel). As already noted, the screening process is intended to be completed in a series of three or four 90-minute sessions when using local residents. In companies, most employees cannot be away from their regular work for long periods of time, and this means planning for more but shorter sessions. Approximately 90 minutes is the maximum one can expect before attention and testing fatigue reduce the effectiveness of the sessions. The sessions are fielded within a limited time period (i.e. consecutive days) so as to reinforce the acquired skill. If the screening is fielded with several days between sessions, subjects tend to forget much of what was learned in the previous session, thus prolonging the screening process.

Based on research performed during the past several decades and literally thousands of tests, certain observations serve as guiding principles. For example, approximately 30% of those who volunteer will not qualify; that is, they will not be able to achieve at least 51% correct matches across 30 trials. What is particularly interesting is that this observation extends across all ages from children ages 8–12 to adults ages 60 years or older and across countries and regions of the world. If one had a smaller pool of people available, one could increase the number of sessions and expect another 10–20% to qualify; however, one soon reaches a decision point to begin testing versus losing those already qualified.

On completion of the screening tests, individuals are categorized according to performance, with those having the greater number of correct scores available for analytical tests (discrimination and descriptive tests). The others could either participate in rescreening or be put in the pool of subjects available for acceptance tests. This latter task requires a different kind of judgment, one that most everyone is capable of providing. We discuss this latter issue in Chapter 7.

Performance from the screening tests is based on the percentage of correct matches across all trials and by test, taking into account the degree of difficulty that would carry more weight. One might have two individuals, both with 70% correct, but one missed the difficult pairs and the other missed the easier pairs. The latter person would be a better choice because he or she appeared to be learning how to take the test based on the easier pairs. In most instances, we use the duo–trio method (see Chapter 5) because it limits product exposure, minimizing sensory fatigue. Thus, any individual who achieves at least 51% correct decisions would be considered as qualified; however, the usual practice has been to select individuals who achieve at least 65% correct. The criterion of 65% is arbitrary; the sensory professional must decide what skill level is appropriate for that product category. In addition, one might select an individual based on a lower percentage correct, such as 55%, because there are an insufficient number who achieved the 65% criterion. The decision is made by the sensory staff, and only after several real tests will the effectiveness of that decision be determined. With the results from a few screening cycles available, the sensory staff can make adjustments leading to a more efficient and effective screening program. We further discuss this topic later in this section.

During screening, it may be necessary to use other methods, such as the paired test or a directional difference test. The choice is made by the sensory staff. It has the added benefit of reminding subjects to follow instructions. This approach also helps to establish whether all methods can be used with a specific company's products, as well as to identify appropriate products to use for screening future subjects.

When selecting product pairs for screening, it is often useful to solicit input from product technologists as to ingredient changes that might serve as variables to be tested. However, there is one caveat. It is recommended to select variables that are easy to prepare because the screening procedure should not require too much effort. After all, if one expects to use these variables at some future time and possibly at a different location, the preparation of the samples cannot be overly difficult. It is here that some common sense is applied, and once the samples have been tested, future selections will be much easier.

Another key point is the use of finished products rather than model systems—for example, aqueous solution of sweet, salt, sour, and bitter or a specialized "reference." Use of model systems can and will create problems in the form of the messages being communicated to subjects. Consider the following example: Subjects were screened based on sensitivity to sweet sour, salty, and bitter solutions. These subjects then participated in a descriptive test, and as a group the first words after tasting a few products were "sweet," "sour," "salty," and "bitter." It took the panel leader considerable time to refocus the subjects to the actual product perceptions of which sour had no role in the product, etc. A panel is more likely to repeat the sensory messages (and experience "phantom perceptions") from the model systems/references rather than what they actually perceived. This occurs because of the context of the various stimuli presented at the start of and during the language development process in descriptive analysis.

For companies that manufacture many different types of products, the selection of appropriate screening products can be challenging. However, there are no shortcuts, and one might find it most appropriate to use the product category for which there is a business priority. This problem cannot be entirely resolved because business priorities can change; however, one could develop groups of subjects screened for specific product categories, provided there is a need. Another option is to use a variety of products during screening. Once again, the decisions about how many people to screen and with what products depend on business priorities and the potential number of people available for testing. In practice, one often finds that subjects sensitive to one product category will do equally well on many other categories. If one is not sure, then the existing qualified subjects can participate in an abbreviated version of screening with the new product category. In this instance, sensory staff can select as few as 8–10 pairs (with a replicate) to verify that the subjects can discriminate differences in this new product category.

There are several key points with regard to the subject qualifying process: (1) familiarity with a product, (2) self-confidence, and (3) the actual testing (i.e. teaching subjects to follow instructions and using their senses to reach a decision). This also increases the sensory staff confidence when reporting results and making recommendations. Over time, it is possible to determine how well the subjects' sensory skills are developed and maintained.

Screening should always involve products of interest to that company. A series of test methods could be used to broaden the individual's experience and to assist sensory evaluation in identifying whether a specific procedure is appropriate for a product. Results should become a part of the individual's performance record and should be kept confidential. At the conclusion of screening, each person should receive a notice acknowledging participation and advising about future testing.

In recent years, there has been an interest in and discussion about the use of standardized materials for screening and training. By standardized, we mean a series of materials one could obtain that would represent, like a standardized test score, stimuli that an individual would have to detect and/or recognize in order to be considered a "qualified" subject. The assumption of identifying those individuals with improved performance using specially prepared samples certainly sounds appealing, almost "too good to be true." The reality is that it is too good to be true primarily because it lacks empirical evidence. Just as threshold testing provided no good indication of product evaluation performance, a similar outcome is likely for use of standardized sets of externally obtained "standards." Each product category has its own sensory properties, and subject screening must be specific to that company's portfolio of products. There is little to be gained and much to lose spending time screening subjects to be generalists or to identify specific aromas and flavors. In any screening process, the use of a stimulus that is not relevant to the company's products will be a distraction and is not recommended. Some subjects might consider these stimuli as having something to do with the products they will be testing. To assume otherwise is naive. Most people are not good at reading the minds of others, and having subjects experience a stimulus, the assumption is that these people will most likely associate that stimulus with their next product test.

As previously mentioned, initial screening may be time-consuming, but once a sufficient pool of subjects is available, screening new subjects can be reduced to once or twice a month depending, of course, on staff availability and the backlog of requests. This organized (and controlled) effort causes the least disruption of regular testing activities and provides for continued but controlled introduction of new subjects. The frequency of the process is dependent on the influx of volunteers; if there is a high rate of change (employee turnover), it will be necessary to determine whether the problem is peculiar to specific areas of the company and plan accordingly. This knowledge, in turn, should be taken into account when seeking new subjects. In practical terms, it should be possible to initiate and complete screening within 1 or 2 weeks and have a panel available for routine testing directly thereafter. If one is using nonemployees, the time requirements can be further reduced.

2.2.8 **Performance monitoring and motivation**

The two additional subject-related activities that warrant special attention are performance monitoring and motivation. Performance monitoring is an integral part of every testing program and requires an accessible and easily understood record system. Each time a subject participates, the date, the product, and the performance measure(s) are recorded. For each subject, there is a file listing performance by test type that can be accessed by name, product category, date, test type, etc. These records constitute the basis for subject selection for subsequent tests. The chapters on specific methods provide more details on the use of performance records. In any operation, record keeping is easily accomplished using any spreadsheet format. The key is for the information to be retained and monitored after each test. Figure 2.8 shows an abbreviated performance record from difference testing for some subjects. The goal is to monitor performance for any atypical patterns that require dialogue with a subject to determine if a "no testing" time is warranted. More about this can be found in the methods chapters. The system needs to be kept current, and performance records older than approximately 3 months should be archived.

Subject motivation is an issue of concern to all sensory professionals. The most common problem of an established subject pool is sustained motivation—that is, maintaining an individual's interest such that the activity is welcomed regardless of the number of times that person has participated. Unfortunately, no foolproof solutions to the problem exist, and an approach that is successful in one situation may prove unsuccessful in another. Because testing is done on a continuing basis, the staff needs to develop a variety of practices to maintain interest. Although the use of electronic noses and tongues is increasing (Bleibaum *et al.*, 2002), they are still in their infancy and will probably have greatest application in environments in which repetitive testing is required (e.g. quality control) rather than be used in place of individual tests.

As with other aspects of testing, there are some general guidelines on motivating subjects. The following guidelines have proved to be especially useful:

1. Subjects should be rewarded for participation, not for correct scores.
2. Subjects should not be given direct monetary rewards, unless they are nonemployees.
3. Subject participation should be acknowledged on a regular basis, directly and indirectly; food and/or beverage rewards (direct) should be available at the conclusion of each test and should be varied (be imaginative, do not serve the same treat every day).
4. Management should visibly recognize sensory evaluation as a contributor to the company's growth as an indirect source of motivation.
5. Subjects should be given "vacations" from testing—for example, for a week every 4 weeks. This is a sensory staff decision.
6. Sensory evaluation should hold an open house during selected holidays for all subjects, using the occasion to acknowledge participation and identifying some special accomplishments. Management participation in these activities is important. In the situation in which nonemployees are the subjects, no management presence is needed.

Discrimination Testing

NUMBER OF SUBJECTS = 30
NUMBER OF EVALUATIONS PER SUBJECT = 20
ACCEPTABLE % = 65
QUESTIONABLE % = 60

Subject	Test 1 R1	R2	Test 2 R1	R2	Test 3 R1	R2	Test 4 R1	R2	Test 5 R1	R2	Test 6 R1	R2	Test 7 R1	R2	Test 8 R1	R2	Test 9 R1	R2	Test 10 R1	R2	# Tests Correct	TOTAL CORRECT	PERCENT CORRECT
1 *																					5	14	70%
2 *																					7	15	75%
3 *																					4	13	65%
4																					2	10	50%
5 *																					4	13	65%
6 *																					5	15	75%
7 *																					3	13	65%
8 *																					9	19	95%
9																					2	11	55%
10 *																					8	18	90%
11 *																					7	17	85%
12 *																					7	15	75%
13 ?																					4	12	60%
14 ?																					3	12	60%
15 *																					6	15	75%
16 *																					7	16	80%
17																					2	10	50%
18 *																					6	16	80%
19 ?																					6	12	60%
20 *																					6	15	75%
21 ?																					6	12	60%
22 *																					5	14	70%
23 ?																					4	12	60%
24																					7	17	85%
25 *																					6	16	80%
26 *																					6	14	70%
27 *																					7	17	85%
28 *																					6	15	75%
29																					2	11	55%
30 *																					8	18	90%
TOTAL CORRECT	26	30	26	25	24	25	24	24	24	22	21	20	19	21	19	19	16	14	15	13			
N	30	30	30	30	30	30	30	30	30	30	30	30	30	30	30	30	30	30	30	30			
PERCENT CORRECT	87%	100%	87%	83%	80%	83%	80%	80%	80%	73%	70%	67%	63%	70%	63%	63%	53%	47%	50%	43%			
POOLED % CORRECT	93%		85%		82%		80%		77%		68%		67%		63%		50%		47%				

* Subjects chosen for panel; a 1=Correct response; 0=Incorrect response

FIGURE 2.8

Example of subject sensory acuity screening results.

7. For employees, memos acknowledging special assistance should be included in the subjects' personnel files.
8. The sensory evaluation staff should exhibit a positive attitude and should display a friendly but professional approach to its work.
9. The sensory professionals should never suggest that sensory testing involves "right" or "wrong" answers.
10. Subjects should be allowed to examine the summary sheet from a discrimination test if it is requested. However, this is not an activity that is encouraged; it should be used in those situations in which staff believe it can be helpful for a subject.

Sustaining subject motivation is especially difficult if the frequency of participation is high, once or twice each day or every day. Developing a sufficiently large pool minimizes reliance on the same group of subjects, boredom is minimized, and for employees, managers will be less concerned about work absence, and so forth. As previously mentioned, sensory staff must develop a balance between frequency of participation, which sharpens sensory skills, and motivation, which tends to decrease with frequency of testing, and the impact of time away from regular work. Of course, using nonemployees eliminates this latter issue.

In some instances, companies have compensated employees for their participation. Although the concept appears to have merit, from a practical viewpoint it leads to a multitude of problems, most of which become evident only after the compensation program is well underway. There is no evidence, to our knowledge, that financially compensated employees perform more efficiently, are more prompt, or express greater interest in their work as panel members. All subjects do not maintain the same skill level all the time; some improve, and some do poorer than chance and should not continue as subjects. Providing brief vacations from testing may still require payment, as will orientation and screening activities. Thus, considerable money is spent without having collected much useful information. Eliminating nonperformers may prove very costly, and a few subjects may create problems, even legal challenges. For example, some subjects may demand to see performance records, complain to the personnel department, and so on. It may be necessary to continue their payments even though they do not participate. Should this occur, it could create motivational problems for the other subjects. The other motivation problem for the sensory staff is the amount and content of the information given to the subjects directly after a test and over the long term. As noted in the guidelines, directly after some tests (e.g. discrimination), it may be useful for each subject to have an opportunity to compare his or her decisions with the sample serving sequence. Note that no one is advised "you were wrong on the first set." This latter situation is unacceptable; it is a value judgment, which will in turn decrease motivation. Subjects should be reminded that there are no "real" correct responses; all information is of equal value.

Over the long term, one can also provide each subject with a performance record, for example, comparing each subject's performance with change performance and assessing the individual's performance relative to that of the panel as a group. Sensory staff should develop specific procedures for displaying performance (see the procedure developed by Swartz and Furia, 1977), especially given the availability of graphics capabilities in today's software environment. ASTM International, Committee E-18, has a task group (E.18.03.09) on panelist feedback.

2.2.9 Requests and reports

When program development evolves to the stage where the various basic elements are in place, it is appropriate to focus on the issue of how tests are initiated and the results communicated to the requestor. There is an anecdote in sensory evaluation that describes the requesting process as follows: Returning from lunch, you find 2 quarts of juice (or whatever product) on your desk with a note that requests that you "do a quick 200-consumer test for my meeting tomorrow morning, and let me know if sample A really is better" or some similar request. This continues to occur albeit less frequently, and it is especially unfortunate that some sensory professionals consider this typical and proceed to do the test. Our attention here should be directed not to the issue of the test but, rather, to the nature of the requesting process. Sensory staff must develop a standard protocol, approved by management, that stipulates that no testing be initiated without receipt of a paper or electronic request and appropriate authorization. Although some individuals may find these requirements to be tedious or unnecessary, the issues are of sufficient importance that they cannot be ignored.

The objective is to obtain a record of the request to make certain that the test purpose is correctly stated, that products are properly identified by code and by location, and that a priority has been assigned. The assignment of priority is not a sensory responsibility; that is, sensory can state the backlog of work but cannot properly assign a priority to a request. The requestors and/or their management must decide on priority. This, of course, assumes that management has agreed on a procedure for assigning priorities.

The form should be relatively simple in its use and should provide space for sensory staff to add sufficient detail so that it can be used in actual test planning and implementation. As shown in Figure 2.9, those requesting a test provide information such as the objective(s), date of request, priority, product source and location (e.g. whether products are commercial or experimental), and who should receive copies of the report. Based on this information, sensory staff record receipt of the request, consider the nature of the request, and, if warranted, schedule a meeting with the requestor.

Generally, several meetings (several test requests) are required to educate requestors on the important issues of test objective and product history. An objective is not "to do a triangle test," "to do a 200-consumer acceptance test," etc. An objective must define the purpose for the test; in scientific terms, what is the hypothesis being tested? Based on the results, what action will be taken? The

```
┌─────────────────────────────────────────────────────┐
│                   REQUEST FORM                      │
│                                                     │
│   To:                          Date:                │
│                                Approval:            │
│   From:                        Priotiry:            │
│                                                     │
│   Test Objective:                                   │
│                                                     │
│   Product with Code:           Project No.:         │
│                                                     │
│   Product Description and History:                  │
│                                                     │
│   Product Location and Amount:                      │
│                                                     │
│   Preparation Procedures:                           │
│                                                     │
│   People to be Excluded:      Report Distribution   │
│                                                     │
│                   For SE Use Only                   │
│                                                     │
│   Test method:_____    Sample quantity:_____ │
│   Suggested test date:_____ Sample temperature_____ │
│   Design:_____       Carrier:_____    │
│   Number and type subject:___ Serving container:_____ │
│   Method of sample presentation:___ Lighting conditions:__ │
│   Number of replications:_____ Other:_____   │
│   Serving conditions:_____ Experimenter comments:___ │
└─────────────────────────────────────────────────────┘
```

FIGURE 2.9

Example of a form for use in requesting a sensory test.

requestor may explain the background leading to the request and in the process help the sensory professional better frame the purpose for the test. For example, the requestor may inform sensory that the company would like to replace a particular flavor (for any of a variety of reasons) used in certain products and minimize the likelihood that the difference will be detected.

In some instances, there may be more than one objective as a result of the brand manager asking questions about sales decline versus whether a difference can be detected. A product has been losing market share for the past 4 months, it's not delivering, improve the flavor ... and so on. By the time such directives reach sensory staff in the form of a request, the test purpose may be vague or contradictory. The sensory staff must recognize the potential risk of unclear requests, and through discussion with the requestor, it must ascertain whether more information about the test is needed and, if so, identify who initiated the request and how the results will be used. Ideally, sensory staff should be involved at the earliest possible time or notified about the situation before actual product work is initiated. This early involvement will facilitate the testing process and, where appropriate, sensory professionals may be able to recommend an alternative test plan as an aid to the requestor. For more on this topic, see Stone and Sidel (2007).

The dialogue between the requestor and the sensory professional is important for both. It minimizes the selection of an inappropriate test, eliminates an unwarranted test, and ensures that products will be available when needed and that priorities will be met. Sensory professionals will establish lines of communication with their customers, who in turn will benefit from an understanding of sensory testing activities and the kinds of information that will be available. Over time, requests will become more informative and discussions will be primarily limited to problems that have a more complex background; however, as we discuss in Chapter 4, the dialogue is never totally eliminated. One additional aspect of this dialogue, important to sensory evaluation, is the financial benefit derived from a test result. If, for example, a test result enables a cost reduction to be made, or minimizes additional testing and so forth, then these benefits must be credited to the sensory program. In this way, the sensory staff can demonstrate to management the benefits that accrue from an organized sensory program.

Reports represent a special opportunity for sensory evaluation, primarily because test results in the form of numbers and statistics may not be as informative and can be easily misunderstood, but tend to form an important basis on which a sensory program is judged. Sensory staff must develop ways of communicating results that make them understood and actionable. Although test results are only one basis for evaluation, they should not detract from the overall accomplishment. Reports are prepared by the appropriate sensory professional and not by the requestor or by some other individual. Allowing others to prepare reports will seriously undermine credibility and lead to other problems. Occasionally one encounters situations in which sensory evaluation provides everything, including a report, but offers no interpretation. This procedure would work only in special situations, such as during the preliminary stages of a research effort, and only if it has been agreed to by the sensory staff.

With regard to the report itself, the first issue is to not write a report as if it were a manuscript for a journal but, rather, to consider it as the recommendation based on the test objective; that is, was the objective satisfied and what is its meaning? What is the next course of action, if any? All numbers, including statistical significance, should be relegated to the back of a report or made available on request. As shown in Figure 2.10, the first section of a report should summarize results and the conclusion/recommendation in a few sentences.

It should identify the products, objective, test request and date, distribution, and author of the report. For results of a discrimination test, one can develop a similar report format. As shown in Figure 2.11, the format closely follows the request format, thus facilitating report preparation by the appropriate sensory staff member. Subsequent pages containing more detail can be added as needed; however, it is unrealistic to expect that a long and tedious report will be read, except by a very

```
┌─────────────────────────────────────────────────────────────┐
│                        REPORT FORM                          │
│   To                               Date:                     │
│                                                              │
│   From                             Project No:               │
│                                                              │
│                                                              │
│   Objective:                                                 │
│                                                              │
│                                                              │
│   Sample Description:                                        │
│                                                              │
│                                                              │
│   Conclusion and Recommendations:                            │
│                                                              │
│                                                              │
│   Results:                                                   │
│                                                              │
│                                                              │
│   Product Preparation:                                       │
│                                                              │
│                                                              │
│   Test Procedures:                                           │
│     Test method:_____      Serving conditions:_____  │
│     Design:_____         Sample quantity:_____   │
│     Number and type subject:____ Sample temperature:_____  │
│     Sample presentation:_____  Carrier:_____     │
│     Replication:_____       Other:_____     │
└─────────────────────────────────────────────────────────────┘
```

FIGURE 2.10

Example of a form for use in reporting test results.

few individuals. Although the technical staff may believe that they require all of the information, including computer printout (if a part of the analyses), sensory evaluation should not confuse this documentation with the purpose of the report (and particularly, the conclusion and recommendation).

A report that successfully summarizes the results in a few sentences is more likely to be read and appreciated. A long and involved discussion should be viewed with concern; there is a problem with the results and the author is trying to explain them. Whereas managers usually welcome brief, concise reports, some frustration may be experienced by the technical staff. Familiarity with the format of scientific journals and a sincere desire to have everything available for examination may prompt the scientist/technologist to request all of the study details. However, this concern for detail should gradually diminish with an increase in sensory evaluation credibility. With today's software and graphics capabilities, it should be easy to create documents that are easy to read and easy to understand.

A standard format should be developed to facilitate report preparation as well as ease of reading. The standard format should be modified for reports that cover a series of tests, for example, for storage or for a research project. For example, each

DIFFERENCE TEST REPORT FORM

To:　　　　　　　　　　　　　　　　　Date:
From:

1. Product Description:

2. Objective:

3. Conclusion and Recommendation:

4. Results:
　　Date of Test _____

	Sample	Number of Judgments	Number of Correct Judgments	Significance

5. Test Procedure:
　　Number of subjects_____ Different sample:_____
　　Subject type_____ Carrier, if any:_____
　　Replication_____ Interval between sets_____
　　Sample temperature _____ Rinse water_____
　　Sample quantity_____ Mouth cleanser_____
　　Container_____ Lighting_____

FIGURE 2.11

Example of a form for reporting results of a difference test.

storage withdrawal should not require a full report, but possibly an interim document, and should incorporate elements of previously reported data where reasonable. Sensory staff should determine the type of report best suited to their particular business activities and, where necessary, obtain management approval for the format, and so forth. In some environments, PowerPoint or Keynote slide presentations are prepared and distributed electronically to everyone involved in a project, thus enabling individuals to examine the details at their leisure.

2.2.10 Operations manual

An operations manual should be available in every sensory department. The document describes all activities of the function, including development of program objectives, how tests are requested and reported, job descriptions for each staff member, examples of scorecards, and subject selection procedures. The manual often includes information on product preparation procedures and amounts required for different test types. Such documents are extremely important in that they serve as a basis for the orientation and training of new staff members, ensure that standardized procedures will not be inadvertently modified, minimize re-creation of procedures already developed, and aid in the efficient use of resources.

Preparation of a manual takes a reasonable amount of time and is modified as new information is developed. We have found that the actual preparation should be undertaken only after sensory staff members have at least 1 year of experience and have sufficient familiarity with all procedures that are likely to be encountered. The important point is the need to document the purpose and procedures of all sensory evaluation activities. In some environments, the document is available to all requestors via the internal electronic communications network.

2.2.11 Planning and research

Planning in sensory evaluation is important to prepare for and/or anticipate special needs, to identify specific gaps such as a declining subject pool, and to propose alternative procedures for overcoming them. Each year, the budget process requires attention so as to identify the needs for the forthcoming year; however, this attention is usually focused on financial rather than operational needs and should therefore be separate from the business planning effort.

The planning document should identify sensory evaluation's accomplishments to date, anticipated skills requirements, the basis for each need, and how these capabilities will enhance the department's activities in the coming years. For the supplier of ingredients (flavors and fragrances), information provided by a potential customer is often limited, as is the time allowed to submit a sample. The supplier's sensory resources may be unable to provide adequate assistance because it did not develop the necessary specialized capabilities. Such procedures are not developed on short notice but, rather, require planning. If one department is providing more than half of all test requests in the most recent 12-month period, then it is desirable to determine if this trend will continue. It will also be useful to meet with potential customers to describe services available to them and assess their interest in these services.

Developing and/or evaluating new methods or the modification of current methods are activities that also need attention. The literature on sensory evaluation appears in a variety of sources. Sensory staff should have funds available in its budget to monitor the literature, to identify new developments, and to provide an assessment of those that have possible application. Most relevant journals are now available electronically along with easy-to-use manuscript libraries. The Internet, with its rapid search capabilities, provides an excellent way of keeping current.

Planning documents should be concise, preferably not more than 4 pages. It may be useful to characterize the previous year's accomplishment—for example, the number of tests by month and by requesting group, total number of responses, and any other pertinent information. Coupled with typical budget requests, the planning document constitutes a business statement about sensory evaluation and its contributions to the company's growth in the marketplace.

2.2.12 Strategic use of research vendors

As mentioned in the beginning of this chapter, organizations are often not wholly self-contained and find a need for outside vendors. They make it possible to hire, on a part-time basis, experts or additional capabilities otherwise unavailable within the company.

2.2.12.1 Advantages of using outside resources

There are some sensory and marketing research agencies with exceptional expertise and success patterns in new product development programs as well as working with flagship products to protect or expand existing market share. The use of such resources can help accomplish a number of objectives, such as the following:

- The sensory activity may be new to the company, and the vendor provides for organizational and method choices.
- Provide a proprietary or unfamiliar technique, which could help research a development issue.
- Act as an extension of the company's existing sensory and marketing research departments.
- Provide a creative resource on an ongoing basis, without disturbing normal operations.
- Act to get more projects moving faster to meet urgent deadlines.
- Maintain a historical database of research used to educate employees (new and existing) on a regular basis (sensory, R&D, and marketing).
- Establish benchmarks for company products versus competition.
- Internal personnel may not have the sensory skills or practical experience. Few companies have specialists in all areas, such as psychology, statistics, focus group moderating, descriptive analysis, consumer testing, and respondent recruiting and scheduling.
- It may be less expensive to use an outside vendor. Professionals who specialize in certain research areas have typically encountered similar problems with other clients. They are therefore more efficient in dealing with complex issues.

Often, outside suppliers have special facilities or competencies such as an established consumer population, a national or regional recruiting and interviewing capability, focus group rooms or suites with one-way mirrors, test kitchens, and special appliances and equipment that would be too costly to duplicate.

Sensory science and consumer research are used increasingly in litigation or in proceedings before regulatory or legislative bodies. The credibility of the findings will generally be enhanced if a respected external supplier conducts the study. In legal issues, this research is subjected to critical questioning and cross-examination and is likely to stand up only if designed to the highest standards, which may exceed those used within the organization for routine decision-making purposes.

To develop a relationship or partnership with a vendor, there are several considerations to get the most of the experience. There are no absolutes, but the following are guidelines:

- Provide as much information as possible concerning the product from R&D and market research.
- Be very specific about objectives of the project, and make sure all components of the company are in agreement on these objectives.

- Do not change objectives during the project without a very good reason, and inform the vendor if you do so because it will impact the research design and subsequent results.
- Ensure that the product, which is supplied to the vendor, is representative and does not vary from one phase of the evaluation to another.
- Be open-minded in trying new methods or abandoning old ones; otherwise, you may not obtain the growth opportunity the vendor may provide.
- Make sure the vendor has the opportunity to interact with the company components at the appropriate times in the project.
- Require a written and/or oral report.

These guidelines will apply to a greater or lesser extent depending on the nature of the project. A simple acceptance test may require only a short written report of results, whereas complex optimization studies require several written and oral reports to different business units of the company. Remember, desired results for a project depend on the interaction of the vendor and the company. It is rare that either one can be held entirely responsible for the results of a project activity.

2.3 Conclusions

In this chapter, we identified and characterized the individual elements that make up a sensory evaluation function in a company. Numerous activities were identified, including management-approved goals and objectives, professional staff, program strategy, facilities, test methods, identification, screening and selection of subjects, subject performance monitoring, test requests and reports, data processing, operations manual, and planning and research activities. Recognition of these specific activities is an integral part of a fully capable sensory evaluation function. This approach assumes a reasonable level of management skill in addition to familiarity with sensory evaluation principles and practices on the part of the sensory staff. It can be argued that this exercise is unnecessary, as well as demanding of the sensory staff; nonetheless, without this background and the individual elements in place, the full potential of sensory evaluation cannot be realized.

Developing sensory resources as described herein can take as much as 2 or more years, based on a plan that is both realistic in terms of staff skills and consistent with the long-range needs of the individual company. This should not be construed to mean that no actionable sensory information will be available for this time period (actually, information can be available within 1 month) but, rather, it takes that long for a program to establish its credibility, to develop a record of successes, and for the sensory professional(s) to know how best to use resources. A first step in the process, after identifying the need for sensory resources, is to establish a dialogue with management and with potential test requestors that identifies how the sensory resources will be a benefit to them. Matching this expressed need with available resources enables sensory evaluation to develop a realistic plan for the company. This approach is consistent with a management-by-objective philosophy; however, there is no evidence that other management approaches would not work equally well. The key is to recognize the need for a formal, business approach to all aspects of a sensory program.

Throughout this discussion, considerable emphasis has been placed on a need for a formal approach to the development of sensory resources, to develop a pool of qualified subjects, use procedures that are scientifically sound, document requests and reports, maintain subject performance records, and establish direct lines of communications with technologists and brand managers. Sensory research provides a unique information source that has significant value in the marketplace. Ultimately, however, the success of a sensory program will depend on the individuals involved, the sensory professionals, and their ability to make meaningful contributions to the decision-making process.

Measurement

3.1 Introduction

The definition of sensory evaluation described in Chapter 1 emphasizes the importance of measurement for treating sensory evaluation as a scientific discipline. Measurement is critical to quantifying responses to stimuli for the purpose of utilizing descriptive and inferential statistics. Such statistics provide a rational basis for decisions about the products that are evaluated and the subjects who did the evaluations. The value of measurement and the requirement for valid scales of measurement are not, however, unique to sensory evaluation. Physics, with its impressive list of achievements, provides an excellent example of what can be accomplished through measurement.

H. Stone, R. Bleibaum, H. A. Thomas: Sensory Evaluation Practices, fourth edition.
DOI: http://dx.doi.org/10.1016/B978-0-12-382086-0.00003-0

Psychology, in its evolution from philosophy to laboratory science, devoted much attention to developing methods and scales for measuring behavior. Many different types of scales were developed, and each was accompanied with controversy regarding the appropriateness of the various scales and how best to measure behavior. This controversy continues to the present, and the interested reader will find the early publications by Boring (1950), Carterette and Friedman (1974), Eisler (1963a,b), Ekman and Sjöberg (1965), and Guilford (1954) most helpful in describing the issues and providing a detailed history of measurement. The publications by Laming (1986), Marks (1974), Nunnally (1978), and Lim (2011) bring this discussion to the current state, but by no means is this the end of the discussion. Guilford (1954), in his text *Psychometric Methods*, credited the early development of behavioral measurement to investigators involved with the two distinctly different research areas of mental testing (e.g. the intelligence quotient) and psychophysics (i.e. the mathematical expression describing the relationship between changes in the physical stimulus and perceived intensity).

Ekman and Sjöberg (1965) also discussed the development of scaling theory and methods as proceeding along these two distinct yet parallel courses of test theory and classic psychophysics. They categorized the two lines of development according to different research interests and the application of scaling to two entirely different kinds of studies—one to measure preference for a product and the other to study the psychophysics of perception. Representative examples of scaling theory related to preference measurement are found in the published works of Thurstone (1959), and examples of psychophysics and scaling theory are found in Stevens (1957, 1962). Although this literature is more than 40 years old, it provides an excellent perspective to current practices. Unfortunately, one still encounters scale use based on faulty or no assumptions and a lack of awareness of the earlier literature, other than a long history of use in a particular company without any evidence of the sensitivity or the reliability of the scale. Part of the problem is the ease with which subjects respond and equating this response as evidence that the scale has merit. The problem is complicated by the unfounded concern that if a different scale is used, it will be difficult to compare results, which itself is a common misconception. Still others search for the universal scale as if it represents some kind of special instrument.

Major developments in scaling from the Thurstonian school included assumptions and procedures related to comparative and categorical judgments, and from Stevens, ratio scaling methods such as magnitude estimation. It is well documented that the data obtained from category and ratio scaling experiments are different (Stevens and Galanter, 1957). However, as Marks (1974) reported, Stevens' belief that only ratio scaling could yield valid scales of sensation has been challenged (Anderson, 1970; Weiss, 1972).

Eisler (1963a,b) suggested that discrimination was the basis of category scale judgments, and because discrimination changed with the magnitude of the stimulus difference, category scale data could be expected to deviate from magnitude scale data. However, output from the two types of scales have been shown to be related, although the relationship is nonlinear. Marks' (1974) view of this relationship was not that one scale was derived from the other (one explanation) but, rather, that both category and ratio scales were valid and also were different types of scales. He postulated that for any given sensory attribute there are two basic underlying scales—a scale of magnitude and a scale of dissimilarity.

The position of the sensory professional regarding this controversy about category versus ratio scales should be pragmatic and eclectic (unless one is engaged in research). Appropriately constructed category scales will enable the sensory professional to determine whether a product is more or less liked or the magnitudes of differences for specific sensory attributes (e.g. color, aroma). To derive a mathematical expression describing the relationship between ingredient concentration and perceived intensity, a ratio scaling procedure such as magnitude estimation might be an appropriate choice. However, that mathematical expression has value only if it has a practical application for the sensory professional. Hence, the selection of a scale will be based on very practical considerations. Philosophical arguments and claims that certain scales are more or less linear compared to one another must be viewed in the context of research hypotheses and not embraced without fully understanding the basis on which such a conclusion was reached. Differences in subjects, instructions, measurement techniques, choice of stimulus, and objective will influence results. Finally, the reader is reminded that studies with simple stimulus systems often do not yield similar conclusions when the stimulus is a more complex product.

In this discussion about measurement, primary emphasis is placed on descriptions of various (but not all) types of scales, suggested applications, and methods for analysis.

3.2 Components of measurement: scales

Selection of a scale for use in a particular test is one of several tasks that need to be completed by the sensory professional before a test can be organized and fielded. Determining test objective, subject qualifications, and product characteristics will have an impact on and should precede method and scale selection. These issues were discussed in Chapter 2 and are considered in more detail in Chapter 4. For purposes of this discussion, it should be kept in mind that these issues are, for the most part, interdependent; that is, a decision on one task will influence the next decision. In the case of scale selection, knowledge of the test objective, who will be the subjects, and the type of information desired must precede choice of scale. Before examining the different types of scales available, it will be useful to review some practical issues. To derive the most value from a response scale, it should be the following:

Meaningful to subjects. The words used for questions and/or to scale the responses must be familiar, easily understood, and unambiguous to the subjects. The words must be readily related to the product and the task, and they must make sense to the subject in how they are applied in the test. Words that have specific and useful meaning to the requester and/or the sensory professional may be much less meaningful to subjects, especially if they are not qualified subjects—that is, typical consumers. Assuming that a consumer will understand terminology that is specific to project team members is, at best, risky. In some situations, it is essential to add an explanation to a particular question as an aid for that consumer. As we will discuss in Chapter 6 on descriptive analysis, word usage by consumers is complicated—using different words to represent the same sensation or the same word to represent different sensations. As a result, sensory professionals must consider providing a context for a word or set of words used on a scorecard.

Uncomplicated to use. Even where questions and words describing the task and response scale are understood and meaningful, the task and scale must be easy to use. If not, it will result in subject frustration, increased measurement error, and provide fewer differences among products. Although a particular scale may be better from a theoretical perspective, it may produce less useful results than a simpler scale that is easier to use. Other issues to consider include the practice of switching scale direction from one question to another, switching number of scale categories for similar scales, or changing scale magnitude. Unfortunately, consumers do not necessarily read every instruction, and switching scale structure and direction without clear delineation will cause problems (e.g. decreased sensitivity) that cannot be corrected after a test is completed. This will be discussed further later in the chapter.

Unbiased. It is critical that results not be an artifact of the scale that was used. Ideally, the scale is a "null" instrument that does not influence the test outcome. Where products are perceived as different, we want to know this; where they are not, we want to know this as well. Unbalanced scales easily bias results because they decrease the expected probability for responses in categories that are underrepresented. They introduce a bias for which no obvious advantage has been demonstrated. The most typical explanation is to learn more about negative responses, as if knowing this enables products to be better liked. Number and word biases have been well documented in the literature, but one continues to encounter their use, particularly in the measurement of quality; for example, "best quality, good quality, poor quality" or 1 is best quality and 5 is worst quality. These latter scales are unique to a company, having been developed years earlier without any research or awareness of the measurement literature, and have taken "a life of their own"— that is, they have been used for so long that no one is able to not use them. It makes it very difficult for the sensory professional to change these practices.

Relevant. This relates to scale validity; that is, the scale should measure that attribute, characteristic, attitude, etc. that it is intended to measure. For example, preference scales should measure preference, and quality scales should measure quality; it is unwise to infer one from the other. Where the subject or the requester does not see the relevance for a particular scale or task, test credibility is lessened for data collection and eventual presentation of results. Simply stated, if the scale is not relevant for the task or issue, do not bother using it. This particular problem extends well beyond the measurement process. It derives in part from a lack of understanding of the perceptual process and a belief that humans will act in certain ways contrary to our knowledge of behavior. The typical situation arises when a request is made to save time by having the subjects respond to questions about differences (the magnitude, the nature of the difference, whether it is good or bad quality, etc.). The subjects are at the test site so why not ask these questions! Unfortunately, the matter of subject qualifications is not appreciated nor is the halo effect of one response on the other, etc.

Sensitive to differences. Not all scales are equally sensitive for measuring differences. Scale length and number of scale categories are major variables that have an effect on scale sensitivity. For example, one continues to encounter disbelief that there are differences in scale sensitivity based solely on the number of categories available. This is purely a mathematical issue without any impact of words or numbers (numerical scales). In effect, a 3-point scale is less sensitive than a 5-point scale (by approximately 30%), and both are less sensitive than a 7-point or 9-point scale. This topic is discussed in more detail later in this chapter.

Provides for a variety of statistical analyses. Statistical analysis of responses is critical to determining whether results are due to chance or to the treatment variables. The more powerful the statistics that can be applied, and the more of them, the greater the opportunity for identifying significant events that have occurred. This does not mean that scales that use less powerful statistical analyses are of no value; they are less flexible and may be less sensitive, thereby making it more difficult to demonstrate their usefulness.

For those situations in which new scales or words for a scale are required, it will be prudent to first do a small pilot test to eliminate potential problems with their use.

Just as there are different types of test methods, there are also different types of scales that provide different kinds of information. For clarification purposes, the classification of scales proposed by Stevens (1951) is followed here. Although Coombs (1964) challenged the classification as being too restrictive, it remains a frequently used system that is relatively easy to follow. Note that Stevens presented the idea that a particular scale determines the permissible mathematical operations for the responses (from that scale). Whether this latter restriction about permissible mathematics must be followed precisely cannot be stated unequivocally, inasmuch as there are scales that are not easily classified or that yield responses consistent with results from scales in other categories. Nonetheless, this system is useful for discussion purposes. The different types of scales are distinguished on the basis of the ordering and distance properties inherent in measurement rules—that is, the property of numbers based on how they are assigned. Stevens postulated four categories of scales:

1. Nominal scales for use in classification or naming
2. Ordinal scales for use in ordering or ranking

3. Interval scales for use in measuring magnitudes, assuming equal distances between points on the scale
4. Ratio scales for use in measuring magnitudes, assuming equality of ratios between points

3.2.1 Nominal scales

In these scales, numbers are used to label, code, or otherwise classify items or responses. The only property assigned to these numbers is that of nonequality; that is, the responses or items placed in one class cannot be placed in another class. Letters or other symbols could be used in place of numbers without any loss of information or alteration of permissible mathematical manipulation.

In sensory evaluation, numbers are frequently used as labels and as classification categories; for example, the three-digit numerical codes are used to keep track of products while masking their true identity. It is important that the product identified by a specific code not be mislabeled or grouped with a different product. It is also important that the many individual servings of a specific product exhibit a reasonable level of consistency if the code is used to represent a group of servings from a single experimental treatment.

An example of a use for a nominal scale is shown in Figure 3.1. In this particular application, no actual product is involved; however, the results would identify the rooms in which air fresheners are used most frequently. This information will be useful for identifying appropriate types of fragrances and alternative positioning statements for further product research.

Nominal scales are also used to classify demographic data about respondents, such as age, gender, and income, as well as to classify product usage behavior. Another feature of nominal scales is the total independence of the order among the various categories. The order can be changed without altering the logic of the question or the treatment of the results.

Name _____ Code _____ Date _____

In which location(s) in your home do you most often use air fresheners? Please check as many as necessary.

❏ Bathroom	❏ Garage
❏ Kitchen	❏ Family room
❏ Bedroom	❏ Dining room
❏ Closet	❏ Living room
❏ Hall	

FIGURE 3.1

Example of a scorecard that uses a nominal scale to obtain information about product usage characteristics.

In general, subjects have little or no difficulty in responding to questions that use nominal scales (this assumes the questions are understood). This is an obvious advantage when a problem has a large number of alternatives—for example, developing a test protocol that best reflects the most frequently used preparation and consumption mode for the product. Or, if one wanted to obtain responses from a large number of respondents without taking a substantial amount of time, the use of a nominal scale might be quite appropriate.

Responses to open-ended questions such as "What did you like about this product?" are used in a post hoc manner to develop a nominal scale, in contrast to *a priori* scale development in which all response categories are present at the beginning of the study.

Once open-ended data have been collected, all responses are read and categories for responses are developed to reflect some minimal number of independently appearing response comments. A frequency count is then obtained to represent the number of times a particular comment was made.

Because the same words may have different meanings and different words may have the same meaning to the respondents and to the experimenter, there exists substantial opportunity for responses to be incorrectly assigned to a category. Because this would be a violation of the restriction for using nominal scales, the mathematical treatment and value of open-ended data is seriously questioned. It has been suggested that one or two individuals could categorize the data independently; however, this approach would neither alter the meanings of the words nor be helpful in situations in which disagreements between the two classifiers existed. Occasionally, the sensory professional may use an open-ended question as an aid in the construction of categories for use in subsequent testing. In this situation, the open-ended question may have application; however, it should not be used as a direct guide for product development. Open-ended questions should not be used as a substitute for focus group or descriptive panel information. Recently, software has been developed that will organize and categorize open-ended comments. Whether this will encourage researchers to use open-ended comments more frequently remains to be seen because the basic problems of deciding which words will be grouped and their interpretation remain to be resolved. However, the key question, previously mentioned, is whether there is any value to the exercise considering the other measurement techniques available.

Mathematics permissible for nominal scale data include frequency counts and distributions, modes (the category containing the most responses), chi-square (χ^2), and a coefficient of contingency. Of the permissible computations, χ^2 is probably the most helpful. It allows for a comparison of frequency distributions to determine whether they are different, comparison of frequencies for data that can be categorized in two or more ways to determine whether the actual responses are different from some expected values, and comparisons of two or more groups relative to a series of categories. Data derived from the scorecard shown in Figure 3.1 would be consistent with this category. For a detailed description of the various applications of χ^2 to nominal responses, the interested reader will find the discussion by McNemar (1969) helpful.

The coefficient of contingency may be viewed as a type of correlation or measure of association between different variables having nominal scale information and is derived from the χ^2 computation. The computation is possible if the same objects have been classified on two variables or attributes, each having two or more categories. For example, to determine whether there is dependence between two income groups and gender of the respondents, the formula is

$$C = \sqrt{\frac{\chi^2}{\chi^2 + N}}$$

and the contingency table would appear as follows:

	Income	
Gender	**Low**	**High**
Male	A	B
Female	C	D

C will have a value <1.0 and will depend on the number of categories involved in the computations. As noted by McNemar, for a two-by-two table, the maximum of C is $\sqrt{1/2}$ or 0.7071, and the closer the computed value to this maximum, the stronger the degree of association.

It is possible to convert nominal scale data by assigning ranks or percentages based on frequency. This conversion permits use of statistical analyses usually restricted to ordinal data and proportions (e.g. t test for proportions). In this case, it would be prudent to identify that scale conversion was done prior to using these inferential analyses.

Although considered "low-order" scales because of the limited permissible computations, nominal scales are a valuable resource to the sensory professional. They are easy to use by the subject, require limited test time, and with limited computations provide rapid results to the requester. Other than the potential for misclassifying responses, the other serious limitation is the ability of respondents to contribute differentially to the database. Some subjects respond to open-ended questions (e.g. "What did you like most about the product?") with many comments compared with other subjects who give terse replies or have difficulty answering. As Payne (1965) noted many years ago, open-ended questions have value in the early phases of research, but closed-ended questions are more informative, and thus serve more useful purposes, in subsequent stages of any testing.

3.2.2 Ordinal scales

Ordinal scales use either numbers or words organized from "high" to "low," "most" to "least," etc., with respect to some attribute of a product set. The categories in an ordinal scale are not interchangeable. No assumptions are made regarding the distance between categories or the magnitude of the attribute represented by a category. Other than direction, all that is assumed is that a category is either greater or less than another category. Ordinal scales are considered to be the first or most basic scale for measuring perceived intensities and as such have more in common with other magnitude scales than with nominal scales.

Ranking is one of the most commonly used types of ordinal scale. It is a relatively easy behavioral task, and a number of procedures have been developed for ranking products. The most direct procedure is to have respondents arrange or sort a set of products so that each succeeding product has more (or less) of an attribute; for example, rank products from most to least sweet or from most to least liked. This procedure works well for products that can be easily manipulated by hand, such as a series of fabrics or a series of bottled liquids. However, for products that are not in closed containers, and especially foods and beverages, the risk of spills may require some modification in the test procedure. For example, having subjects list the products by their codes rather than rearranging the products would be an acceptable step, as shown in Figure 3.2.

The paired-comparison test is a special use of the rank-order test, as are the directional discrimination (e.g. which sample is sweeter) and the paired-preference tests. Chapter 5 is devoted exclusively to the discrimination test and its applications in sensory evaluation. For purposes of this discussion, attention is directed to the binary form of the data derived from these two-product rank tests. In Guilford's (1954) discussion about paired tests, there are procedures for transforming binary data to interval data. In multidimensional scaling, Shepard (1966) described procedures for deriving "partially metric" data from multiple paired comparisons. Such approaches take advantage of the fact that there is a constant interval or distance between the first and second rank in a two-sample, forced choice situation. These latter computations are not typical of the use of ranking in a laboratory sensory test. Nonetheless, they offer opportunities where there is a large array of stimuli (e.g. products or concept statements), all stimuli are compared with one another, subjects are limited to forced choice responses, and results are described in terms of interval data. This methodology is more likely to be encountered in consumer research when there is interest in determining product preference under different purchase options, for example.

With simultaneous product presentation, ranking is considered a direct method; it does not depend on memory. The products serve as their own frame of reference before any response is obtained, and they require uncomplicated statistical assumptions and applications. Beyond ranking, the subject does not provide a number or score for each product or mark a word scale. Both of these tasks have some bias associated with them. However, there are limitations to the more widespread use of ranking in product evaluation, and only the paired comparison for directional difference or preference finds much use in sensory evaluation. Limitations include the following:

1. All products in a multiproduct ranking test must be considered before a judgment is made. This can easily result in sensory fatigue and interactions, a

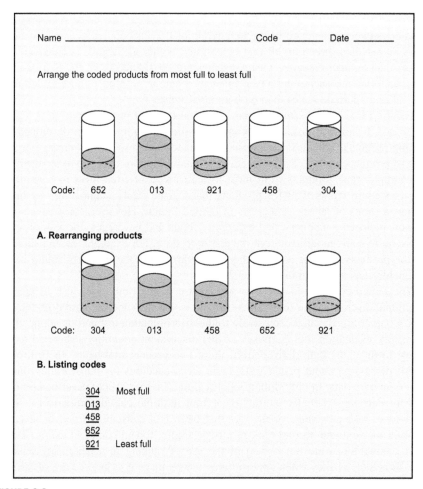

FIGURE 3.2

Examples of a direct ranking test in which the respondents can (A) rearrange the products or (B) list the codes. In the former procedure, the products are moved; in the latter, the subject records the order and no product movement is required.

problem that is most acute with products having a lingering flavor or odor or when a large number of products are to be evaluated. In paired-comparison tests, the number of pairs increases at a rate of n^2 for each additional product beyond the first two. Of course, in a visual test, sensory fatigue is not an issue.

2. Because all rank tests are directional, it is necessary to specify the characteristics and direction for the ranking. For example, ranking flavor intensity for a set of products, from most intense to least intense, assumes that all subjects are familiar with the specific flavor (the characteristic for which the judgments are provided). If the subjects are not trained or qualified to judge the specific

characteristic, there is no assurance that they actually perceived that characteristic in making their decisions. Although it may be easy for the sensory professional or a trained subject to perceive this characteristic, untrained subjects (the typical consumer meeting demographic criteria) may not understand the specified characteristic unless it has been demonstrated. This problem is not unique to ranking; it can occur whenever a descriptive characteristic is included in a scorecard used by untrained subjects.

3. The data provide no indication of the overall location (high or low) of products on the attribute rated and no measure of the magnitude of difference between products.

It is probably the latter limitation that has resulted in the infrequent use of ranking in sensory evaluation. However, it would be unrealistic to disregard ranking entirely—for example, when there is a large array of products and when time constraints make it unrealistic to use either paired-comparison or a scoring procedure, or if one were seeking a new fragrance option and there were as many as 50 submissions. Other than an individual arbitrarily eliminating submissions, the most reasonable approach would be a rank test, using an incomplete block design (see Chapter 4 for a discussion about test designs in sensory evaluation). Each subject evaluates a subset (e.g. 8 of 16 products) and ranks them based on the appropriate criterion. The product concept statement would be an appropriate criterion. In this way, a rank order can be achieved, and those products that equal or surpass a specified value would be subject to further evaluation. We have employed this specific approach quite successfully and hence our insistence that ranking should not be disregarded as a test method. It is most helpful for screening a large array of products to a smaller, more manageable product subset. In a "round-robin" procedure, in which an incomplete block has been used, only the highest ranked product(s) from each segment is selected for inclusion in follow-up testing. Informal ranking procedures are used for benchtop screening to reduce the number of product alternatives submitted for sensory tests.

Analysis of rank data can be accomplished by several different methods, including those appropriate for nominal scales and particularly those referred to as nonparametric methods. Methods that will be helpful include Wilcoxon signed ranks test, Mann–Whitney, Kruskal–Wallis, Friedman two-way analysis of variance, x^2, and Kendall's coefficient of concordance. A detailed description and worked examples of the various tests can be found in Daniel (1978), Hollander and Wolfe (1973), and O'Mahony (1986). Kramer (1960, 1963) also developed a set of tables for ease in determining whether there was a significant difference in the ranks for a set of products; however, some errors were identified in those tables, and the publications by Joanes (1985) and by Newell and MacFarlane (1987) provide more precise directions and analyses for ranked data.

An alternative to the limited information obtained from direct ranking is provided by the use of rating scales. These scales provide subjects with an unbroken continuum or with ordered categories along a continuum. As a group, they are perhaps among the most widely used and oldest scales of measurement in sensory evaluation. This durability is attributable primarily to the ease with which they can be formulated and administered, the large number of statistical tests that can be used to analyze results, and empirical evidence that they work.

Numerous examples of rating scales can be found in the literature as evidence of their diversity and application (Amerine *et al.*, 1965; Baten, 1946; Ellis, 1966; Hall, 1958; Lawless and Heymann, 2010). Scale categories varied from as few as 5 to as many as 12, although the majority of the scales had 8 or 10 response categories. Some scales had a word and/or number for every scale category, whereas others were anchored only at the extremes. The most difficult of issues appeared to be the number of categories and the specific words used to anchor them. Cloninger *et al.* (1976) used results from a series of rating scales and after applying various normalization and transformation techniques concluded that a 5-point scale was more suitable than scales with more categories.

Contrary to this conclusion, there is extensive literature on scaling and information theory that supports the 9-point rating scale as being more useful and optimal for information transmission (Bendig and Hughes, 1953; Cox, 1980; Garner, 1960; Garner and Hake, 1951). Two examples of ordinal scales are shown in Figure 3.3, both representing methods that use words, numbers, and/or categories for the measurement of intensity. The first example (A) represents a type of hybrid consisting of 5 word and 10 numerical categories. Obviously, more weight is given to some categories (associated with the word "strong" are 3 numerical categories) than to others ("none" has only 1 numerical category). The second example (B) represents a less complicated scale with no numbers and only 2 word anchors. The use of fewer words is intended to minimize bias. As mentioned previously, one can easily demonstrate an advantage in sensitivity for the scale with more categories. Taken to an extreme, one could envision greatest sensitivity to a scale with 100 or more categories; however, the reality is quite different (as reported by Bendig and Hughes and Garner and Hake). As the number of categories increases from 2 to approximately 10, sensitivity increases, reaching an optimum at approximately 9 or 10, and then decreases as the number of categories increases (beyond 10). This inverted "U" is explained by having either too few or too many categories; either leads to reduced sensitivity and lack of product differentiation.

Selection of the word anchors for rating scales often appears too arbitrary, providing opportunities and especially pitfalls. By opportunities, we refer to the use of words that are meaningful and unambiguous to the subjects relative to the specific scale. An example of ambiguous words would be a scale of overall reaction to a product having anchors of excellent to poor, good and bad quality, best tasting ever and worst tasting ever, etc. These are general quality terms, not personal preferences, and do connote different perceptual meanings to different people; opportunities for confusion in scoring (and a concomitant loss of sensitivity) increase dramatically when such words are used. Finally, measures of product quality might not be equivalent to specific product differences; that is, there could be perceived preference differences between two products, but the judgments on the "quality"

Name _____ Code _____ Date _____

Check one of the boxes that represents your opinion about the taste intensity of the product you are evaluating.

		Product	
Intensity of taste		487	924
		Taste	Taste
None	10		
Slight	9		
	8		
Moderate	7		
	6		
	5		
Strong	4		
	3		
Extreme	2		
	1		

A

Name _____ Code _____ Date _____

Check the box that represents the relative intensity for that characteristic you are evaluating.

Characteristic A

      Light   Dark

☐   ☐   ☐   ☐   ☐   ☐   ☐   ☐   ☐

Characteristic B

      Weak   Strong

☐   ☐   ☐   ☐   ☐   ☐   ☐   ☐   ☐

B

FIGURE 3.3

Two examples of ordinal-type rating scales that have been used in sensory evaluation. The first (A) represents a structured scale that contains both numerical and word categories, some of which have been weighted. The second (B) is a less complicated scale with no numerical values and only two word anchors.

scales may not be sufficient to yield a significant score difference. It is possible to have different preferences for products that are perceptually different but of equal quality.

The use of scoring is intended to determine the magnitudes of the differences between products. If products are being evaluated based on quality, it will be quite difficult to determine in what ways a product should be modified. In addition, it would be quite risky to have a sensory panel of 10–20 provide judgments of product quality, a task for which it may not be suitable (Sidel *et al.*, 1981, 1983). A more detailed discussion about the use of sensory evaluation for measuring quality is presented in Chapter 8; however, our interest here is in the use of word anchors that are least likely to be misinterpreted and to not use words that connote quality. As we will show in the discussion on descriptive analysis, the use of intensity measures, usually from low to high combined with word anchors that can be demonstrated to subjects (given examples of products that represent those sensory measures), is a very successful procedure in the sense that one can achieve optimal sensitivity and minimal variability without extensive effort devoted to training of subjects. This does not mean that there is only one ordinal scale or that scales that do not follow this pattern of development will not be useful. As with the use of any scale, it is a question of risk on the part of the sensory professional considering the problem, the products, and the extent to which the subjects are familiar with the use of the scale.

In addition to responsibility for selection and/or development of a specific scale, it should be kept in mind that ordinal scale data may exhibit interval properties. In fact, the same scale under different operations may exhibit more or less equality of intervals between scale points. The degree to which the distance between intervals is equal has some bearing on the risk involved with using various statistical techniques to analyze results. However, we agree with Guilford (1954), Nunnally (1978), and Labovitz (1970) that violation of the assumption of equality of intervals between points on these rating scales usually is sufficiently tolerable to have minimal effect on the use of parametric statistical analyses of these data. Later sensory evaluation literature (Land and Shepherd, 1988; McBride, 1983) provides further support for this conclusion. However, the reader is cautioned that such violations can be quite risky if the rules are stretched to excess. Although it may be difficult to specify what is "to excess," some degree of protection is afforded if a conservative approach is taken in scale construction and selection and data interpretation. There are operational (Anderson, 1970) and mathematical (Guilford, 1954) procedures for producing intervals sufficiently equal to be treated as equal interval data. For the sensory professional, little is to be gained and much is to be lost by following an unnecessarily restrictive policy that would classify all rating or category scales as ordinal scales, limit their analyses to nonparametric techniques, and sacrifice any internal quality that they contained. Although we recommend a more flexible point of view than that of O'Mahony (1982), this should not be interpreted as disregarding the order and interval requirements consistent with the use of parametric statistics. Rather, it is to allow the professional to take full advantage of the interval component of properly constructed and used rating scales.

According to Nunnally (1978), "when rating scales are used to obtain interval responses ... they are said to constitute the method of equal-appearing intervals." Furthermore, Guilford (1954) indicated that the task of sorting stimuli into equal-appearing intervals produces category values as interval scale values, which then can be treated statistically as such. It is this procedure with its theoretical and mathematical foundations attributable to Thurstone that has produced useful scales such as the 9-point hedonic scale (Jones *et al.*, 1955). Because of its widespread use, this particular scale is discussed in a separate section of this chapter.

Analysis of ordinal and rating scale data falls into two broad categories—parametric and nonparametric. The application of the latter was described previously. The parametric methods are applicable given adequate equality of intervals of the scale data and assuming the results are consistent with a normal distribution. For parametric data, there are numerous methods for analysis, including t test, analysis of variance, and correlation, as well as typical summary statistical measures such as mean and standard deviation. These tests and suggested references are described in more detail in the statistics discussion in Chapter 4.

No discussion on ordinal scales would be complete without some comments about the relative sensitivity of paired-comparison versus rating methods. The statement is often made that the paired-preference test is the most sensitive method for measuring consumer acceptance-preference attitudes. This belief may be supported in part by the psychophysical axiom that states that man is a better discriminator than a judge of the absolute. It is also believed that presenting the consumer with both products simultaneously makes the choice decision easier because the respondent has simultaneous access to both products. In sensory evaluation, however, few responses are absolute, even if they involve a single product, because product memory plays an important role when no other product is available. The ability to "go back and forth" between products, when served simultaneously, is certainly not an advantage for products that have strong aroma and flavor characteristics. This technique maximizes the potential for sensory fatigue and increases the likelihood of a loss in differentiation between products, which is the most probable outcome when product differences are relatively small.

As Seaton (1974) concluded, in a review of the comparative merits of the two procedures, rating and comparison methods were comparable; however, the former offered substantial additional information not possible with the latter, the comparative procedure. In particular, he was referring to the score for each product, which provides a measure of location on the scale, a measure of the magnitude of difference between the products, as well as the opportunity to convert the responses to ranks and proceed with an analysis of the comparative information as was used in the direct paired comparison. Also, obtaining scaled responses from products served monadically is more typical of consumer behavior—that is, evaluating one product at a time. These are most useful measures that are not directly obtainable with a comparative method. Although Laue *et al.* (1954) concluded that direct comparative methods were more sensitive to small differences when the dimensions of difference were known to the subjects, this is most unlikely in a consumer test. Considering the potential for sensory fatigue and sensory interaction and the limited output of information, we see no advantage or demonstrable evidence for the paired comparison, and we recommend the use of rating scales for measuring product acceptance-preference. However, there can be situations such as an advertising challenge in which the message is based on the direct comparison, in which case the paired method would be appropriate. Children older than a certain age (usually 7 or 8 years) also find it easy to use a scale for stating their reaction to a product. As discussed in Chapter 7, Kroll (1990) and Popper and Kroll (2003) found that children used rating scales as effectively as they did paired comparison. This has been our experience as well; however, there will be situations in which the paired-comparison procedure would be the method of choice—for example, with children whose cognitive skills were not sufficiently developed to understand the scaling concept. These issues are discussed in Chapter 7.

3.2.3 Interval scales

An interval scale is one in which the interval or distance between points on the scale is assumed to be equal and the scale has an arbitrary zero point, thereby making no claims about the "absolute" magnitude of the attribute measured. Interval scales may be constructed from paired-comparison, rank, or rating scale procedures or by the method of bisection, equal sense distances, and equal-appearing categories. For a description of each of these procedures, see Guilford (1954).

An example of an interval scale is the monthly calendar, in which each day constitutes an equal interval of time. A true or rational zero is not necessary for effective use of the calendar, and the interval between days is independent of whether that interval occurs early or late in that month. For example, the interval between the third and fifth day of the month is the same as that between the 13th and 15th day. Any *x* day interval is equivalent to any other *x* day interval.

In the foregoing discussion, we made note of the equal-appearing intervals with some ordinal scales and the need to be cautious about always assuming that any ordinal rating scale is an interval scale. Relatively few interval scales have been developed by directly setting out to formulate a scale with equal intervals. The two interval scales with which most sensory professionals should be familiar are the 9-point hedonic scale and the graphic rating scale. The hedonic scale is considered later in this chapter.

The graphic rating scale (sometimes referred to as a line scale) was developed from the work of Anderson (1970, 1974), utilizing a procedure described as functional measurement. In this procedure, the subjects are exposed to the stimuli they will measure in pretest sessions and are provided practice with stimulus end anchors—that is, as examples of scale extremes. These two steps when coupled

Characteristic

Weak   Strong

FIGURE 3.4

An example of a line scale–graphic rating scale. The subject places a vertical line across the horizontal line at that place that best reflects the intensity of that characteristic. Typically, the two anchors reflect a continuum from weak to strong intensity.

with a line scale result in response behavior that can be stated mathematically as equal interval. In descriptive analysis, use of a line scale has proven to be very effective (Stone and Sidel, 1998; Stone *et al.*, 1974), and their use in descriptive analysis is discussed more extensively in Chapter 6. Analyses of hundreds of tests using this type of scale have made clear the equal interval nature of the scale. With untrained subjects, Lawless (1989) and Lawless and Malone (1986a,b) found line scales to be at approximate parity with other standard scales used in sensory evaluation. Because best use of the scale requires subjects experience using the scale, we would expect it to be more sensitive when used by experienced subjects. An example of a line scale is shown in Figure 3.4. One distinct advantage of the line scale is the absence of any numerical value associated with the response plus the limited use of words to minimize word bias. Measuring the distance from the left end of the line to the vertical line yields a numerical value for computational purposes.

Interval scales are considered to be truly quantitative scales, and most statistical procedures can be used for their analysis; these include means, standard deviations, *t* tests, analysis of variance, multiple range tests, product–moment correlation, factor analysis, and regression. Numerical responses may also be converted to ranks, and standard rank order statistics may be applied to the data.

3.2.4 **Ratio scales**

Ratio scale data exhibit the same properties as interval scale data, and in addition, there is a constant ratio between points and an absolute zero. Stevens (1951, 1957) described four operational procedures for developing psychophysical scales having ratio properties: magnitude estimation, magnitude production, ratio estimation, and ratio production. Of the four, magnitude estimation is most frequently used for developing ratio scale data. This is primarily because of organizational issues— that is, the relative ease with which the experimenter can organize the test and the absence of an elaborate scorecard. In addition, minimal amounts of product are required compared with the methods of magnitude production and ratio production. In a magnitude estimation experiment, the respondent assigns a numerical value (neither less than zero nor a fraction) to each stimulus. This numerical value should represent the perceived intensity for that stimulus or more specified attributes (e.g. loudness, brightness, sweetness, and odor strength). When presenting subjects with a series of different stimulus concentrations, using any of the ratio-scaling procedures described previously together with a specific method for treating the obtained responses, researchers found that equal stimulus ratios produced equal response ratios. Stevens (1957) called this the "psychophysical law" and expressed it mathematically as

$$\psi = ks^n$$

where ψ is the geometric mean response to a stimulus, k is a constant, s is the concentration of the stimulus, and n is the exponent of the function, equivalent to the slope of the line. Engen (1971) and others refer to this equation as the power law or Stevens' power law. When data from ratio-scaling experiments are plotted on log–log coordinates, a linear relationship is obtained between stimulus concentration and perceived intensity. From this background, Stevens (1957) concluded that scales other than ratio were biased and should not be used for measuring prothetic continua. By prothetic continua, we are referring to stimuli that are additive, such as loudness and brightness. Metathetic continua are those stimuli that involve substitution or change, such as location between two stimuli. These concepts relate to scaling and scaling theory; a discussion on the subject is provided by Stevens and Galanter (1957).

Ratio scaling has had a significant impact on psychophysics, but there is considerable controversy concerning its role in measurement and especially in relation to its claimed superiority to other scales. The interested reader will find the discussions by Anderson (1970), Carterette and Friedman (1974; see the chapters by Anderson, Jones, and Stevens), Nunnally (1978), and Birnbaum (1982) especially helpful in characterizing the various issues associated with this controversy about scale superiority. The literature (Birnbaum, 1982; Land and Shepherd, 1988) of the past 20 years suggests no such superiority. For the sensory professional, the impact of ratio scaling and especially the use of magnitude estimation in sensory evaluation has been controversial and almost chaotic. Proponents of the method (Moskowitz, 1975) emphasized advantages in the use of magnitude estimation, including a means of circumventing the problem of word selection associated with most other scales, and provided direct numerical measures in response to products. It also was claimed that task instructions were so easy as to make it possible for any individual to participate as a subject without benefit of any detailed instructions, and that the elimination of numerical restrictions would further reduce number biases. Since the early 1970s, few publications on sensory evaluation have failed to make use or mention of the methodology. However, it was soon realized that subjects, especially consumers, required more instruction, that the procedures were no easier to learn than other methods, and that for some consumers it took more than the usual amount of time to learn the task. In addition, the use of numbers was not as anticipated. For example, some individuals will use a relatively narrow range of numbers (1–10, in equal increments), whereas others will use a very wide range (1–1000). In effect, each subject may settle into a specific pattern in the use of numbers, and the issue of bias has not been removed at all.

The issue of scale superiority, as noted previously, has never been demonstrated satisfactorily insofar as sensory evaluation of products is concerned. Studies involving different products have not shown any substantive advantage compared with standard sensory test methods (Giovanni and Pangborn, 1983; Lawless and Malone, 1986a,b; Moskowitz and Sidel, 1971; Vickers, 1983; Warren *et al.*, 1982). The fact that ratio scaling does not exhibit superiority is not surprising. It was developed in the context of inquiry into the measurement process, which in turn it was hoped would better explain the stimulus–response relationship. The fact that the shape of a psychophysical function relating responses to a series of stimuli could be stated mathematically opened new research vistas. However, it tells us nothing about the validity of that function (Shephard, 1981), and it tells us nothing unique about similarities and differences for complex products, which are of primary concern to the sensory professional. The fact that responses are numerical does not alter the situation because all other sensory rating methods also yield numerical values. Thus, it is not surprising that the use of ratio scaling has not proven to be superior. It is important for the sensory professional to be aware of the scaling option afforded by ratio scales; however, there should be no expectation that it will result in more or better results.

Traditional analysis of ratio-scaled data involved computation of an exponent, or slope of the line, representing the increase in sensory magnitude as a function of an increase in stimulus magnitude. Numerous examples of ratio-scaled data are available in the literature. The publications by Stone and Oliver (1969) and Engen (1971) are especially clear in describing the analysis of magnitude estimation ratio data.

Data analysis is usually accomplished by first normalizing the obtained responses to eliminate inter- and intrasubject variation. The most frequently used normalization procedure requires initial transformation of raw stimulus and response values to logs, where the mean of the logarithms for each subject on each sample is equivalent to a geometric mean. Log transformations compress the range of data values, and it could be inferred that extreme positive skewness is expected. To avoid this criticism, Powers *et al.* (1981) described normalization procedures that do not require the log transformation. In either case, the concept of normalization is consistent with the view held by Stevens (1951, 1957) that variability represents measurement error and as such should be eliminated from the analysis. Fortunately, not all researchers hold this view, and many choose to utilize variance measures in establishing confidence levels for the responses obtained. This is an extremely important point not to be overlooked. Products vary as do subjects, and our reason for using a panel of subjects and evaluating an array of products is intended to help quantify responses to variables of interest and to better understand and account for variability that may be inherent in the test or non-test variables.

Once the entire data matrix from a magnitude estimation experiment has been transformed and normalized, curve fitting commences. The method of least squares is used to determine the line of best fit. The resulting equation takes the form of

$$\text{Log response} = \text{log intercept} + (\text{slope} \times \text{log stimulus})$$

The slope is equivalent to the exponent as described previously where $\psi = ks^n$ and n is the exponent of the function. The data are then plotted on log–log coordinates, where the ordinate represents the log mean response and the abscissa the log of the stimulus.

Analysis of variance (ANOVA) and other statistical analyses as described for nominal, ordinal, and interval scales may be applied to ratio-scale data. However, responses obtained using a magnitude estimation procedure present practical problems when models such as ANOVA are used to determine statistical significance. Raw magnitude responses generally are positively skewed, which can translate into very large variance measures as mean intensity scores increase. In the example reported by Engen (1971, see p. 76), the highest concentration received raw scores ranging from 7.5 to 150, with a mean of 45.95 and a standard deviation (SD) of 38.66. The lowest concentration sample had raw scores ranging from 0.5 to 75, with a mean of 6.68 and an SD of 13.6. This type of situation is susceptible to significant violations of the homogeneity and distribution assumptions on which ANOVA models are based. In addition, the large standard deviations that are possible with the high mean values can result in large error terms, increasing the likelihood of Type 2 errors.

To eliminate the effects of inter- and intrasubject variability, the data can be normalized prior to using the ANOVA. This will result automatically in nonsignificant interactions between subjects and products, a measure that could be critical in making a decision regarding which product formulation warrants further attention. Thus, normalization prior to use of an ANOVA will weaken the analysis of the data; without it, however, some critical information could be lost. By weaken, we mean the aforementioned interactions will be lost, as will the information it tells us about the products, as perceived by the subjects. Of course, this dilemma can be avoided by using any other scoring method. Ratio scaling methods may be well suited to determining the form of a relationship between perceived and stimulus magnitude (which can be predicted); however, they are not well suited for measuring differences among products that vary according to different sensory modalities and attributes. Additional discussion about psychophysics and scaling in sensory evaluation is found in Land and Shepherd (1988), Frijters (1988), Lawless and Heymann (2010), and Lim (2011).

The labeled magnitude scale described by Green et al. (1993) is considered a hybrid, having characteristics of both a labeled category scale and a ratio scale. Labeled categories are not automatically spaced at equal intervals, a contrast to how one constructs a traditional category scale. Each category is labeled, and spacing for the individual categories is determined from previously collected ratio-scaling data. The end anchors use extreme statements such as "strongest imaginable" and "not at all detectable."

3.3 **Selected measurement techniques**

In this section, we describe some scales that have special appeal. This appeal may be based on ease of use, popularity, or simply habit, without an appreciation for why it is used, which can be risky.

3.3.1 **Hedonic scale**

Of all scales and tests methods, the 9-point hedonic scale occupies a unique niche in terms of its general applicability to the measurement of product acceptance/preference. The scale was developed and is described in detail by Jones *et al.* (1955) and by Peryam and Pilgrim (1957). As part of a larger effort to assess the acceptability of military foods, these investigators studied a number of different scales of varying length and number of categories as well as selection of most appropriate words used as the anchors for each category. This research yielded a scale with nine points or categories and nine statements. As shown in Figure 3.5, the hedonic scale is simple to describe, and as it turned out, it is equally easy to use. We believe that this latter feature is a major reason for its general usefulness in assessing product likes and dislikes for all types of foods, beverages, cosmetics, paper products, and so on and why it is used on a worldwide basis (when translated).

The scale was developed to assess acceptability of several hundred food items (Peryam *et al.*, 1960), and since then it has been reconfirmed by further studies of

Name _____ Code _____ Date _____

Please circle the term that best reflects your attitude about the product whose code matches the code on this scorecard.

Like extremely
Like very much
Like moderately
Like slightly
Neither like or dislike
Dislike slightly
Dislike moderately
Dislike very much
Dislike extremely

FIGURE 3.5

An example of the 9-point hedonic scale. The subject's task is to circle the term that best represents his or her attitude about the product. Boxes adjacent to the terms could also be used. The responses are converted to numerical values for computational purposes: like extremely, 9; dislike extremely, 1.

foods served to the military (Meiselman *et al.*, 1974). These investigations demonstrated the reliability and the validity of the scale to a degree that has been especially satisfying. Of particular value has been the stability of responses and the extent to which such data can be used as a sensory benchmark for any particular product category. A product may have a mean score and standard deviation of 6.47 ± 1.20; tests with an array of competitive products will typically yield an ordering of the products with mean values within this range that is quite stable— that is, independent of panel size and region of the country. There is no question that for some products, a subset of the population of consumers may alter the ordering; however, the usefulness of the benchmark is not lost. This degree of stability is especially important for companies that seek to develop a database for their own products as well as to have a means for rapid assessment of formulation changes and/or to track competition. In addition, knowing that a particular product category has an average score of 6.02 ± 1.50 provides a frame of reference as to what scores might be possible. This is especially useful if management has an expectation that the product should receive a score >7.5. Alternatively, marketing guidelines might require a particular product score (e.g. 7.0) for the project to proceed. In this latter instance, the sensory database could be used as a warning system that the action standard might not be met. Although the sensory acceptance test result is not intended for use in this kind of a situation, it is not surprising that it is often the only acceptance information available. The usefulness of the method with employees is discussed in more detail in Chapter 7. Parametric statistical analysis, such as ANOVA, of 9-point hedonic scale data can provide useful information about product differences, and data from this scale should not be assumed to violate the normality assumption (contrary to O'Mahony (1982) and Vie *et al.* (1991)). Figure 3.6 shows the results from an acceptance study in which 222 consumers evaluated 12 products using the 9-point hedonic scale. The sigmoid shape of the curve indicates that the scores are normally distributed. In many other tests involving thousands of consumers, the method has proven to be effective in ordering of preferences and the scale comes as close as one would like to being an equal interval scale.

Where there is interest in converting hedonic scale data to ranks or paired preference data, this too is readily achieved. The implied preference technique only requires counting the number of subjects scoring one product higher than the other and analyzing the result using a $p = 1/2$, or binomial distribution, as discussed in Chapter 7.

Periodic efforts to modify the scale by eliminating the midpoint (the category of "neither like nor dislike") or some other categories ("like moderately" and "dislike moderately" have been suggested) have generally proven to be unsuccessful or of no practical value, even where children are test subjects (Kroll, 1990). A frequent comment has been that there is avoidance of the midpoint; however, there has been no systematic study demonstrating the existence of such a bias or that it results in a loss of significance between products. Similar arguments about the moderate categories have been equally unsuccessful. Still another concern for some sensory professionals is the bipolarity of the scale from the experimenter's viewpoint and its

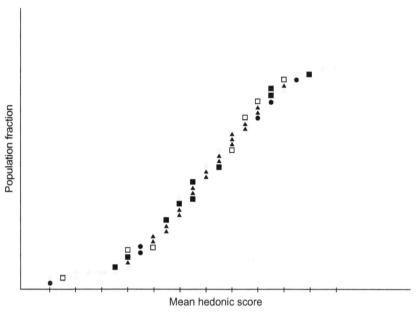

FIGURE 3.6

Results for 12 products, each evaluated by 222 consumers using a 9-point hedonic scale. Note that the y-axis is the consumer population fraction in cumulative percentage, and the x-axis is that portion of the scale from 2.0 to 7.5 in 0.5 units.

treatment mathematically as unidirectional. Whether consumers consider it bipolar cannot be easily determined nor should it necessarily be so. Empirically, consumers respond in ways that make clear they are using it in a way one can describe as equal interval. Although it is reasonable to expect subjects to experience difficulty with bipolar scales (generally more variability because of the avoidance of the neutral or midpoint of such a scale), there is no adequate evidence of this problem from our experience (after using the scale in thousands of tests with a wide range of populations in more than 25 countries). The computational issue seems to be equally trivial; the numbers used are of less importance than the significance or lack of significance in the difference in scores.

Efforts to demonstrate that magnitude estimation is a more useful scale for measuring product acceptance-preference have also proven to be less than successful. The earliest comparative study by Moskowitz and Sidel (1971) concluded that magnitude estimation was not a superior test method, as does the research by Pearce *et al.* (1986) and Pangborn *et al.* (1989). The authors of the latter study concluded that magnitude estimation may be inappropriate for scaling liking. McDaniel and Sawyer (1981) provided a contrasting conclusion; however, their study had design flaws that make it difficult to conclude much about the question.

In conclusion, it appears that the 9-point hedonic scale is a unique scale, providing results that are reliable and valid. Efforts to either directly replace or improve this scale have been unsuccessful, and it should continue to be used with confidence. In recent years, research by Schutz and Cardello (Cardello and Schutz, 1996; Schutz and Cardello, 2001) on extensions to the scale have proven to be very promising, and are discussed in detail later in this chapter.

3.3.2 Face scales

These scales were primarily intended for use with children and those with limited reading and/or comprehension skills. They can be described as a series of line drawings of facial expressions ordered in a sequence from a smile to a frown, as shown in Figure 3.7, or they may depict a popular cartoon character. The facial expression may be accompanied by a descriptive phrase and may have five, seven, or nine categories. For computational purposes, these facial expressions are converted to their numerical counterparts and treated statistically, as in any other rating scale. Little basic information on the origins and development of the face scale is available. A sensory test guide prepared by Ellis (1966) provided an example of a face scale similar to one shown in Figure 3.7. It was identified by the author as having been used successfully; however, no details were provided.

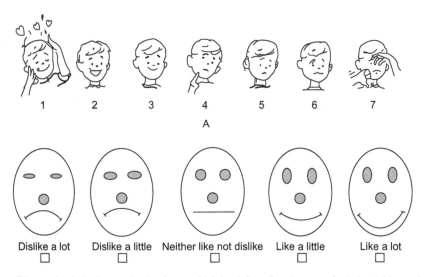

Please check the box under the figure which best describes how you feel about this product.

B

FIGURE 3.7

Two examples of face scales that can be found in the literature and appear to have been used for measuring children's responses to products.

The face scale is the type of scale that is frequently used and would be expected to have considerable merit; however, it will create more problems than it will solve. Very young children (6 years or younger) can be distracted by the pictures and can even be disturbed by the mean look of the frowning face. The scales may add undesirable and possibly complex visual and conceptual variables to the test situation. Matching a product to a face representing the respondent's attitude is a complex cognitive task for a child, and it may in fact be more complicated than some other more typical scaling procedures. For example, in a study of flavorings for use with children's medication, it was observed that the children tended to use the happy smile portion of the scale because they thought that they should feel better after taking the medication. This information was derived from post-test interviews necessitated by the lack of differentiation of the products and a desire by the investigators to reformulate the products. As a result, it was necessary not to reformulate the product but, rather, to develop a scale that was not subject to misinterpretation by the children.

There is no question that children's testing is challenging. The ability to read and to comprehend test instructions is not uniform among children of the same age. This does not necessarily mean that typical scales cannot be used; rather, it suggests that some changes may be necessary in the test protocol and especially with the oral instructions given at the time of the test. It is interesting that the face scale would be proposed for use with individuals having limited reading and comprehension skills when one of the basic requirements of its use is the ability to interpret reaction to a product represented by a face. There is no question that with instructions, some children can learn the task. However, this would defeat the claimed, primary advantage for the scale—the ease with which it can be used by the child. Finally, if one must train an individual to use a scale, it would be more reasonable to work with a scale that does not require transformations. In our experience working with children 8 years or older, we obtain reliable acceptance information using the 9-point hedonic scale provided all the children can read and, most important, can understand the meaning of the words. It should not be a surprise that many adults also do not understand the meaning of all of the statements. However, this is overcome through an appropriate orientation such that the children (and adults) develop an understanding of the scale's direction and what will be their task. Alternatively, if there are doubts about the use of individual experimenters or the subjects' ability to follow instructions, then it is recommended that some version of the paired preference model be used. As noted elsewhere (see Chapter 7), this is a very basic task and minimal reading or comprehension skills are required.

Although there exists considerable anecdotal information about the use of face scales, little research appears to have been done with them (or little has been published). In her study with children, Kroll (1990) found no advantage for face scales over other category scales. Until more positive evidence is available, we do not recommend use of face scales. Sensory professionals should give consideration to modifying test instructions and using more typical measurement techniques. Chapter 7 contains additional discussion about children as subjects.

3.3.3 **Labeled affective magnitude scale**

The labeled affective magnitude (LAM) scale is one of the newly developed scales that effectively measures liking for products or stimuli (Schutz and Cardello, 2001). It is a category scale that has ratio scale properties such that liking phrases are placed along a line in a ratio relationship. The development of the scale used magnitude estimation scaling in which extreme terms for the highest and lowest ratings provided labels beyond the current standard 9-point hedonic scale labels. Reliability and validity of the scale were tested, and results indicated that the scale was reliable, discriminated well or better than other scales among highly liked foods, was highly correlated with the hedonic and magnitude scale, and was user friendly. As a ratio scale, the LAM scale meets the assumptions of statistical inference analysis and allows ratio-type statements concerning the results (e.g. one product can be said to be twice as liked if it rates at twice the scale units).

The LAM scale has precise locations for the scale points and their corresponding labels (Schutz and Cardello, 2003). Figure 3.8 shows the LAM scale as it would be presented to the consumer for evaluation. Note that there are no numerical anchors on this scale, and "it would not be prudent to place numerical anchors on the scale that do not correspond to the ratios of semantic meanings among the verbal labels" (Cardello and Schutz, 2003, p. 345). The transformation of the scale for analysis purposes can be accomplished and is often converted from zero for greatest imaginable dislike to 100 for greatest imaginable like. Analysis of this scale is similar to the 9-point hedonic scale using the parametric statistical analysis, such as ANOVA. Although it was previously noted that the benefit of benchmark or historical data is of upmost importance, this fact should not discourage the researcher from using the LAM scale for current or future testing. For example, if the company historical data are based on 9-point hedonic research, then the LAM scale results can be converted to represent expected means ranging from 1 to 9. In addition, as with the common practice of calculating the frequency of responses to the 9-point hedonic category labels, so too the LAM scale lends itself to this type of summary output (Table 3.1).

3.3.4 **Just-about-right scale**

The just-about-right (JAR) scale is one of the most frequently encountered in larger scale consumer testing. These bipolar scales, as shown in Figure 3.9, have three or five categories, usually anchored with statements of too much, too little, and just about right for each product attribute.

We do not recommend this type of scale for sensory evaluation tests. JAR scales are championed as a diagnostic tool for consumer tests but are an ineffective substitute for designed experiments (e.g. DOE) or good sensory descriptive data. Reliance on these scales is usually an indication of limited resources or limited knowledge about sensory descriptive methods or both. These scales combine (or, more correctly, confound) attribute intensity and preference in a single response, and they are highly susceptible to interpretive and/or semantic errors because the

Please taste as much of the food as needed to form an opinion. Mark a "/" that intersects the line on the range that best describes your like/dislike of the food you just tasted.

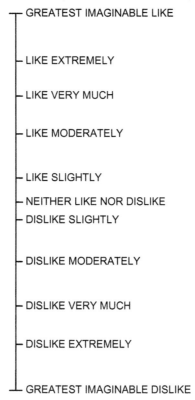

GREATEST IMAGINABLE LIKE

LIKE EXTREMELY

LIKE VERY MUCH

LIKE MODERATELY

LIKE SLIGHTLY

NEITHER LIKE NOR DISLIKE

DISLIKE SLIGHTLY

DISLIKE MODERATELY

DISLIKE VERY MUCH

DISLIKE EXTREMELY

GREATEST IMAGINABLE DISLIKE

FIGURE 3.8

Example of LAM scale scorecard for consumer acceptance testing.

product attribute to be measured is given a name. This particular risk is common to any scale that uses word anchors; however, consumers are especially vulnerable. Even if the consumer does not understand the particular descriptive word, a response is still obtained. As a result, there is a preponderance of judgments placed in the middle category of the scale. Enterprising investigators have proceeded from this experience to formulate these scales with five or even seven categories. The difficulty arises when trying to anchor each category: Leaving blanks typically results in consumers avoiding the unlabeled categories.

Analysis of data from these scales also presents numerous problems. Frequently, only the percentage responding in each category is reported without any rule for determining how much of a difference between percentages is to be considered significant. Although we do not advocate use of these scales, data from them may be treated as follows.

Table 3.1 Distribution Table Used for Calculating Frequencies for the LAM Scale[a]

	High End	Label	Low End	Point Range
Greatest imaginable like		100	94	7
Like extremely	93	87	83	11
Like very much	82	78	73	10
Like moderately	72	68	62	11
Like slightly	61	56	53	9
Neither like nor dislike	52	50	48	5
Dislike slightly	47	45	40	8
Dislike moderately	39	34	28	12
Dislike very much	27	22	17	11
Dislike extremely	16	12	6	11
Greatest imaginable dislike	5	0		6
				101

[a]There are 101 possible values that a consumer can choose. If the value is 94–100, then the count is included in the "greatest imaginable like" category. If the value is 83–93, then the count is included in the "like extremely" category.

Name _____ Code _____ Date _____

Make a mark in the box that represents your reaction to the product.

Aroma
- ❑ Too strong
- ❑ Just about right
- ❑ Too weak

Sweetness
- ❑ Much too strong
- ❑ Strong
- ❑ Just about right
- ❑ Weak
- ❑ Much too weak

FIGURE 3.9

Two examples of just-about-right scales. Both types of scales would not be placed on the same scorecard. They are presented here for illustrative purposes.

3.3.4.1 Option A

Establish an agreed-on "norm" or "minimum" response percentage for the "just about right" category (a reasonable number may be 65% just about right). When the minimum is achieved, ignore the other responses. If the minimum is not achieved, use a table for $p = 1/2$ to determine whether there is a significant difference between responses in the remaining two categories. Do not treat the "just about right" response as tie scores; the total number of observations is the sum of the two extreme categories only. Obviously, a large proportional difference will be necessary for significance when total N is small (as it should be).

3.3.4.2 Option B

First combine the number of responses in the extreme categories, and using a $p=1/2$, determine whether this number is significantly different from the number of responses in the "just about right" category. State whether the "just about right" responses are significantly larger (i.e. more) than the combined data from the extreme categories. Next, compare (again using $p=1/2$) the extreme categories with one another to determine whether a significant product defect exists (and, if so, its direction). Once again, do not consider the "just about right" responses as tie scores because this would be an unacceptable way to increase the N, thereby making it easier to obtain a significant product defect.

Other single-sample analyses may be attempted; however, they require assumptions about the estimated frequency of response in each category (e.g. χ^2). We have not found any acceptable criteria on which to base such estimated or expected frequencies.

The task of finding an appropriate analysis for determining statistical significance between related samples is even more difficult. In our opinion, the bipolar character of the JAR scale limits the value of standard two-sample analysis methods. Finally, analyses that compare the distributions between samples tell us little about which distribution is the more desirable.

3.3.4.3 Option C

This analysis is based on serial use of the Stuart–Maxwell and the McNemar tests as described by Fleiss (1981). The test is for matched products in which there are more than two scale categories. The Stuart–Maxwell test is used to determine whether there is a significant difference in the distribution of responses for the products. Where a significant difference is obtained, the data matrix is collapsed into a series of matrices and the McNemar test is used to determine individual scale categories for which differences are significant.

Table 3.2 contains the appropriate sorting of data from 100 consumers who evaluated two products using the three-category JAR scale. To construct this table, it was necessary to determine the number of too sweet, not sweet enough, and just about right responses assigned to product A when product B was scored too sweet, not sweet enough, and just about right. Obviously, this categorization of responses should be planned in advance to minimize repetitive handling of the original database. The Stuart–Maxwell statistic, as described by Fleiss (1981) for a three-category classification, is as follows:

$$\chi^2 = \frac{\bar{n}_{23}d_1^2 + \bar{n}_{13}d_2^2 + \bar{n}_{12}d_3^2}{2(\bar{n}_{12}\bar{n}_{13} + \bar{n}_{12}\bar{n}_{23} + \bar{n}_{13}\bar{n}_{23})}$$

Table 3.2 Hypothetical Data to Illustrate the Stuart–Maxwell Test[a]

| Product B | Product A | | | |
	Too Sweet	Not Sweet Enough	Just about Right	Total
Too sweet	20	10	20	50
Not sweet enough	7	20	3	30
Just about right	5	10	5	20
Total	32	40	28	100

[a]The entries are the responses to both products for each subject. For example, of the 32 subjects who indicated product A was too sweet, 20 indicated product B was too sweet, 7 said it was not sweet enough, and 5 said it was just about right. Of the 40 subjects who said product A was not sweet enough, 10 said B was too sweet, 20 said B was not sweet enough, and 10 said B was about right.

where

$$d_1 = (n_{1.} - n_{.1}), d_2 = (n_{2.} - n_{.2}) .... d_k = (n_{k.} - n_{.k})$$

$$\bar{n}_{ij} = \frac{n_{ij} + n_{ji}}{2}$$

and $n_{..}$ is the total number of matched pairs, $n_{1.}$ is the number of paired responses in row 1, and $n_{.1}$ is the number of paired responses in column 1. Applying this formula to the data in Table 3.2 yields the following value for χ^2:

$$\chi^2 = \frac{\dfrac{3+10}{2}(50-32)^2 + \dfrac{20+5}{2}(30-40)^2 + \dfrac{10+7}{2}(20-28)^2}{2\left(\dfrac{10+7}{2} \times \dfrac{20+5}{2} + \dfrac{10+7}{2} \times \dfrac{3+10}{2} + \dfrac{20+5}{2} \times \dfrac{3+10}{2}\right)}$$

$$= \frac{3900}{485.50} = 8.03$$

which with 2 degrees of freedom is significant at the 0.02 level. We conclude that the distribution of responses for product A is different from that for product B.

Because the distributions have been found to be different, it is necessary to determine those categories or combinations of categories that are significantly different. Fleiss (1981) correctly warns us that a control is needed to minimize the chances of incorrectly declaring a difference as significant when a number of tests are applied to the same data. The suggested control is to use the table value of χ^2 with $k - 1$ degrees of freedom (where k is the number of scale categories).

Using the McNemar test, where

$$\chi^2 = \frac{[(b-c) - 1]^2}{b+c}$$

Table 3.3 Two-by-Two Table for Comparing Ratings of "Too Sweet" for Products A and B[a]

Product B	Product A		
	Too Sweet	Other	Total
Too sweet	20	30	50
Other	12	38	50
Total	32	68	100

[a]Note that the entries were derived from Table 3.2 by combining categories.

with the data in Table 3.3, we obtain

$$\chi^2 = \frac{[(30 - 12) - 1]^2}{30 + 12} = \frac{17^2}{42} = 6.88$$

The critical value of χ^2 with two degrees of freedom is 5.99. Because the obtained value of McNemar's χ^2 is larger than the table value, we may conclude that product B has significantly more "too sweet" responses than does product A.

Due to the increased use of the JAR scale by both sensory and marketing research, the industry continues to develop and modify a variety of analysis approaches in an attempt to make the scale more useful for product development. The most frequently used strategy is penalty analysis (PA). The analysis boasts its ability to provide product improvement direction using descriptive JAR scales when the data are collected in conjunction with liking measurements. In other words, the analysis allows prioritization of the product characteristics that most heavily penalize the consumer liking result. The analysis procedure requires basic grouping of consumers based on their JAR scale responses and then recalculation of their liking data. First, consumer data are sorted and grouped into three categories: those consumers who indicated the product was too strong, those who indicated it was too weak, and those who considered the product just right. Using the three categories, the corresponding liking means are calculated, and then the "too strong" and "too weak" means are subtracted from the "just right" mean. The resulting means are considered the penalties in liking that resulted from the product being considered too strong or too weak by consumers.

The PA output is often taken a further step and presented graphically as a mean drop plot. Figure 3.10 illustrates a typical display format for the mean drop plot. Common practice has been to not display mean drops for those categories with less than 20% of consumers indicating the attribute was too weak or too strong. Weighted or net penalties are often calculated in an attempt to give more or less weight to the penalty, depending on the portion of consumers indicating the product was not just right. This calculation is the penalty score multiplied by the percentage or proportion of consumers who scored the product too weak or too strong.

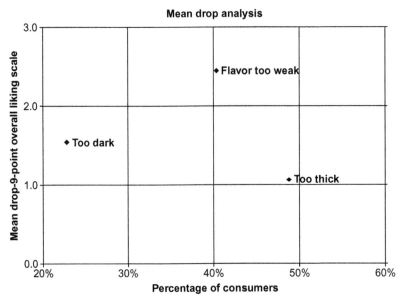

FIGURE 3.10

Illustration of mean drop analysis.

It is important to stress that although the number of approaches to the analysis of JAR scales continues to grow and greater emphasis is being placed on its ability to drive product improvement efforts for increased product liking, it is not and should not ever be considered a replacement for quality sensory descriptive information.

3.3.5 Other scales of interest

In addition to the previously discussed scaling techniques, another family of scales is of interest to the sensory professional. In particular, we refer to scales such as semantic differential, appropriateness measures, and Likert or summative scales. These scales are used primarily by market research to measure consumer behavior as it relates to product image, social issues, sentiments, beliefs, and attitudes. They impact sensory evaluation when results are used to direct product formulation efforts or when results are compared with those from a sensory test. However, when combined with appropriate sensory analysis data, the relationships between the two types of information have major benefits for a company. Sensory professionals should be familiar with them and their relationship to sensory testing activities. They are not a substitute for sensory data; however, they extend the information base and lead to more informed product business decisions.

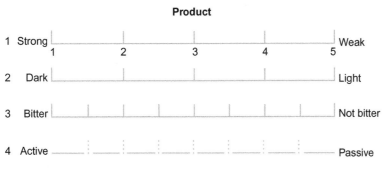

FIGURE 3.11

Examples of possible semantic differential scales. Note that the selection of word pairs and type of scale are the responsibility of the experimenter.

Semantic differential scales may be characterized as a series of bipolar scales with as many as 30 scales anchored at the extremes with word pairs that are antonyms. The scales are often referred to as summative because the scores can be summed across scales. Examples of some alternative formats for this scale are shown in Figure 3.11, and as can be seen, there are numerous variations. As Hughes (1974) observed, there are five basic issues that are addressed in the preparation of semantic differential scales: balanced or unbalanced categories, types of categories (numerical, graphical, and verbal), number of categories, forced choice or not, and selection of word pairs. In each instance, the experimenter has an option to establish which particular scale format will be used. Although this can be considered advantageous, it also assigns some risk to the experimenter. For example, if the word pairs are inappropriate or are misinterpreted by the subjects or the experimenter, this will introduce problems in interpretation of the results.

Likert scales are also used regularly by market research, and like semantic differential scales, they can take different forms. One such measurement is of agreement (or disagreement) with a particular statement, as shown in the examples in Figure 3.12. Importance ratings are also seen regularly in market research studies and are designed to assess the weight or value given to various product characteristics or benefits during a consumer's purchase decision.

Total unduplicated reach and frequency (TURF), a method of analysis for bipolar scales related to purchase intent and nominal scales that address expected purchase intent behavior of the consumer, is a common practice among marketing research professionals. Adapted from media research, the method's primary objectives are stated to include optimizing a line of products, whether they are flavor lines, colors, etc.; identifying the combinations that will have the most appeal; and identifying the incremental value of adding to a line of products (Cohen, 1993). Table 3.4 illustrates a typical TURF summary for a SKU line by flavor.

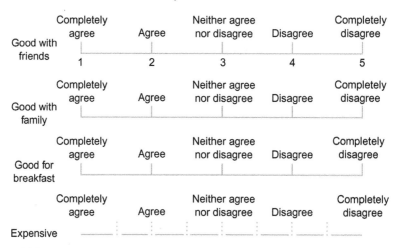

FIGURE 3.12

Example of Likert scale; the subject's task is to mark the scale at that point which reflects degree of agreement.

Table 3.4 TURF Analysis Using "Definitely/Probably Would Purchase"	Incremental Reach (%)	Cumulative Reach (%)
Flavor:		
Current in-market	73	73
Additions:		
Burst	15	87
Fizz	4	91
Ice	3	94
Original	1	95
Cosmic	1	96
Total reach		96

There is no question that these scales provide useful information about products, whether it is during concept development, when assessing the potential impact of advertising, or when linking the imagery and the sensory information. We discuss this further in Chapter 6. More discussion about these techniques can be found in Hughes (1974) and Nunnally (1978).

3.4 **Conclusion**

In this chapter, we described the four basic types of measurement techniques; nominal, ordinal, interval, and ratio. We identified the properties of each scale type and provided examples of how they are applied to the sensory evaluation of products. There is no question that these measurement techniques are a necessary part of sensory evaluation. It is useful, at this point, if we give some consideration to the question of whether one of these scales is best. Although having just one scale might appear to be most helpful, current evidence supports a position that is exactly the opposite. For any particular problem, more than one scale can be appropriate, just as more than one test method can be applicable.

It is reasonable to expect that behavioral research will result in the development of new measurement techniques and this will be of benefit to sensory evaluation. However, the sensory professional should not assume that what is new is necessarily better. In the past two and a half decades since the first edition of this text, there have been numerous developments in sensory evaluation. Much of this has been associated with the ease of data capture and real-time analyses and the application of methods or adaptations rather than with enhanced methods of measurement. It must be kept in mind that selection of a scale should always occur after a problem has been defined and the objective stated. In proceeding from nominal to ratio scales, the instructional set becomes more complicated and demands on the subject increase in terms of the type of responses that they provide. Thus, each scale offers advantages and disadvantages, and it is only after the problem is defined that the sensory professional can make a decision regarding which scale is best for that specific test.

Suggested methods for analysis according to the specific measurement technique were also listed. Additional detail about these procedures is provided in Chapter 4 and in the three chapters on sensory test methods (Chapters 5–7). Those chapters clarify the relationship between the test objective, experimental design and analysis, test method, and measurement technique in achieving reliable and valid product information.

Test Strategy and the Design of Experiments

H. Stone, R. Bleibaum, H. A. Thomas: Sensory Evaluation Practices, fourth edition.
DOI: http://dx.doi.org/10.1016/B978-0-12-382086-0.00004-2

4.1 Introduction

Every test, beginning with the request and ending with the recommendation(s), must reflect a strategy that addresses the key questions being asked—that is, the test's objective(s) and how the results will be used. Strategy is used here to mean a plan that enables the sensory professional to determine the basis for the request, how the test will be designed to answer the question(s), and how the results will be communicated. A test objective may have several parts and may not be entirely clear (i.e. looks can be deceiving); nonetheless, it is essential that the sensory professional establish the primary purpose or purposes for the request before any test strategy can be formulated. Once the objective has been established and there is agreement regarding the specific question(s) to be answered, the test strategy can be formulated. This includes selecting an appropriate test method, determining which qualified subjects are available, selecting an appropriate experimental design and method for analysis, determining when and where the products were prepared (i.e. pilot plant or production, etc.), and determining when and how they will arrive at the test site. Although these activities will require additional time before the test is implemented, it is critical that the objective(s) be clearly understood. A test with one objective is often the exception; in many situations, there will be several questions needing answers and the professional must engage the requestor in a dialogue to be sure that all relevant information has been obtained. Otherwise, the sensory professional may have proceeded with a particular strategy that yields less useful results. For example, the discrimination test provides information regarding whether there is a perceived difference between products, but it cannot be used to determine the basis for the difference. Only through a dialogue with the requestor can the sensory professional conclude that the discrimination model is appropriate. Further dialogue and bench screening of the products may make it clear that the product differences are easy to detect and a descriptive test is a more appropriate choice. In practice, requestors often want several kinds of product information, of which preference is usually somewhere in their discussion. Obviously, a single test cannot satisfy both the difference and the preference question, and some discussion is necessary to separate the issues and establish priorities. For example, establishing that preference information will not directly determine that products are different or the basis for a difference is a sensory responsibility; however, the sensory professional must make this point clear to the requestor before the test is initiated. These alternatives are discussed in more detail in the chapters on test methods (Chapters 5–7); however, our interest at this point is to emphasize their importance to the sensory professional in formulating an appropriate test strategy. From a sensory standpoint, the issue is not a simple matter of selecting a "right or wrong test" but, rather, a matter of selecting a test plan that takes into account all reasonable alternatives and, most important, the business objective. Test methods are not interchangeable nor should an attempt be made to answer different questions within the same test method. Not only is it wrong for that request but also it will have longer term implications, particularly with regard to sensory professional's reputation of independence and professionalism. Thus, test strategy involves integration of several elements into a cohesive plan that is most likely to satisfy the objective(s) and the requestor as well.

As a further complication, a strategy must take into account the impact of the product in relation to the psychological errors of testing and the product source as well. Regarding the former, we refer to testing errors such as first-sample effects, contrast and convergence, and the error of central tendency (to name the most common), all of which will influence the outcome of a test (Amerine *et al.*, 1965; Guilford, 1954; Lawless and Heymann, 2010). Product source also impacts strategy, particularly with regard to recommendations. Results of a test with products prepared at the bench will not likely have the same effects as those prepared at a pilot plant or at a full-scale production facility. Furthermore, knowledge as to when and how products were prepared must be known; otherwise, recommendations need to be qualified. Often, different results are obtained based solely on source of manufacture, and this information is not generally shared with all parties to a project. As a result, decisions are made that are not supported in subsequent tests, leading to credibility issues. Product source is discussed further in the methods chapters.

Although sensory professionals generally are aware of testing errors, it is not clear that there is as much appreciation for their effects in the design of a test as there should be, and this leads to confusion not only in test planning and design but also in interpreting results. To those not familiar with testing errors, the situation can be very confusing, if not troubling, in the sense that one can "manipulate" the serving order to achieve a particular result. For those unfamiliar with testing errors, this has the effect of clouding the scientific nature of the process. Nonetheless, it is important to share this information with a requestor when a particular array of products has the potential of causing such errors to occur. It is equally important to understand that not all testing errors occur in a test nor are biased responses easily identified as such. It is not easy to demonstrate that using qualified subjects minimizes Type 2 decision errors, that one- or two-digit codes are more susceptible to number biases than are three-digit codes, and so forth. The effects of these errors are not easily recognized nor do they occur all of the time and to the same extent, and they are not easy to demonstrate to those who question their impact. Even the more obvious use of red lights (in an attempt to mask color differences not intended as part of the test) continues to create confusion in decision making, but many still do not grasp the implications of this atypical testing environment on response behavior. The problems caused by a disregard for these errors occur much later when business decisions are found to result in no sales improvement or consumer complaints that products thought to not be perceived as different are, in fact, perceived and consumers express their unhappiness through declining sales and/or complaints to the company.

This chapter focuses on test strategy in relation to experimental design, taking into account the request for assistance (and the requestor), product criteria, and the impact of psychological errors on response behavior. Experimental design is also considered with a sensory orientation. The reader is assumed to have a working knowledge of statistics and to be seeking a guide to the design subtleties of a sensory test and thereby developing a greater appreciation for response behavior beyond their numerical representations. This latter issue is especially critical because, as in a detective story, this behavior is an important clue. By and large, subjects are telling us something about products and we must be able to identify that "something" without hiding behind a screen of numbers.

4.2 Test request and objective

In Chapter 2, test strategy was discussed in the context of the overall positioning of a sensory program within the research and development structure. We now focus on strategy as it is applied to the individual request for assistance. The overall goal is essentially unchanged—that is, to formulate and implement a test plan that will satisfy the stated objective. Specific issues that need to be considered before a plan is completed, and the test implemented, are in addition to the more obvious considerations of test priority, objective, and product identification and availability. In particular, appropriateness of the products relative to the objective and how the results will be used and also the possible effects of psychological errors on response behavior must be taken into account before the sensory professional can proceed with preparation of the actual experimental design, the scorecard, and so on.

Although test objective was discussed previously (see Chapter 2), its importance to formulating a strategy demands that we include it in this discussion. As a preliminary step, the sensory professional has to ensure that all the necessary information is provided and that the request is reasonable—reasonable in the sense that what is being requested is consistent with sensory goals and objectives and with corporate goals and objectives, and that it warrants the expenditure of effort. For example, measuring purchase intent or attempting to represent unique population segments are unreasonable design variables in a sensory test involving only 40–50 subjects. Although the issue of what is reasonable can be a source of disagreement between sensory and requestor, it must be addressed. Not only does it require direct discussion between the two but also it further increases the likelihood that the best possible plan will be formulated. In most situations, products will be examined to ensure they are appropriate. To the extent possible, such meetings should provide for a thorough exchange of information regarding the products and possible methodological approaches. Questions about an objective should also be thorough to avoid problems after a test is completed. Occasionally, a requestor will withhold information unknowingly or in the belief that this information is not relevant or will create experimenter bias. This is more likely to lead to an unsatisfactory test. For example, if the objective is to compare several competitive products with each other, the design strategy will be different from a test in which the objective is to compare products to a market leader but not to each other. Both tests will have some common design features, but there could be differences in the product serving order, the analyses, and any conclusions drawn from the results. Failure to obtain a clear understanding of a test's objective increases the likelihood of dissatisfaction after that test is finished, and over the long term it will erode the mutual trust between the sensory program and the requestor. Questionable results are remembered long after confused objectives are forgotten. Over time, requestors will become more sophisticated in the preparation of requests and discussions about objectives and consideration of reasonableness will be minimized, leading to a more efficient and productive evaluation process.

4.3 **Product criteria**

Test strategy should include the possibility of an initial rejection of a request if the products are inappropriate or the basis for the request cannot be justified. Receipt of a request should not automatically result in a test or in rejection of the request. Previous results or other product information may eliminate the need for a test. For example, product instability or variability may be greater than a contemplated ingredient change, and sensory evaluation's responsibility is to provide such an explanation with a rejection.

Rejecting products because they are not appropriate is a necessary part of test strategy and one that is likely to provoke considerable dialogue, at least initially. Little is gained from a discrimination test that yields 90–100% correct decisions or from a test involving one or more easily recognized products. A test of very different products typically results in biased response patterns for some of the products. Later in this chapter, we discuss contrast and convergence effects that can occur in such tests. Initially, requestors may consider this approach as a lack of cooperation or as an attempt to exert too much influence over their work. However, it is important that the sensory staff be very rigorous in considering product appropriateness. To an extent, the sensory professional is educating the requestor with regard to the many problems that can arise because of indiscriminate testing or testing products that are either not ready or a test is not warranted. It is not a question of a subject's ability to complete a scorecard but, rather, the difficulty in assessing response bias and the concomitant risk in the decision-making process. There are, of course, guidelines that can be used as an aid in reaching a decision about appropriateness; however, each product category will have its own unique characteristics, and the sensory professional must identify these to requestors to minimize conflict and enable them to be more selective in their product submissions. General guidelines include the following: (1) Products obviously different from each other should not be tested in a discrimination test; (2) easily recognized product (e.g. branding on the product) will significantly alter response patterns to it and other products in a test; and (3) comparing experimental products with competitive products, especially during the early stages of development, will yield atypical results.

Although assessing product appropriateness would appear to be time-consuming and potentially disruptive, its importance far outweighs any additional time or delay in testing. Obviously, not every request or every product will require such scrutiny. With experience, this increased vigilance is more than rewarded by a much greater level of awareness by all professionals involved with the sensory program. The reader should keep in mind that concern about product does not end with the initial review. In fact, vigilance is essential just prior to the actual test to be sure that the products are the same as what was observed when the request was first discussed. In some instances, the sensory staff are surprised when the products are prepared for the test and are found to be different from the original plan. This calls for immediate action—for example, contacting the requestor and determining whether the test will proceed as planned. In our experience, product is often the most critical element in a sensory test (besides subjects). The simple question of asking which product is control and where was it produced often yields a very interesting dialogue that helps to explain unexpected results; hence our emphasis on product. We now turn our attention to an associated issue, psychological errors in testing.

4.4 Psychological errors

One of the more intriguing aspects of a sensory test is the extent to which psychological errors influence individual responses. What Guilford (1954) has referred to as "peculiarities of human judgment, particularly as they affect the operation of certain methods of measurement" is the topic of interest. With an awareness of the peculiarities, we need not be puzzled by atypical response patterns, and in many instances we can organize and design a test to minimize the impact of these errors or at least to enable them to affect all products equally. For example, the first product in a multiproduct preference test typically is assigned a higher score compared with the score it will receive in subsequent serving positions. This phenomenon is referred to as a first-sample effect, which falls within the much broader classification of a time-order error. To minimize the impact of this error and ensure that it has an equal opportunity to affect all products, the products must be served equally often in the first position. In addition, data analysis must include provision for examination of response patterns by serving order. Of course, it also is true that some individuals will purposely request a single product or monadic test, knowing that it could result in a higher score. In effect, one is taking advantage of specific knowledge about human behavior. Single product tests are very popular when a company has established action standards for a product to move into a larger scale market test that includes, among other criteria, a specific preference score such as a 6.5 on a 9-point hedonic scale. By moving to a larger test, the appearance of progress has been realized and there is always the hope (mistaken) that product acceptance will improve!

A more comprehensive discussion about the various types of errors is provided in the psychological literature (see Guilford, 1954). Our intention is to discuss these errors in relation to their impact on the product evaluation process. In general, many of these errors are more noticeable when subjects are naive and unfamiliar with the test method and/or the particular product or product category. Obviously, the impact of these errors can be reduced if one works with experienced and trained subjects, following the recommended procedures and practices described in Chapter 2. However, there are psychological errors that are more directly related to the specific products being tested; for example, contrast and convergence, and other context effects, can be minimized but never totally eliminated through appropriate experimental design considerations. The key issues for the sensory professional are awareness of these errors and the importance of designing each study to minimize their effects and also to ensure that their effects will have equal opportunity to occur to all products.

4.4.1 Error of central tendency

This error is characterized by subjects scoring products in the midrange of a scale, avoiding the extremes (or extreme numbers) and having the effect of making products seem more similar. This scoring error is more likely to occur if subjects are unfamiliar with a test method or with the products, or they are relatively naive with regard to the testing process. The more experienced (and qualified) subject is less likely to exhibit this scoring error.

Avoidance of scale extremes is a particularly interesting issue. For example, it is occasionally suggested that the 9-point hedonic scale (Jones *et al.*, 1955; Peryam and Pilgrim, 1957) and similar category-type scales are in effect more restricted; that is, the extremes are avoided, and in the case of the 9-point scale one actually has a 7-point scale. Guilford (1954) and Amerine *et al.* (1965) recommended that this avoidance of extremes could be offset by changing the words or their meaning or by spacing the terms farther apart. This approach is more applicable to scales other than the 9-point hedonic scale, which does not produce this error when product serving order is balanced and subjects are familiarized with the scale and the products prior to a test. Schutz and Cardello (2001) introduced a "labeled affective magnitude scale" to improve spacing between scale categories. Although more research is being conducted to assess the latter scale, current summaries indicate it has improved discrimination among well-liked products. One of the more illogical extensions to the avoidance of the extremes is to propose dropping two of the 9-point scale categories and using a 7-point scale. Unfortunately, this does not alter the error and therefore one now has a 5-point scale that is significantly less sensitive than a 9-point scale. Avoidance of extremes will also occur with subjects who have been trained (directly or indirectly) to think that there are correct responses for perceived attribute intensities. This is likely to occur when using references (or standards) and subjects are advised to always assign specified value(s) for those standards (e.g. the scale midpoint in difference from reference judgments). The problem here is the inherent variability in standards and in the subjects themselves, a topic discussed more extensively in Chapter 3. Nevertheless, this error of central tendency can be minimized through standard sensory procedures and practices. In tests with naive consumers, demonstrating scale use before the test has a very beneficial effect and minimizes this error.

4.4.2 Time-order error

This error is referred to as an order effect, first-sample, or position error. This error in scoring results in the first product (of a set of products) being scored higher than expected, regardless of the product. The error can create difficulties in interpretation of results if the serving order is not balanced and results are not discussed as a function of serving order. Although it is possible to statistically measure the effect of order through the use of a Latin-square design (Winer, 1971), these designs are seldom used in sensory evaluation, primarily because of the unlikely assumption that interaction will be negligible. Interactions between subjects and products are not unusual, nor are interactions resulting solely from various treatment combinations. However, examination of response patterns by order of presentation as a means of assessing the error will provide very useful information about product similarities and differences.

The series of examples shown in Table 4.1 is intended to demonstrate this point as well as to illustrate the close relationship between the impact of a psychological error and product differences as perceived by a panel (when is the difference attributable to the order error and when to the products?). Example A (in Table 4.1) demonstrates the typical product served first effect; both products were scored higher by an average of 0.3 scale units. The conclusion derived from these results is that both products were liked equally and the time-order error affected both products

Table 4.1 Examples of Scoring Patterns from a Series of Two-Product Tests Displayed by Serving Order as a Demonstration of Order Effects

		Product		
Example	Serving Order	X	Y	
A	First	6.4[a]	6.3	6.3
	Second	6.1	6.0	6.1
	Pooled	6.3	6.1	
B	First	6.5	6.4	6.5
	Second	6.7	6.1	6.4
	Pooled	6.6	6.2	
C	First	6.3	6.5	6.4
	Second	6.2	5.9	6.1
	Pooled	6.3	6.2	
D	First	6.4	6.5	6.5
	Second	7.0	6.5	6.7
	Pooled	6.7	6.5	

[a]Entries are mean values based on the use of the 9-point hedonic scale (1, dislike extremely; 9, like extremely), N = 50 in each case except for pooled values, with a monadic-sequential serving order. See text for further explanation.

equally. Example B provides a more complicated possibility; product X was liked more when evaluated after product Y. When served first, product X obtained a score of 6.5, and this score increased to 6.7 when it was served second. The reverse was obtained for product Y; when served first, the score was 6.4, and it decreased to 6.1 when it was served second. The net result of this would be a recommendation that product X was preferred over product Y. It is possible that the 6.6 versus the 6.2 pooled result was sufficient to reach statistical significance at the 95% confidence level; nonetheless, the more telling issue was the 0.6 scale units of difference between products in the second position. If my product were Y and I were interested in attracting consumers of product X, these results would let me know that I would not be very successful from a product perspective!

Example C provides another possible outcome, except both products show the decline as occurred in example A; however, the magnitude of change was greater with Y (suggesting possible vulnerability of Y or opportunity for X). Example D provides another outcome in which X increased and Y remained essentially unchanged. Once again, a possible explanation for the increased preference for X over Y (7.0 compared with 6.5) in the second position would be that Y has a sensory deficiency that was only apparent after first evaluating Y (and then X).

Obviously, the sensory professional cannot afford to look at these scoring patterns solely in the context of the time-order error but, rather, must consider the product scoring changes relative to each other and relative to the test objective. Simply assigning the change in a mean score to a time-order error is quite risky at best, as is the reporting of only results from the first position. In addition to the issue of selective reporting, the loss of comparative information about the products should be a very significant concern to the sensory professional and to the test requestor.

4.4.3 Error of expectation

This error arises from a subject's knowledge about product and is manifested in the expectation for specific attributes or differences based on that knowledge. For example, in a discrimination test, the subject is likely to report differences at much smaller concentrations than expected. Conversely, the subject may overlook differences that are not expected in that type of product. In effect, the subject may have too much knowledge about products and should not have been selected to participate in the test. In descriptive testing, procedures that expose subjects to a broad array of irrelevant references will produce an error of expectation observed as a "phantom attribute" representing an attribute not in, or not at a perceptible level in, the test products. Such phantom perceptions represent "false hits" for the presence of that attribute. It is reasonable for subjects to expect an attribute if the panel leader has *a priori* provided them a reference for that attribute.

4.4.4 Error of habituation and of anticipation

These errors arise from the subject providing the same response to a series of products or questions. For example, in threshold tests in which products are systematically changed in small increments by increasing or decreasing stimulus concentration, the subject may report "not perceived" beyond the concentration at which the stimulus actually is perceived (habituation). An example of anticipation error is when the subject reports perception of the stimulus before a threshold concentration is actually achieved, but is anticipated at any time because there has been a series of "not perceived" responses. To some extent, these errors are similar to the error of expectation.

4.4.5 Stimulus error

This error occurs when subjects have (or think they have) prior knowledge about products (i.e. stimuli) in a test and as a result will assign scores based on knowledge of the physical stimulus rather than based on their perception of the stimulus or will find differences that are unexpected. Of course, this latter situation can be especially difficult to address because non-product variables may arise from the normal production cycle or if the product was formulated in a way that was nonrepresentative. To reduce this risk, the sensory professional must be very thorough in examining products before a test and make sure there is sufficient product available to avoid using one or two containers of product and assume they are representative. If one follows test procedures as described in this book, then this particular error should not be of major concern. It is also one of the reasons why individuals who are directly involved in formulation of a product should not be subjects in a test of that product.

4.4.6 Logical error and leniency error

These errors are likely to occur with naive subjects who are not given precise instructions as to their task. The logical error occurs when subjects are unfamiliar with the specific task and follow a logical but self-determined process in deciding how the products will be rated. The leniency error occurs when subjects rate products based on their feelings about the experimenter and, in effect, ignore product differences. These errors can be minimized by giving subjects specific test instructions and by using experienced subjects; the use of qualified and trained subjects is required for analytical tests.

4.4.7 Halo effect

The effect of one response on succeeding responses has been associated by Guilford (1954) with stimulus error in the sense that it involves criteria not of direct interest to the particular experiment. One example of a halo effect is a test in which consumers provide responses to product preference followed by responses to a series of related product attributes. In this situation, the initial response sets the stage and all subsequent responses justify it; alternatively, asking preference last will not change or eliminate the effect. Halo effects are more often observed in tests involving consumers and particularly in those situations in which there are many questions to be answered.

With many questions, product re-tasting becomes necessary, which in turn leads to physiological fatigue, further exacerbating the situation and leading to errors of central tendency. To gain insight into this problem, one can examine response patterns and variance measures within and across scorecards. To overcome or circumvent halo effects, some investigators will require test participants to start (their evaluations) at a different place on a scorecard or to change the order in which the questions are listed. Although such approaches have surface appeal, the problem still remains but now it affects responses to different questions. In addition, there is an increase in the amount of time required to complete a scorecard and an increase in the number of incomplete scorecards. Although it is not our intent, at this point, to analyze the relative merits of lengthy consumer scorecards (this will be discussed in Chapter 7), it is necessary to make the reader aware of this particular effect because it will have direct implications regarding scorecard preparation and business decisions based on results from such tests. Also, it is unreasonable to expect that one can design out of a study the halo effect in its entirety. One must keep the effect in mind when examining results and providing an interpretation.

4.4.8 Proximity error

In this error, adjacent characteristics tend to be rated more similar than those that are farther apart. Thus, the correlations between adjacent pairs may be higher than if they were separated by other attributes, and as Guilford (1954) suggested, this would be one reason to not put great emphasis on these correlations. It has been suggested that this error could be minimized by separating similar attributes, by rating attributes on repeated sets of products, or by randomizing the attributes on the scorecard. Unfortunately, each of these suggestions raises questions concerning their practicality and usefulness in a business environment. Perhaps more important is the question of whether this error can be demonstrated in most consumer products (foods, beverages, cosmetics, and so forth). Research on the error appears to be associated with social studies (i.e. personality traits) and not with consumer products. Nonetheless, some problems associated with the proposed solutions warrant consideration.

In descriptive tests, the subjects decide on the sequence in which the attributes are perceived and scored (usually in the order of their perceived occurrence), and efforts to randomize the sequence create confusion (and increases variability). For example, subjects allowed to score attributes on a random basis demonstrate three effects that do not appear to lessen with time: an increase in variability, an increase in test time (by as much as 50%), and a tendency to miss some attributes entirely. We are not aware of situations in which the proximity error has been demonstrated in sensory tests.

4.4.9 Contrast and convergence errors

Probably the most difficult errors to deal with are those of contrast and convergence. Their impact on evaluation of foods was described in detail by Kamenetzky

(1959). These errors are product related and often occur together. The contrast error is characterized by two products scored as being very different from each other and the magnitude of the difference being much greater than expected. This could occur when a product of "poorer" quality was followed by a product of "higher" quality. The difference was exaggerated, and the higher quality product was scored much higher than if it were preceded by a product closer in quality. Convergence is the opposite effect, usually brought about by contrast between two (or more) products masking or overshadowing smaller differences between one of these and other products in the test. In tests involving three or more products, the existence of one effect most often is accompanied by the other. These two effects are depicted in Figure 4.1; the examples at the top of the figure represent the impact on intensity measurement, and the examples at the bottom are for preference. Obviously, this exaggeration of the difference for some products and compression of the difference for other products can have serious consequences. Although it is reasonable to acknowledge that this problem can be minimized through proper product selection, it is quite another matter to delineate "proper product selection." In the

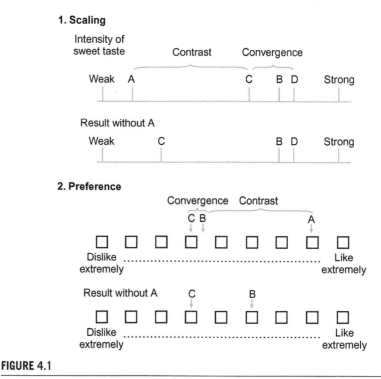

FIGURE 4.1

Examples of scoring and preference patterns demonstrating contrast and convergence effects. For purposes of display only, the products appear to be scored on a single scale. In actual tests, each product would have its own scorecard.

first instance, the two errors are more likely to be encountered when one or more of the products are recognizable, or several of the products are experimental and one is commercially available (such as a "gold standard"). Of course, one may still find it necessary to include such products, in which case experimental design and examination of responses by serving sequence become particularly important. For example, with a balanced design in which every product appears in every serving position, results can be examined when a particular product is served last and therefore has no effect on the preceding products. In this way, estimates can be obtained for the effects of contrast and convergence. There can also be situations in which a product is always served in the last position so as to eliminate at least one of these two effects. To the extent possible, the sensory specialist must be able to assess products, well in advance of a test, to be able to determine the potential for contrast and convergence before deciding on a test design.

However, the question of reporting these results remains. One cannot ignore the errors and report the results or simply state that the particular testing errors had a significant effect on responses and therefore no conclusion can be reached. Somewhere between these two extremes exists a reasonable approach to reporting. The most reasonable approach is to describe the results in relation to the product(s) serving order. Consider the preference data in Figure 4.1 (lower portion of the figure), where product B was more preferred than product C (when A was not a factor in the decision). This approach serves two purposes: (1) It satisfies the requestor's objective, and (2) it makes the requestor aware of the consequences of including products in a test that are predisposed to these errors.

In this discussion about psychological errors, the intent has been to alert the sensory professional to the potential impact of these errors on response behavior. For the most part, careful product selection and the use of qualified subjects will minimize the impact of many of these errors but will not eliminate them entirely. Results must always be reviewed for the presence of these errors, particularly when there is a concern based on a pretest examination of the products.

4.5 **Statistical considerations**

Sensory evaluation uses statistics to determine whether the responses from a group of subjects are sufficiently similar or represent a random occurrence. Knowing that there is this degree of similarity enables the sensory professional to make a decision about the products being tested with some stated measure of confidence in the context of that population of subjects and, where appropriate, to subjects in general. The remainder of this chapter focuses on statistical issues that sensory professionals consider (or should consider) when determining the most appropriate experimental design and analysis for a particular problem. The reader is assumed to have a reasonable knowledge of statistics and be familiar with commonly used summary statistical procedures, including the analysis of variance (ANOVA). [For a review of these methods as they relate to sensory evaluation, see Sidel and Stone (1976), O'Mahony (1982),

Gacula and Singh (1984), Gacula (1993), and Korth (1982).] Although we do not provide detailed mathematical steps for the various methods, some attention is given to methods of particular value to sensory evaluation, such as the analysis of data derived from split-plot designs and other related designs. The latter provide the means by which one can analyze results from different groups of subjects who have evaluated a set of products at different locations and/or at different times. The chapters on specific sensory test methods (Chapters 5–7) also include recommended designs and analyses appropriate for those methods. However, a more thorough exposition on statistics can be found in Alder and Roessler (1977), Bock and Jones (1968), Box et al. (1978), Box and Draper (1987), Bruning and Kintz (1977), Cochran and Cox (1957), Cohen and Cohen (1983), Cohen (1977), Guilford (1954), Hays (1973), Krzanowski (1988), McCall (1975), McNemar (1969), Morrison (1990), Phillips (1973), Smith (1988), Winer (1971), and Yandell (1997).

The design of a test, the selection of a statistical method for analysis, and the interpretation of results are responsibilities of the sensory professional. Delegating this responsibility to a statistician or to some other individual is not recommended. A statistician cannot be expected to be cognizant of the subtleties of sensory evaluation, to be aware of the kinds of subject-by-product interrelationships described earlier in this chapter, or to easily accept some of the trade-offs that occur when the sensory requirements of a problem take precedence over the mathematical and statistical requirements. This should not be misconstrued as a recommendation to avoid statisticians; rather, it is an acknowledgment that the experienced sensory professional must have sufficient knowledge about statistics to recognize when and what type of assistance is required.

In many sensory tests, the statistical design requirements cannot be completely satisfied and compromises must be made. For example, one uses a prescreened number of subjects who meet specific sensory requirements (i.e. they are not a random selection of people), who will most likely evaluate each product on a repeated basis, and who will use a scale that may not be linear throughout its entire continuum. The products are not selected randomly. The subject's responses will most likely be dependent, not independent, and in most instances the number of subjects will be less than 20. For many statistical methods, it is easy to cite examples of a "failure to meet" one or more of the requirements based on these criteria; however, this should not constitute a basis for modifying a test plan or for not using a particular statistical analysis. It must be kept in mind that statistical models are based on stated assumptions about the response behavior of the individuals in a test; however, these assumptions are not absolute—they are not laws. In fact, the average score will likely not represent any one subject's score. When one or two assumptions are violated, there is a potential risk of reaching an incorrect conclusion. As noted by Hays (1973),

> *Statistics is limited in its ability to legislate experimental practice. Statistical assumptions must not be turned into prohibitions against particular kinds of experiments, although these assumptions must be borne in mind by the*

experimenter exercising thought and care in matching the particular experimental situation with an appropriate form of analysis.

For some tests, statistical risk can be determined in relatively precise terms, such as in the case of multiple *t*-test comparisons. When one is deciding that differences between products are significant, the error rate increases very rapidly in direct relation to the number of comparisons. As discussed by Ryan (1959, 1960), if one were comparing 10 means at the 0.01 level of statistical significance, the error rate per experiment would be equal to 0.45. In fact, multiple *t*-tests should be avoided because of this problem. There are also examples in which a precise statement of risk is less clear. For a discrimination test, a precise computation of risk is difficult because assumptions must be made regarding the true magnitude of difference between products, the actual sensitivity of the subjects, and the use of replication, which is a necessary part of the discrimination test. These issues are discussed in more detail in Chapter 5 as they relate to the discrimination process. However, it should be mentioned that one of the consequences of repeated trials testing is thought to be an increased potential for finding differences. Because we want to minimize unanticipated changes (i.e. minimize β risk), this particular risk would seem to have a positive effect. It would be foolish to not obtain replicate responses within a test just because there is a risk that the responses may be dependent. The mathematical and particularly the behavioral implications of this risk need more study. Our purpose here is to provide a frame of reference for the role of statistics in the overall scheme of sensory evaluation. After all, statistics, in and of itself, is not perfect in its analysis of responses, and new procedures continue to be developed. Sensory evaluation uses statistics as one of several tools in helping to solve a problem. This tool may not be entirely satisfactory, primarily because it was not designed exclusively for that purpose (sensory evaluation); nonetheless, we use it just as we make use of scales and subjects that are not perfect. They are perfect enough to enable actionable product decisions to be made with minimal risk.

In the following discussion, attention is given primarily to parametric, compared with nonparametric, statistical methods. By parametric, we refer to methods that assume data that are normally distributed, whereas nonparametric methods are distribution free—that is, no assumptions are made about the distribution from which the responses were obtained. Interest in nonparametric methods was very strong in their early development (during the mid-1940s) because the methods were quite easy to use (e.g. Wilcoxon signed rank test and Mann–Whitney) and enabled the experimenter to utilize proportions or other similar kinds of data that could be categorized in contingency tables. It should be clear that nonparametric tests provide the sensory professional with additional tools for data analysis, when there are reasons to justify their use. As O'Mahony (1982) noted, parametric methods are preferred because they use scaled data as they were obtained from the subjects. Using scaled data directly as obtained from the subjects is very important because it provides an unobstructed view of the responses with no modifications or transformations that are mathematically correct but destroy individual differences. It is the latter that often provide important clues regarding the nature of product similarities and differences. However, there can be situations in which a few subjects may score a product very different from all the other subjects, and those results would be sufficient to change a difference, for example, from significant to nonsignificant. In this situation, converting to ranks eliminates the magnitude effect, and an appropriate nonparametric procedure can be used. For more detail on nonparametric methods, the reader is referred to Hollander and Wolfe (1973) and Daniel (1978).

Before describing some statistical tests, it is useful to briefly discuss issues of a more general nature, such as reliability and validity, replication, independence versus dependence of judgments, the concept of random selection of subjects, and Type 1 and Type 2 errors and risk in the decision-making process. Some of these issues are also considered in the discussions about the analysis of data obtained using specific test methods.

4.5.1 Reliability and validity

These two related elements (reliability and validity) should be of concern to every sensory professional because they have a direct effect on each test and on program credibility. By reliability, we mean the extent to which subjects will provide similar responses to similar products on repeated occasions. The closer this agreement, the higher the degree of reliability and, therefore, the greater confidence there will be in the potential validity of the results. However, a high degree of reliability is no guarantee of validity, and with low reliability there can be no consideration of measuring validity. For example, if a group of subjects replicate their judgments but are assigning responses opposite to the known or expected direction, there would be a high degree of reliability but no validity. This could occur if subjects were scoring a series of products containing known amounts of some ingredient such as an added flavor. If the subjects were confused about what they were measuring or how to use the particular scale, one could achieve a high degree of reliability and equally high variability but with no guarantee of validity. In the discussion on analysis of descriptive data (Chapter 6), a method is described for establishing a numerical measure of reliability for each subject. There are other procedures for estimating reliability, and these can be found in the sensory literature (Bressan and Behling, 1977; Sawyer et al., 1962; Zook and Wessman, 1977).

For analytical tests (discrimination and descriptive), the need for reliability cannot be underestimated. These tests typically rely on approximately 10–20 subjects, which is a relatively small number of people on which to make an actionable recommendation. If the recommendations and the results are to be accepted as valid, then the experimenter must have a high degree of confidence in those subjects, and such confidence is derived in part from the data—that is, from the repeated measures, which provide the basis for computing a reliability estimate. Even if one were to increase the number of subjects in these tests, the trade-off would not be equivalent. For example, results from a discrimination test using 50 subjects would not carry the same "weight" as the results from 25 subjects with a replicate (in both instances, the $N = 50$). Just as any physical or chemical test would be done more than once, so too would a sensory analytical test. What is surprising is how little attention is given to this issue of replication yet is fundamental to minimizing risk in decision making.

Validity is more difficult to deal with because it is not directly measured and there are several kinds of validity (for which researchers disagree on the types and definitions). In its broadest sense, it may be considered as the degree to which results are consistent with the facts. Of the different kinds of validity, we are most interested in face validity and external validity. We state that there is face validity when we measure what we intended to measure and results are consistent with expectation. For example, if we know that product A was formulated to contain more sweetener and the results show that this product was scored as more sweet than other products in a test, then we can conclude with confidence that the results have face validity. However, this observation by itself would not necessarily be sufficient to conclude that the results are therefore generalizable. External validity would need to be considered. By external validity, we mean that given a different and/or larger group of respondents, will the same conclusion be reached? How confident are we that the conclusion reached in test A will be repeated in test B? For many sensory problems, determining that there is face validity, or external validity in particular, is usually more difficult than the relatively simple sweetener example. The effects of ingredient or processing changes are not always known or easily measured, and they are not usually very direct as stated in the sweetener example. That is, the addition of any ingredient is likely to alter sensory properties in an indirect manner. A process variable might alter product appearance, or the change may be too small to be detected. Therefore, one's expectations about the validity of results will require more investigation.

Dialogue with the requestor may provide further information about product formulation that would explain observed response patterns and, in turn, this might be sufficient to substantiate the face validity. However, the question of external validity cannot be completely and directly addressed through technology; rather, it must also be addressed through testing in the marketplace. Are consistent results obtained when the same products are evaluated by two groups of consumers? Note that we talk in terms of consistency rather than in terms of an exact agreement of the numerical values. In any comparative study, one expects the results for two (or more) tests to be consistent; however, the expectation that the numerical values will be exactly the same is quite unrealistic. When considering the external validity of test results, it is important to remember that the basis for determining that this validity exists is whether the conclusions we reach are the same and therefore can be generalized.

From a purely sensory perspective, a concerted effort should be made to develop evidence for the validity of sensory test results. Because there is usually a bias associated with the use of employees for sensory tests, in general, the issue of validity, and especially external validity, becomes very important (see Chapter 2 for more discussion on this problem of use of employees versus nonemployees in a test).

More dialogue about measuring validity is needed, and a stronger case should be made for the importance of validity in the long-term development of a sensory program's credibility.

4.5.2 Replication

In the previous section, the importance of replication in assessing subject and panel reliability was noted as input to the determination of validity of the results. It is useful if we also consider replication in the context of the test itself and the trade-off between subjects, products, and the practicalities of repeated testing in a single session and over a period of several days.

Replication is absolutely essential for all sensory analytical tests—that is, for discrimination and descriptive tests—because of the very nature of the tests and the limited number of subjects in each test. In the first instance, analytical tests provide precise information about product differences and the specific attributes that form the basis for those differences. Like physical and chemical analyses, there must be more than a single analysis, and the analysis is done using more than a single sample of a product. As described in Chapter 2, subjects are selected based on their sensory abilities as determined from a series of screening tests. By eliminating individuals who are insensitive or are inconsistent, and by selecting those who are most sensitive, a more precise sensory instrument is created. This enables one to rely on fewer (and more precise) subjects. However, one must obtain replicate judgments from each subject to be sure that they are consistent in detecting differences. In addition, because there are a limited number of subjects, each one's contribution to the database is relatively large and a few atypical responses could shift a result from significance to nonsignificance or the reverse. The use of replication provides insight into individual response behavior. Given several opportunities to score products or to select the one most similar to an identified reference, how similar are the responses? If they are not similar, then one must determine their impact on test conclusions and decide whether a qualified recommendation is appropriate. Such attention to detail is critical to the successful use of sensory evaluation small-panel testing, which probably represents as much as 75% of the activity of most sensory evaluation programs in business today.

Thus far, no attempt has been made to state how much replication is necessary. Obviously, a replicate is superior to none, but at what point does it no longer become useful and additional responses have no impact on the conclusions? In addition, as the total number of judgments increases, the greater the likelihood of finding statistically significant differences of no practical value. For discrimination testing, we have observed that a single replicate (each subject evaluates each product twice) has proven to be sufficient, providing the experimenter with sufficient information to monitor discrimination ability and to allow a conclusion about the products with minimal risk. With a replicate, each subject will have sampled six products within a 20- to 30-minute time period, which usually is sufficient without experiencing any sensory fatigue. For descriptive tests, three or four replicates are usually sufficient to assess subject reliability and to reach conclusions about the products. Empirically, we have observed that variability decreases somewhat as one proceeds from the first to the second replicate and levels at the third replicate, but subsequent replications do not exhibit any further decrease in variability. Of course, having a measure of within-subject variability improves the sensitivity of a statistical test and increases the likelihood of finding a difference. A more detailed discussion about the use of replication within the context of specific test methods is provided in Chapters 5 and 6.

There are practical as well as statistical considerations involved in the planning of an experiment and a decision regarding the extent of the replication necessary. Practical considerations include product availability, preparation requirements, subject availability in terms of numbers and frequency, and the nature of the request (early stages of a project or just prior to test market). If insufficient product is available or the preparation procedures are so demanding as to cause concern for product stability, then a design that includes testing on successive days may not be realistic without finding larger than expected variability. In this instance, replication would have to be achieved within a single day by having several sessions.

To some extent, additional subjects can be used in place of replication; however, it should be approached with an appreciation for the risks involved: As noted previously, the exchange is not equivalent. With replication, a subject has additional opportunity to score products, which in turn provides a measure of within-subject variability and leads to an increased likelihood of finding significant differences (i.e. statistical power is increased). The expectation that increasing the number of subjects will compensate for any advantage gained through replication is not necessarily true in every instance. Empirically, it can be shown that variability in scoring decreases with replication and the largest decreases occur from first to second and from second to third replicates, respectively. Of course, if a project is in its early stages and several experimental versions of a product are under consideration, then a single replicate may be sufficient. On the other hand, as a project progresses and there are fewer experimental products and differences are smaller, the need for more replication becomes greater. The sensory professional must consider these issues when planning a test. However, it is important to keep in mind that without replication, risk in decision making increases, and no amount of additional subjects can compensate for it.

A statistical approach to the problem of replication is also possible. By selecting appropriate mean and variability (e.g. standard deviation) values and the probability value for accepting or rejecting the null hypothesis, one can compute the number of judgments and then divide by the number of subjects to determine the extent of replication that is necessary. Obviously, this assumes that a database about the products and/or the subjects is available.

In affective testing, replication is atypical because of the more subjective nature of the judgmental process and because of the manner in which the results are used (i.e. to estimate market potential, purchase intent, etc.). Such broadly based objectives require larger numbers of respondents, particularly if one is dealing with, and representing, unique population segments. With a larger number of qualified (demographic and product usage) consumers, the various statistical measures become better estimates of the target population and thus the generalizations have greater validity. A broad representation of potential consumers is desirable for determining if there are any product deficiencies evident to a subset of the population and the extent to which those deficiencies represent business problems (or opportunities). If one intends to segment results based on typical demographic criteria, such as age, gender, and income, or based on other criteria (e.g. preference clustering), then the importance of a large database is more meaningful. Affective testing is discussed in greater detail in Chapter 7.

The issue of how much replication is necessary, like many other aspects of sensory evaluation, is not and perhaps should not be mandated. Each problem has a unique set of requirements, just as each product has a unique set of properties, and therefore the exact amount of replication cannot always be stipulated.

4.5.3 Independence and dependence of judgments

For sensory evaluation, the concept of independent judgments is not realistic, at least not in a literal sense. All individuals participating in tests have experience as product users (a primary screening criterion), and to that extent, their responses will reflect this dependence along with prior test participation. From a sensory design standpoint, the use of a balanced serving order and other design considerations enable the experimenter to minimize the impact of this dependence, or at least to more evenly distribute its effect across all products. As a further consideration, a monadic sequential serving order is most appropriate; that is, only one product and a scorecard are served and then both are removed before the next product and scorecard are served. In that sense, the evaluations are made independently; however, the responses still have some dependency. The impact of dependence on the results and conclusions of a test is not entirely clear. Although some statistical models (e.g. one-way ANOVA and t-test for independent groups) assume independence of judgments as a requirement, the risk is not well-defined. These models require more subjects than their respective dependent models (e.g. two-way ANOVA and dependent t-test), and this introduces risk of finding more differences.

4.5.4 Random selection of subjects

In sensory evaluation, the random selection of subjects is rarely achieved or rarely, if ever, desired. Sensory tests involve a limited number of subjects who are selected from a larger group of screened and qualified individuals. Selection of those subjects will be based on immediate past performance and the frequency with which these individuals have participated in tests. Test design reflects this selection, as will be described in the section on experimental design.

4.5.5 Risk in the decision-making process: Type 1 and Type 2 errors

An important consideration in the statistical analysis of results is the probabilistic statement that is made about those results and the risk implicit in that statement.

The question most frequently asked is whether there is a difference between the products, if the difference is statistically significant (discussed at length in Chapter 5), and what the probability is that the result is real and not due to chance. In sensory evaluation, it is accepted practice to consider a difference as statistically significant when the value for α is ≤ 0.05. In effect, when α is ≤ 0.05, we are concluding that there is a difference between products A and B, and we are confident enough to conclude that there is only 1 chance in 20 that this result was due to chance. If there is reluctance to accept this degree of risk, then the criteria for statistical significance can be changed; for example, setting $\alpha \leq 0.01$ decreases the probability of wrongly concluding there is a difference. From a practical standpoint, however, changing the decision criteria has its own risk. Others may view this changing of the decision criteria as using statistics to suit the particular situation, particularly to demonstrate that statistical significance was obtained. On the other hand, marketing and consumer insights may be using an α of 0.20 (or some other level), which also creates the potential for conflict. The magnitude of difference necessary to achieve significance at an α of 0.20 is smaller than that for an α of 0.05, and consumer insights could easily conclude that its test found differences missed by sensory evaluation. Of course, the choice of α in this instance could be based on knowledge that consumers are less sensitive and/or product is inherently variable, so a less restrictive significance criterion might be appropriate. This particular problem is reconsidered in Chapter 7 as part of a larger discussion on consumer testing; it is mentioned here as it relates to the need of the sensory professional to have a clear understanding of the level of statistical significance being used by others and its associated risk in the decision-making process. It is not a matter of one level being more or less correct or of everyone necessarily operating at the same level; rather, the decision criteria and risk should be defined to prevent misunderstandings between the different business units. It is very possible that the significance level necessary to initiate the next action could be different for different product categories, and this would not be wrong.

Taking risk into account acknowledges that the possibility of an incorrect decision still exists. For example, if we were to reject the null hypothesis when it is true, we will have committed an error referred to as a Type 1 error; this has a probability equal to α (which is the significance level). In practical terms, we might have a situation in which we are comparing an experimental product with a target product, and the hypothesis is that there is "no difference" between the products (referred to as the null hypothesis and expressed mathematically as $H_0: \bar{x}_1 = \bar{x}_2$). If we concluded from the results that there was a difference when in fact it was not true, then a Type 1 error was committed; the probability of such an error is equal to α. On the other hand, if we concluded that there was no difference when in fact there was a difference, then a Type 2 error was committed, and the probability of this error is equal to β. There is a mathematical relationship between α and β; when one decreases, the other increases. However, this relationship is determined by the number of judgments, the magnitude of the difference between products, and the sensitivity and reliability of the subjects.

This particular relationship is discussed in detail in Chapter 5 on discrimination testing along with worked examples. However, its importance obviously extends to all sensory tests as well as to decisions about the selection of particular statistical tests for use after the analysis of variance when there is a significant difference among the means. For example, the Newman–Keuls and Duncan multiple range tests are more protective with regard to Type 2 error compared with the Scheffé and Tukey tests, which are more protective with regard to Type 1 error. In tests after the analysis of variance, these alternatives present the sensory professional with additional considerations regarding which approach to use. Some worked examples are presented in Chapter 6 and are intended to demonstrate these risks and their effects.

As an additional consideration, the statistical power of the test should also be taken into account (Cohen, 1977; Lawless and Heyman, 2010; Thorngate, 1995). Cohen (1977) described statistical power as follows:

> *The power of a statistical test of a null hypothesis is the probability that it will lead to the rejection of the null hypothesis, i.e. the probability that it will result in the conclusion that the phenomenon exists.*

Thus, by setting α, both β and the power are determined; however, the problem is somewhat more complicated. For example, if α is set very small, statistical power will be small and β risk will be high. As Cohen noted, the issue of statistical significance involves a series of trade-offs between the three factors. For example, with α set at 0.001, the statistical power might be equal to 0.10 and β to 0.90 (where $\beta = 1 - \text{power}$), which would be risky if one were attempting to minimize the prospects of committing a Type 2 error. This is a particularly important issue inasmuch as most product testing is focused on making changes that are least likely to be detected by the fewest people. We wish to minimize the risk of saying the products are the same when they are not.

As previously noted, there are several elements of a test that influence risk and statistical power: the number of subjects and/or responses, their sensitivity and reliability, the magnitude of the difference between products (the effect size), and the testing process. Obviously, a very large number of responses (e.g. several hundred or more) increases the likelihood that a given difference will be detected and β risk minimized. However, the use of large numbers of responses in a test as a means of minimizing β risk and of decreasing statistical risk carries with it the problem of obtaining differences that have no practical value.

Products are also a source of differences independent of the variable being tested. In practice, one often discovers that products are a greater source of variability than are subjects. Whether one is obtaining products from a production line or from a pilot plant, it is not unusual to discover that there are easily discernable differences among products within a production run or comparing production products with those from a pilot plant. Unfortunately, many of those differences are not controllable, and the sensory professional must take this into account when designing each test. It should be kept in mind that the use of replication in a sensory test is essential to obtaining a measure of product variability; that is, by using more product, the responses represent the range of variability within the product being tested. In addition, as Cohen noted (1977), any procedures and/or controls that can be applied to eliminate non-test variables will increase power. This means strict adherence to test protocol (proper preparation and serving), use of an appropriate test site with environmental controls, and so on. Finally, it must be re-emphasized that the use of qualified subjects is, in all likelihood, the most important element in minimizing risk.

In recent years, some researchers have taken another look at the Type 1 and Type 2 errors; however, no significant changes in thinking have occurred, other than to note the availability of tables for calculating the β risk (Schlich, 1993) and further speculation about the process (Smith, 1981). Additional study will be useful because most sensory tests involve limited numbers of subjects and a database of not more than 50 responses, whereas acceptance tests and consumer insight tests use much larger numbers of participants. We would be remiss if we did not emphasize that these small panel tests are done with qualified subjects experienced with the test procedure, and each provides repeated responses, so there are some clear benefits to sensory tests that use replication as an integral part of every analytical test.

4.5.6 **Statistical measures**

We now direct our attention to a discussion about some statistical measures that may be considered basic to all sensory tests. Many of these measures should be obtained regardless of other analyses contemplated. First is the examination of responses by product and by order of presentation for each product and independent of product. Frequency distributions such as these provide for an assessment of any skewed response patterns, as well as an overall judgment regarding the quality of the database. Second is the examination of means and standard deviations for the products, also by order of presentation. These latter measures provide for an assessment of order effects. It should always be kept in mind that these are summary statistics measures; that is, information is being summarized and the question of representativeness is addressed by observation of the frequency distributions.

In Figure 4.2, results are presented from a group of 40 subjects, each of whom scored each product on one occasion (i.e. no replication). The products were served in a balanced order, one at a time. For each product, the responses were plotted as a series of histograms; responses were pooled independent of serving order and whether served first or second. Also included were the means, SDs, N, and measures of the shapes of the distributions and the signs $(+, -)$ of these latter measures, which provides the direction and magnitude relative to a bell-shaped or normal distribution. A distribution is considered skewed when the value for skewness (Sk) exceeds ± 1. In the current example, none of the values approached this, ranging from $+0.0314$ to -0.6377. The measurement of kertosis (K) would yield a value of 3 for a normal distribution; the current results provide evidence for the more normal distribution for the responses to product B and the flatter distribution for product A. For more discussion about the measures of skewness and kertosis, see Duncan (1974). It is a matter of opinion whether one makes the decision to assess these

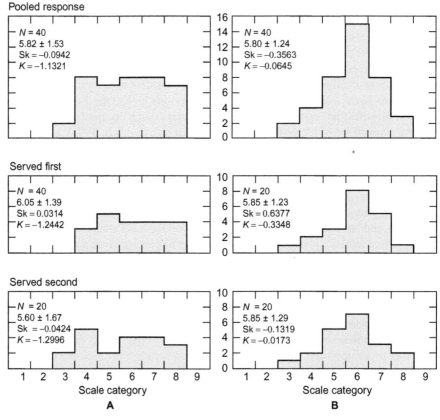

FIGURE 4.2

Histograms for two products, A and B, scored on a 9-point intensity scale. Results are tabulated for both serving orders and combined for each product. Also presented are the number of responses (N), mean ± SD, skewness (Sk), and kertosis (K).

measures or whether a simple perusal of the histograms is satisfactory. However, the important issue is that such an assessment is made because all subsequent tests will reflect the nature of the database. Note that the histograms for each product are quite similar (served first and second compared with pooled), and therefore they provide no more information than can be gained from the pooled results. The relatively flat distribution for A means that about the same number of subjects scored it as 4, 5, 6, 7, or 8, whereas the distribution for B has a peak at the scale value of 6 and is more bell shaped with fewer judgments on either side.

In effect, this result could be interpreted as a reflection of greater agreement by these subjects concerning the score for product B and more disagreement concerning the score for product A. In this situation, the similarity in the means and standard deviations would be less informative or misleading. For the formulation specialist, such a similarity could be most frustrating if there are no other clues as to product change. Of course, the usefulness of these measures is dependent on the total number of responses (N). When N is relatively small, such as 30 or fewer judgments per product, then segmentation will yield smaller numbers for each subset and one must use caution when reaching any generalized conclusions. This would also be true if results were compared from different studies using different scales of measurement and the responses were standardized. Scales can be transformed on some common basis and appropriately tested for agreement (for more detail, see McCall (1975) or McNemar (1969)). In each instance, however, it is very important that the original database be reviewed not only before but also after any transformations to ensure that the conclusions are consistent with that original database. Tests can also be made for homogeneity of the variance and for comparisons of variances (correlated or independent) as further steps in the examination of data.

The most important concern is for the sensory professional to recognize the need to assess the quality of each data set. The assessment should be more than just a perusal of an analysis output; it should be sufficiently extensive to determine its quality. Finally, note that no one measure can be expected to provide a complete picture of a database, just as no one statistical analysis can provide all of the answers to a problem.

4.5.6.1 The t-test

The t-test (or Student's t distribution) has been a very useful statistical procedure, although its popularity appears to have declined with the increase in multiproduct tests (i.e. tests involving three or more products) and accessibility to ANOVA models using standard statistical packages. Nonetheless, the t-test does have utility, especially when one is working with a relatively small database (e.g. $N \leq 30$) and only two products are being tested. If more than two products are being tested, a single comparison of any two products is possible; however, the reader is cautioned that such a test is easily misused. In fact, an inappropriate and common error occurs when multiple t-tests are made from the same database. This incorrect use of the t statistic will result in a substantial and changing Type 1 error, an issue that is described by Alder and Roessler (1977) and Ryan (1959, 1960). Similarly, if large-sample statistical procedures are used for analysis of data that are appropriately analyzed by the t-test, the risk of a Type 1 error is increased. Additional discussion of this issue can be found in Alder and Roessler.

Selection of the appropriate t statistic and degrees of freedom for a two-product analysis will depend on a number of factors, including paired or unpaired variates, equal or unequal numbers of judgments per cell, and equality or inequality of variances. Each alternative entails some computational differences, and the sensory professional should be familiar with them.

Most sensory tests use the dependent t formula (paired-variate procedure) because the same subjects evaluated both products. However, if there is interest in determining whether there was a significant order effect for a product, then an independent t statistic would be used in an unreplicated design, as shown in the following scenario. For this computation, responses from those subjects who evaluated product A in the first position would be compared with responses to product A served in the second position (obtained from the other subjects). If comparing product A with B in the first position only is of interest, then an independent t statistic is appropriate. Similarly, a comparison of the products in the second position also would be made with the independent t statistic.

Obviously, the t statistic can be used in a number of ways to analyze results from a test. It is necessary that the researcher be clear as to the alternative(s) to be used, and care is taken that the data are consistent with the model. The relative ease with which the computations are made has made the method popular, and it should continue to be used.

4.5.6.2 Analysis of variance

Of the many summary statistical procedures, ANOVA is probably the most useful as well as the most frequently used method for data analysis in sensory evaluation. It is especially useful in analyzing multiproduct tests, and it can utilize data from a variety of experimental situations without serious compromise or risk in decision making.

The analysis, as its name implies, is a statistical procedure designed to partition all the sources of variability in a test, thus providing a more precise estimate of the variable being studied. It might be considered an extension of the t-test, which is a two-product test (or a procedure that provides for the comparison of two means). The relationship between the two procedures can be shown in the one-way ANOVA and the independent t-test, in which the F value is equal to the t value squared. For a detailed discussion and description of these computations, see Alder and Roessler (1977), Bruning and Kintz (1997), and McCall (1975).

There are many different types of ANOVAs, depending on the nature of the problem, whether the variable is expected to have more than a single effect, whether the subjects might be expected to respond differently to the different products, whether subjects evaluate each product on more than a single occasion, whether different subjects evaluate each product on more than a single occasion, or whether different subjects evaluate the products at different times. For each experimental design, the particular ANOVA model will have certain distinguishing features. In most instances, these features are associated with partitioning the sources of variability and in determining appropriate error terms to derive the F value (the ratio of variances) for estimating statistical significance.

Before proceeding with a discussion about the different ANOVA designs, some of the common features of ANOVA are considered. As noted at the outset, the analysis involves a series of computations (in which scores are summed and squared or vice versa) to yield total sums of squares, treatment sums of squares, and error sums of squares. With unequal numbers of subjects, it is necessary that the sensory professional account for the missing observations. One possible alternative is to use a general linear model hypothesis. With replication, decisions will also need to be made, before any computations, regarding the most useful way to partition or group the treatment and error sums of squares. Replication is extremely important not only because it provides a measure of subject reliability but also because it increases the statistical power of the test—that is, it increases the likelihood of finding a difference. The expectation is that within-subject variability will be smaller than the between-subject variability. For a trained panel test, how else can one directly demonstrate within-subject and panel reliability? The next computational step is the comparison of the various sums of squares (after dividing each value by its associated degrees of freedom), the assumption being that if the responses (to all the products) are from the same population, then the ratio will be unity. However, the more this ratio deviates from unity (and is greater than unity), the more likely we are to consider the responses as due to a true treatment effect. The ratio, referred to as the F ratio, is then compared with tabular values for determining statistical significance. It is common practice to summarize the computations using a source of variance table and listing the sources of variation, degrees of freedom, sums of squares, mean squares, F values, and associated probability values. A description of these computations with worked examples can be found in Alder and Roessler (1977), Bruning and Kintz (1997), Gacula and Singh (1984), McCall (1975), and O'Mahony (1986). Most statistical packages also provide the user with details regarding the output, and the sensory professional should read the information carefully.

The sensory professional must be familiar with the various types of ANOVA designs and how to read and understand results. As we will demonstrate in the next section, with studies involving more than a single treatment, the proper partitioning of the sources of variability becomes a critical issue in terms of identifying and isolating sources such as interaction that could mask a true treatment effect. Finally, the increasing reliance on statistical packages with default systems is both an advantage and a disadvantage. The use of these statistical packages is advantageous in that analyses are readily available to the researcher; however, this can be a disadvantage if the statistical procedures are not clearly stated, or the package default analysis is not appropriate for the sensory test. For example, only one type of ANOVA model may be available (fixed effects), and this will lead to erroneous conclusions about differences if, for example, the particular design used was a mixed design (i.e. fixed and random effects). Split-plot designs and other more complex designs require even more specific information from the sensory professional to ensure that the computations and the error terms selected for tests for significance will be appropriate. Attention to details is important, and one can never assume that the default system has done the thinking for him or her.

A brief comment about tests after the F test is also in order. ANOVA provides evidence that there is a statistically significant treatment effect, but it does not indicate specifically how the treatments differ. The answer requires the use of one of several tests that are referred to as range tests or tests after the F test. The seven most frequently used tests after the F test are Fisher's LSD (least significant difference) or t-test and Duncan's, Newman–Keuls, Tukey (a), Tukey (b), and the Scheffé and Bonferroni tests. An eighth test, the Dunnett test, can also be used, specifically when one is comparing products not to each other but, rather, only to a control. In the discussion on descriptive analysis, the use of these tests and especially the former seven is described in conjunction with several worked examples. Briefly, these tests provide more or less protection against a decision error (Type 1 and Type 2 errors), and depending on the problem, one test or another would be selected. The reader should be aware that the tests are not interchangeable and that these tests are applicable only if there is a statistically significant F value. Failure to observe this rule can lead to embarrassing consequences, such as reporting significant differences between certain products, when an F test would yield no significant difference.

4.5.6.3 Multiple-factor designs and analyses

The treatments-by-subjects (T × S) design is one of the most frequently used designs in sensory evaluation. It is a straightforward design, with each product evaluated once by each subject. For example, if scoring four competitive products by 40 subjects and determining whether there is a significant difference between any of the products is of interest, then a T × S design is appropriate. The products will be served in a balanced monadic sequential order one at a time to each subject, until all four products are scored. This design provides an assessment of each product relative to the others and a basis for determining the effect of one product on (the score for) another. Typically, these effects are product dependent, hence the need to separate the effect from the variable of primary interest. This design is different from a completely randomized block design in which each subject evaluates only one product (40 responses per product in a four-product test will require 160 subjects). The former is preferred because it provides a more effective use of subjects and yields a separate subject term that can be partitioned from the error. It can also be expected to increase the likelihood for finding a difference because extracting the variance for subject differences reduces the residual error term, hence the value of the T × S design.

Table 4.2 is a source of variance table for the T × S design. In this design, the error term is represented by the T × S interaction value (AS), which might be considered unsatisfactory even though it is appropriate. It is unsatisfactory in the sense that this source of variability contains more than interaction (variability that cannot be assigned elsewhere); however, the data cannot be partitioned any further. Keep in mind that the mean square interaction value (MS_{AS}) is used in the computation for F as the denominator, and the smaller this value, the more likely that F will be significant.

As shown in the source of variance table (Table 4.3), this error term can be improved with the addition of replication. If interaction is nonsignificant, it need not be used as the error term for testing treatment effects. Testing the treatment effect may be achieved by using the residual or, in the situation of a nonsignificant interaction, by pooling the residual with the interaction sum of squares and the degrees of freedom (df) and using the resulting error term. However, it should be kept in mind that when the interaction is significant, it is the appropriate term for testing the treatment effect. Also, a significant subject effect is neither unusual

Table 4.2 Source of Variance Table for a Treatments X Subjects Design[a]

Source of Variation	df	Sum of Squares	Mean Square	F
Treatments (A)	$a-1$	$ss_A = \dfrac{\sum^a(\sum^s x)^2}{s} - \dfrac{(\sum^{as} x)^2}{as}$	$ms_A = ss_A/ \times (a-1)$	ms_A/ms_{AS}
Subjects (S)	$s-1$	$ss_S = \dfrac{\sum^s(\sum^a x)^2}{a} - \dfrac{(\sum^{as} x)^2}{as}$	$ms_S = ss_S/ \times (s-1)$	ms_S/ms_{AS}
Residual (AS)	$(a-1) \times (s-1)$	$ss_{AS} = ss_T - ss_A - ss_S$	$ms_{AS} = ss_{AS}/ \times (a-1)(s-1)$	
Total	$as-1$	$ss_T = \sum^{as} x^2 - \dfrac{(\sum^{as} x)^2}{as}$		

[a]Design plan: all subjects evaluate all products.

Table 4.3 Source of Variance Table for a Treatments X Subjects Design with Replications[a]

Source of Variation	df	Sums of Squares	Mean Square	F
Treatments (A)	$a-1$	$ss_A = \dfrac{\sum^a(\sum^{sr} x)^2}{sr} - \dfrac{(\sum^{asr} x)^2}{asr}$	$ss_A/(a-1)$	
Subjects (S)	$s-1$	$ss_S = \dfrac{\sum^s(\sum^{ar} x)^2}{ar} - \dfrac{(\sum^{asr} x)^2}{asr}$	$ss_S/(s-1)$	
Treatment x subjects (AS)	$(a-1) \times (s-1)$	$ss_{AS} = \dfrac{\sum^{as}(\sum^r x)^2}{r} - \dfrac{\sum^a(\sum^{sr} x)^2}{sr} - \dfrac{\sum^s(\sum^{ar} x)^2}{ar} + \dfrac{(\sum^{asr} x)^2}{asr}$	$ss_{AS}/(a-1) \times (s-1)$	ms_{AS}/ ms_R
Residual (R)	$as(r-1)$	$ss_R = ss_T - ss_A - ss_S - ss_{AS}$	$ss_R/as(r-1)$	
Total	$asr-1$	$ss_T = \sum^{asr} x^2 - \dfrac{(\sum^{asr} x)^2}{asr}$		

[a]Design plan: all subjects evaluate all products.

nor of much consequence. It indicates that the subjects used different parts of the scale when scoring the products and should not be misconstrued as a problem with the subjects. Individual differences in scoring are to be expected, as was noted in Chapter 3.

When the research problem must accommodate two treatment conditions— for example, in evaluating a carbonated beverage product in which there are levels of carbonation and levels of flavor—then a treatment-by-treatment-by-subject (T × T × S) design will be most appropriate. In this situation, the requestor may be seeking direction in formulation of a new product. Not only is it more test efficient to include an array of products that bear a relationship to each other but also the effect of one variable on another can be measured. When each subject evaluates each product, further partitioning is possible using the various treatment interaction effects. As shown in the source of variance table (Table 4.4), the partitioning of the sources of variability is considerably more extensive, thus providing for more precise estimates for F. Although it would be possible to analyze this test as a treatment-by-subject design, the interaction between the two treatments is best analyzed with the T × T × S design. Similar to the unreplicated T × S design, each main treatment effect must be tested using its respective interaction value as the appropriate error term—for example, MS_A/MS_{AS} and MS_B/MS_{BS}. (See Lindquist (1953, pp. 237–239) for additional information on this particular use of the interaction terms.) Figure 4.3 illustrates the usefulness of the T × T × S design with the easily identifiable effects of flavor, separate from carbonation level, on panel perception of tingling mouthfeel.

The next three designs, sometimes referred to as "split-plot" or "mixed" designs, represent useful models for product evaluation that extend the study of treatment interactions, providing for evaluation of matrices of products (i.e. multi-product tests) and the use of different groups of subjects. The latter practice is typical in evaluating the stability of products in storage. The designation of "split-plot" or "mixed" refers to the subplots of other less complicated designs that make up each design—for example, T × S and randomized blocks.

Table 4.5 is a source of variance table for a "one-within and one-between" split-plot design. This particular design would be most appropriate in situations in which, for example, different groups of subjects at different locations evaluated the same set of products or different groups of subjects evaluated a product set representing different plants. Another application would be a test in which different subjects are presented with only one of several alternative concepts, but all subjects evaluate all products. The latter situation enables the sensory professional to measure the effect of the concept on scores for the products, in effect, to determine whether responses to one or all products are more strongly influenced by one concept than by another (concept). Storage tests are especially well suited to these designs because different subjects evaluate each test product (the "within" variable) and the products change over time (the "between" variable).

The particular example in Table 4.5 yields a separate error term for evaluating the two main treatment effects ($MS_{error(b)}$ and $MS_{error(w)}$). The products are the "within" variable, and the production plants are the "between" variable. The interaction term (A × B) is treated as the "within" variable in this design. If there is a choice, the variable of most importance should be used as the "within" variable.

A "two-within and one-between" split-plot design is a T × T × S design achieved with different subjects segmented according to a third test variable. The third test variable might be a production facility, a concept, a treatment variable

Table 4.4 Source of Variance Table for a Treatments X Treatments X Subjects Design[a]

Source of Variation	df	Sum of Squares	Mean Square	F
Flavor concentrations (A)	$a-1$	$SS_A = \dfrac{\sum^a(\sum^{bs}x)^2}{bs} - \dfrac{(\sum^{abs}x)^2}{abs}$	$SS_A/(a-1)$	ms_A/ms_{AS}
Carbonation levels (B)	$b-1$	$SS_B = \dfrac{\sum^b(\sum^{as}x)^2}{as} - \dfrac{(\sum^{abs}x)^2}{abs}$	$SS_B/(b-1)$	ms_B/ms_{BS}
Subjects (S)	$s-1$	$SS_S = \dfrac{\sum^s(\sum^{ab}x)^2}{ab} - \dfrac{(\sum^{abs}x)^2}{abs}$	$SS_S/(s-1)$	
A × B (AB)	$(a-1)(b-1)$	$SS_{AB} = \dfrac{\sum^{ab}(\sum^s x)^2}{s} - \dfrac{\sum^a(\sum^{bs}x)^2}{bs} - \dfrac{\sum^b(\sum^{as}x)^2}{as} + \dfrac{(\sum^{abs}x)^2}{abs}$	$SS_{AB}/(a-1)$ $\times (b-1)$	ms_{AB}/ms_{ABS}
A × S (AS)	$(a-1)(s-1)$	$SS_{AS} = \dfrac{\sum^{as}(\sum^b x)^2}{b} - \dfrac{\sum^a(\sum^{bs}x)^2}{bs} - \dfrac{\sum^s(\sum^{ab}x)^2}{ab} + \dfrac{(\sum^{abs}x)^2}{abs}$	$SS_{AS}/(a-1)$ $\times (s-1)$	
B × S (BS)	$(b-1)(s-1)$	$SS_{BS} = \dfrac{\sum^{bs}(\sum^a x)^2}{a} - \dfrac{\sum^b(\sum^{as}x)^2}{as} - \dfrac{\sum^s(\sum^{ab}x)^2}{ab} + \dfrac{(\sum^{abs}x)^2}{abs}$	$SS_{BS}/(b-1)$ $\times (s-1)$	
Residual (ABS)	$(a-1)(b-1)$ $\times (s-1)$	$SS_{ABS} = SS_T - SS_A - SS_B - SS_S - SS_{AB} - SS_{AS} - SS_{BS}$	$SS_{ABS}/(a-1)$ $\times (b-1)(s-1)$	
Total	$abs-1$	$SS_T = \sum^{abs}x^2 - \dfrac{(\sum^{abs}x)^2}{abs}$		

[a]Design plan: all subjects evaluate all levels of two treatment conditions.

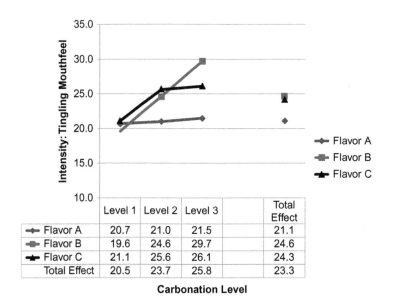

	Level 1	Level 2	Level 3		Total Effect
Flavor A	20.7	21.0	21.5		21.1
Flavor B	19.6	24.6	29.7		24.6
Flavor C	21.1	25.6	26.1		24.3
Total Effect	20.5	23.7	25.8		23.3

Carbonation Level

FIGURE 4.3

Results of a treatment-by-treatment-by-subject design testing flavor and carbonation level.

such as flavoring options, or a demographic classification such as age or gender. The source of variance table for this design is Table 4.6. As was noted for the T × T × S design, special attention must be given to testing each main effect and interaction with the appropriate error term for testing the "between" treatment main effect and the three separate error terms for testing the "within" treatments main effects and interactions. In this particular example, the "within" variable is the flavor concentration and the two "between" variables are carbonation and sucrose; that is, a set of products has been prepared using different levels of carbonation, with all other formulae variables held constant, and another set of products has been prepared using different levels of sucrose. Both sets of products would have two different levels of added flavor.

Finally, Table 4.7 is a source of variance table for a "one-within and two-between" design. Note that the variables are the same as those listed in the previous design. However, the analysis of these variables is determined by which variables were evaluated by all subjects and which variables were evaluated by different groups of subjects. A typical layout to facilitate planning a design and subsequent analysis is shown in Figure 4.4. The "within" variables are placed across the top of the matrix, and the "between" variables are listed down the side.

Numerous other multiple-variable designs have application in sensory evaluation. Most of these require a reasonable effort on the part of the sensory professional to delineate the sources of variability, selecting the appropriate error term for

Table 4.5 Source of Variance Table for a Split-Plot One-Within and One-Between Design[a]

Source of Variation	df	Sum of Squares	Mean Square	F
Between-subjects		Between $SS_T = \dfrac{\sum^{bn}(\sum^a x)^2}{a} - \dfrac{(\sum^{abn} x)^2}{abn}$		
Production plants (B)	$b - 1$	$SS_B = \dfrac{\sum^b(\sum^{an} x)^2}{an} - \dfrac{(\sum^{abn} x)^2}{abn}$	$SS_B/(b-1)$	$ms_B/ms_{error(b)}$
Error (b)	$b(n-1)$	$SS_{error(b)} = \dfrac{\sum^{bn}(\sum^a x)^2}{a} - \dfrac{\sum^b(\sum^{an} x)^2}{an}$	$SS_{error(b)}/b(n-1)$	
Within-subjects	$bn(a-1)$	Within $SS_T = \sum^{abn} x^2 - \dfrac{\sum^{bn}(\sum^a x)^2}{a}$		
Products (A)	$a - 1$	$SS_A = \dfrac{\sum^a(\sum^{bn} x)^2}{bn} - \dfrac{(\sum^{abn} x)^2}{abn}$	$SS_A/(a-1)$	$ms_A/ms_{error(w)}$
A × B	$(a-1)$ $(b-1)$	$SS_{AB} = \dfrac{\sum^{ab}(\sum^n x)^2}{n} - \dfrac{\sum^a(\sum^{bn} x)^2}{bn} - \dfrac{\sum^b(\sum^{an} x)^2}{an} + \dfrac{(\sum^{abn} x)^2}{abn}$	$SS_{AB}/(a-1)$ $\times(b-1)$	$ms_{AB}/$ $ms_{error(w)}$
Error (w)	$b(n-1)$ $\times(a-1)$	$SS_{error(w)} = \sum^{abn} x^2 - \dfrac{\sum^{ab}(\sum^n x)^2}{n} - \dfrac{\sum^{bn}(\sum^a x)^2}{a} + \dfrac{\sum^b(\sum^{an} x)^2}{an}$	$SS_{error(w)}/b(n-1)$ $\times(a-1)$	
Total	$abn - 1$	$SS_T = \sum^{abn} x^2 - \dfrac{(\sum^{abn} x)^2}{abn}$		

[a]Design plan: each different group of "B" subjects evaluate all "A" products.

Table 4.6 Source of Variance Table for a Split-Plot Two-Within and One-Between Design[a]

Source of Variation	df	Sum of Squares	Mean Square	F
Between-subjects	$cn - 1$	Between $SS_T = \sum^{cn}\dfrac{(\sum^{ab}x)^2}{ab} - \dfrac{(\sum^{abcn}x)^2}{abcn}$		
Carbonation (C)	$c - 1$	$SS_C = \dfrac{\sum^c(\sum^{abn}x)^2}{abn} - \dfrac{(\sum^{abcn}x)^2}{abcn}$	$SS_C/(c-1)$	$ms_C/ms_{error(b)}$
Error (b)	$c(n-1)$	$SS_{error(b)} = \dfrac{\sum^{cn}(\sum^{ab}x)^2}{ab} - \dfrac{\sum^c(\sum^{abn}x)^2}{abn}$	$SS_{error(b)}/c(n-1)$	
Within-subjects	$cn(ab - 1)$	Within $SS_T = \sum^{abcn}x^2 - \dfrac{\sum^{cn}(\sum^{ab}x)^2}{ab}$		
Flavor concentration (A)	$a - 1$	$SS_A = \dfrac{\sum^a(\sum^{bcn}x)^2}{bcn} - \dfrac{(\sum^{abcn}x)^2}{abcn}$	$SS_A/(a-1)$	$ms_A/ms_{error1(w)}$
Sucrose level (B)	$b - 1$	$SS_B = \dfrac{\sum^b(\sum^{acn}x)^2}{acn} - \dfrac{(\sum^{abcn}x)^2}{abcn}$	$SS_B/(b-1)$	$ms_B/ms_{error2(w)}$
A × B	$(a-1)(b-1)$	$SS_{AB} = \dfrac{\sum^{ab}(\sum^{cn}x)^2}{cn} - \dfrac{\sum^a(\sum^{bcn}x)^2}{bcn} - \dfrac{\sum^b(\sum^{acn}x)^2}{acn} + \dfrac{(\sum^{abcn}x)^2}{abcn}$	$SS_{AB}/(a-1)(b-1)$	$ms_{AB}/ms_{error3(w)}$
A × C	$(a-1)(c-1)$	$SS_{AC} = \dfrac{\sum^{ac}(\sum^{bn}x)^2}{bn} - \dfrac{\sum^a(\sum^{bcn}x)^2}{bcn} - \dfrac{\sum^c(\sum^{abn}x)^2}{abn} + \dfrac{(\sum^{abcn}x)^2}{abcn}$	$SS_{AC}/(a-1)(c-1)$	$ms_{AC}/ms_{error1(w)}$
B × C	$(b-1)(c-1)$	$SS_{BC} = \dfrac{\sum^{bc}(\sum^{an}x)^2}{an} - \dfrac{\sum^b(\sum^{acn}x)^2}{acn} - \dfrac{\sum^c(\sum^{abn}x)^2}{abn} + \dfrac{(\sum^{abcn}x)^2}{abcn}$	$SS_{BC}/(b-1)(c-1)$	$ms_{BC}/ms_{error2(w)}$

Source	df	SS	MS	F
A × B × C	$(a-1)(b-1)$ $\times (c-1)$	$SS_{ABC} = \dfrac{\sum^{abc}(\sum^{n}x)^2}{n} - \dfrac{\sum^{ab}(\sum^{cn}x)^2}{cn}$ $-\dfrac{\sum^{ac}(\sum^{bn}x)^2}{bn} - \dfrac{\sum^{bc}(\sum^{an}x)^2}{an}$ $+\dfrac{\sum^{a}(\sum^{bcn}x)^2}{bcn} + \dfrac{\sum^{b}(\sum^{acn}x)^2}{acn}$ $+\dfrac{\sum^{c}(\sum^{abn}x)^2}{abn} - \dfrac{(\sum^{abcn}x)^2}{abcn}$	$SS_{ABC}/(a-1)(b-1)$ $\times (c-1)$	$mS_{ABC}/mS_{error3(w)}$
Error₁(w)	$c(a-1)(n-1)$	$SS_{error_1(w)} = \dfrac{\sum^{acn}(\sum^{b}x)^2}{b} - \dfrac{\sum^{ac}(\sum^{bn}x)^2}{bn}$ $-\dfrac{\sum^{cn}(\sum^{ab}x)^2}{ab} + \dfrac{\sum^{c}(\sum^{abn}x)^2}{abn}$	$SS_{error_1(w)}/c(a-1)$ $\times (n-1)$	
Error₂(w)	$c(b-1)(n-1)$	$SS_{error_2(w)} = \dfrac{\sum^{bcn}(\sum^{a}x)^2}{a} - \dfrac{\sum^{bc}(\sum^{an}x)^2}{an}$ $-\dfrac{\sum^{cn}(\sum^{ab}x)^2}{ab} + \dfrac{\sum^{c}(\sum^{abn}x)^2}{abn}$	$SS_{error_2(w)}/c(b-1)$ $\times (n-1)$	
Error₃(w)	$c(a-1)(b-1)$ $\times (n-1)$	$SS_{error_3(w)} = \sum^{abcn}x^2 - \dfrac{\sum^{abc}(\sum^{n}x)^2}{n}$ $-\dfrac{\sum^{acn}(\sum^{b}x)^2}{b} - \dfrac{\sum^{bcn}(\sum^{a}x)^2}{a}$ $+\dfrac{\sum^{ac}(\sum^{bn}x)^2}{bn} + \dfrac{\sum^{bc}(\sum^{an}x)^2}{an}$ $+\dfrac{\sum^{cn}(\sum^{ab}x)^2}{ab} - \dfrac{\sum^{c}(\sum^{abn}x)^2}{abn}$	$SS_{error_3(w)}/c(a-1)$ $\times (b-1)(n-1)$	
Total	$abcn - 1$	$SS_T = \sum^{abcn}x^2 - \dfrac{(\sum^{abcn}x)^2}{abcn}$		

[a]Design plan: each different group of "C" subjects evaluate all levels of two treatment conditions.

Table 4.7 Source of Variance Table for a Split-Plot One-Within and Two-Between Design[a]

Source of Variation	df	Sum of Squares	Mean Square	F
Between-subjects	$bcn - 1$	Between $SS_T = \dfrac{\sum^{bcn}(\sum^a x)^2}{a} - \dfrac{(\sum^{abcn} x)^2}{abcn}$		
Carbonation level (B)	$b - 1$	$SS_B = \dfrac{\sum^b(\sum^{acn} x)^2}{acn} - \dfrac{(\sum^{abcn} x)^2}{abcn}$	$SS_B/(b-1)$	$ms_B/ms_{error(b)}$
Sucrose level (C)	$c - 1$	$SS_C = \dfrac{\sum^c(\sum^{abn} x)^2}{abn} - \dfrac{(\sum^{abcn} x)^2}{abcn}$	$SS_C/(c-1)$	$ms_C/ms_{error(b)}$
B × C	$(b-1)(c-1)$	$SS_{BC} = \dfrac{\sum^{bc}(\sum^{an} x)^2}{an} - \dfrac{\sum^b(\sum^{acn} x)^2}{acn} - \dfrac{\sum^c(\sum^{abn} x)^2}{abn} + \dfrac{(\sum^{abcn} x)^2}{abcn}$	$SS_{BC}/(b-1)(c-1)$	$ms_{BC}/ms_{error(b)}$
Error (b)	$bc(n-1)$	$SS_{error(b)} = \dfrac{\sum^{bcn}(\sum^a x)^2}{a} - \dfrac{\sum^{bc}(\sum^{an} x)^2}{an}$	$SS_{error(b)}/bc(n-1)$	
Within-subjects	$bcn(a-1)$	Within $SS_T = \sum^{abcn} x^2 - \dfrac{\sum^{bcn}(\sum^a x)^2}{a}$		
Flavor concentration (A)	$(a-1)$	$SS_A = \dfrac{\sum^a(\sum^{bcn} x)^2}{bcn} - \dfrac{(\sum^{abcn} x)^2}{abcn}$	$SS_A/(a-1)$	$ms_A/ms_{error(w)}$
A × B	$(a-1)$ $(b-1)$	$SS_{AB} = \dfrac{\sum^{ab}(\sum^{cn} x)^2}{cn} - \dfrac{\sum^a(\sum^{bcn} x)^2}{bcn} - \dfrac{\sum^b(\sum^{acn} x)^2}{acn} + \dfrac{(\sum^{abcn} x)^2}{abcn}$	$SS_{AB}/(a-1)(b-1)$	$ms_{AB}/ms_{error(w)}$
A × C	$(a-1)$ $(c-1)$	$SS_{AC} = \dfrac{\sum^{ac}(\sum^{bn} x)^2}{bn} - \dfrac{\sum^a(\sum^{bcn} x)^2}{bcn} - \dfrac{\sum^c(\sum^{abn} x)^2}{abn} + \dfrac{(\sum^{abcn} x)^2}{abcn}$	$SS_{AC}/(a-1)(c-1)$	$ms_{AC}/ms_{error(w)}$
A × B × C	$(a-1)$ $(b-1)$ $(c-1)$	$SS_{ABC} = \dfrac{\sum^{abc}(\sum^n x)^2}{n} - \dfrac{\sum^{ab}(\sum^{cn} x)^2}{cn} - \dfrac{\sum^{ac}(\sum^{bn} x)^2}{bn} - \dfrac{\sum^{bc}(\sum^{an} x)^2}{an} + \dfrac{\sum^a(\sum^{bcn} x)^2}{bcn} + \dfrac{\sum^b(\sum^{acn} x)^2}{acn} + \dfrac{\sum^c(\sum^{abn} x)^2}{abn} - \dfrac{(\sum^{abcn} x)^2}{abcn}$	$SS_{ABC}/(a-1)(b-1)(c-1)$	$ms_{ABC}/ms_{error(w)}$
Error (w)	$bc(a-1)$ $(n-1)$	$SS_{error(w)} = \sum^{abcn} x^2 - \dfrac{\sum^{abc}(\sum^n x)^2}{n}$	$SS_{error(w)}/bc(a-1)(n-1)$	
Total	$abcn - 1$	$SS_T = \sum^{abcn} x^2 - \dfrac{(\sum^{abcn} x)^2}{abcn}$		

[a]Design plan: each different group of "BC" subjects evaluate all "A" products.

(a) Layout for a Split-Plot with One-Within and One-Between Variable

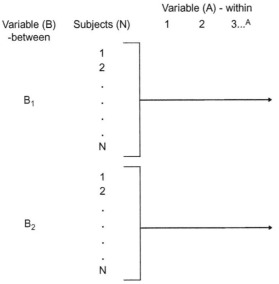

^{A}This design is equivalent to "stacking" two T x S designs on one another.

(b) Layout for a Split-Plot One-Within and Two-Between Design

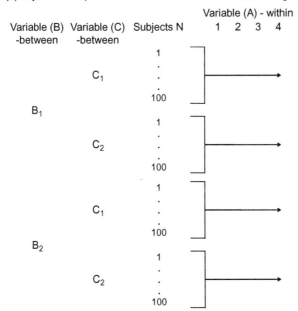

FIGURE 4.4

Layouts for design planning.

testing the different effects, and to decide which comparisons are most appropriate, but this should not be an impediment to their use. Their use allows for the evaluation of arrays of products that might not otherwise be considered feasible. The sensory professional must take the time to become familiar with the range of possibilities and, if needed, to seek assistance from the specialist in statistics. This is the only approach that offers hope for achieving maximum information from every test. In the section on Optimization, in Chapter 8, we discuss some additional experimental designs (e.g. response surface) that are used in sensory evaluation.

4.6 **Experimental design considerations**

In preparing an experimental design—that is, the written plan that designates product serving order—the sensory evaluation professional will have addressed all of the issues considered in the preceding sections (objective, product appropriateness, etc.). These latter issues may be considered as sensory input to test design as distinct from mathematical or statistical input. A considerable number of texts on experimental design exist (Box *et al.*, 1978; Cochran and Cox, 1957; McNemar, 1969; Winer, 1971). Nevertheless, the unique requirements of sensory evaluation necessitate additional discussion and selected examples (of designs) to reflect these requirements.

Experimental design and statistical analysis act in concert to substantively influence the outcome of a test. The design dictates the serving order of the products as well as the most appropriate analysis to satisfy the specific test objective. One may still be able to analyze the results of a poorly designed study, but the variables of interest are likely to be confounded and therefore inseparable from other effects so that any clear statement about the products may be quite risky. On the other hand, neither the experimental design nor the analysis can compensate for poor test execution or an ill-defined objective.

With these thoughts in mind, we direct our attention to consideration of the sensory requirements in experimental design, followed by selected examples of experimental designs and a discussion on the application of more sophisticated designs in sensory evaluation.

The following design guidelines reflect the aforementioned issues as well as the practicalities of the test as one of several sources of business information used to make a recommendation within the briefest possible time period:

1. Responses of a subject are easily influenced by familiarity with the products and by previous test experience. Use balanced-block designs that minimize, or evenly distribute, bias from the interactions of subject, products, and time.
2. When reasonable, all products should be evaluated equally often by each subject. By reasonable, we mean tests involving four or fewer products. For tests involving five or more products, the number of permutations precludes strict subject adherence to this requirement. Alternatives are discussed in guidelines 5 and 6.

3. All products should be served equally often before and after every other product and in all positions equally.

4. With balanced-block designs, the complete permutation is served before repeating a serving order. This procedure helps to minimize any time order effects.

5. Completely randomized designs are considerably less useful than balanced-block designs because some serving orders will occur infrequently, whereas others will occur too frequently. With large numbers of respondents, such as 100 or more, these deficiencies may be overcome. However, most sensory tests involve less than 50 subjects, which could introduce considerable order bias. Table 4.8 demonstrates such a problem with a four-product random plan (from Sidel and Stone, 1976) for 24 subjects; Section B shows that each product is not served in each position equally often. The problem is further exacerbated if one considers the serving order based on adjacent pairs (see Section C). For example, the sequence of product 1 followed by product 2 is served a total of eight times—four in the first and second positions, once in the second and third positions, and three times in the third and fourth positions. At the other extreme, the sequence of product 2 followed by product 3 occurs only three times—once in the 1–2 serving position, twice in the 2–3 position, and not at all in the 3–4 position. Such an imbalance is avoided with the balanced-block designs.

6. If all products cannot be evaluated by each subject within a single session, the order is balanced across the panel within each session. Balanced-block designs are preferred to balanced-incomplete-block designs because they allow for all possible comparisons by each subject and also use the fewest number of subjects. With several sessions, it is possible for each subject to evaluate all products and avoid the use of an incomplete design. Incomplete-block designs provide different product context and subtleties to the subjects. This recommendation should not be misconstrued to mean that incomplete-block designs should not be used but, rather, that they should be considered only after a decision is reached that the balanced-block design approach is not possible. The literature contains numerous examples of incomplete designs (Cochran and Cox, 1957) and more continue to be published (Gacula, 1978). In defense of incomplete designs, however, it should be pointed out that their greatest value arises when one is looking at a large array of products requiring extensive time to evaluate. For example, one might be evaluating an array of 20 skin lotions and the test protocol requires a 2- or 3-day use. Data collection would be extended 60 days, which might be quite impractical to achieve and therefore an incomplete design might be preferred. "Mixed" designs (split-plot designs) might also be appropriate (Lindquist, 1953); however, the sensory evaluation professional should consider the problem in detail before reaching a final decision. In our own work, we avoid incomplete-block designs in which each subject evaluates less than 65–70% of the product set.

7. If it is not feasible to have all subjects evaluate all products, serving order balance is maintained across the panel. For example, in tests involving five products, the number of serving order permutations is 120. The determination of

Table 4.8 Serving Order for a Four-Product Random Test Plan (A) and the Frequency of Product Serving by Position (B) and the Pairing Sequence (C)

A.	Subject	Serving Order	Subject	Serving Order
	A	4 3 1 2	M	1 3 2 4
	B	1 2 4 3	N	4 3 2 1
	C	2 1 3 4	O	2 1 4 3
	D	3 2 1 4	P	3 2 1 4
	E	4 3 2 1	Q	2 4 3 1
	F	1 2 3 4	R	3 1 2 4
	G	2 4 1 3	S	4 2 1 3
	H	3 4 1 2	T	3 1 4 2
	I	2 1 3 4	U	1 2 3 4
	J	2 4 3 1	V	2 3 4 1
	K	1 2 4 3	W	1 3 4 2
	L	4 3 1 2	X	4 1 3 2

B.	Product	Serving Position First	Second	Third	Fourth	Total
	1	6	6	7	5	24
	2	7	7	4	6	24
	3	5	7	7	5	24
	4	6	4	6	8	24
Total		24	24	24	24	

C.	Pairs	Serving Position 1–2	2–3	3–4	Total
	1 2	4	1	3	8
	1 3	2	3	3	8
	1 4	0	2	2	4
	2 1	3	3	2	8
	2 3	1	2	0	3
	2 4	3	3	2	7
	3 1	2	2	2	6
	3 2	2	3	1	6
	3 4	1	2	3	6
	4 1	1	2	1	4
	4 2	1	0	2	3
	4 3	4	2	3	9

the permutations is based on the formula $P_{n,n} = n!$ (from Alder and Roessler, 1977), where n,n refers to the number of products taken in n sets, and $n!$ refers to n factorial: $(n)(n-1)(n-2)$, and so on. For five products taken in sets of 5, $P_{5,5} = 5 \times 4 \times 3 \times 2 \times 1 = 120$ orders of presentation of the five products. In many situations, all 120 orders of presentation will not be served, and it is the responsibility of the sensory professional to select serving orders that are within reasonable balance. Examples are given in Section 4.7.

8. A monadic sequential serving order provides for greater control of the time interval between products—hence control of sensory interaction and sensory fatigue—and provides for a more independent response to each product.

Although it may be argued that these guidelines are obvious to the experienced sensory professional and the statistician, the evidence to date suggests that considerable confusion continues to exist concerning the concept of experimental design in relation to sensory evaluation test activities. The literature provides excellent references on the subject of experimental design, including Lindquist (1953), Cochran and Cox (1957), and Winer (1971). However, the more practical sensory issues remain relatively untouched. Although publications by Gacula (1978) and Mullen and Ennis (1979) represent efforts to correct this deficiency, much more work will be needed to improve the quality of sensory information that is so heavily dependent on design. This effort is especially important with the systems for the optimization of products and the development of predictive models for product acceptance. These programs reflect the competitive forces of the marketplace where manufacturers seek to identify and develop products with unique characteristics. In Chapter 8, we discuss some of these optimization programs and the role of sensory evaluation in making them successful.

4.7 Selected product designs

To supplement the guidelines, we have prepared a series of designs reflecting the sensory evaluation requirements of testing. These particular designs are presented here for illustrative purposes and should not be followed without consideration for the test objective. In large tests, those involving five or more products, the particular design presented here is one of many possible designs and is presented solely to demonstrate that each sensory professional must develop his or her personal file of experimental designs. The sensory professional will be able to handle test requests with minimal delay and will have one less source of error.

An experimental design for a three-product test is presented in Table 4.9. The six orders of presentation for the three products (3 factorial, 3!) are taken in sets of three, so the number of subjects should be in multiples of 6—that is, 6, 12, 18, 24, and so on—if the balanced order is to be maintained. In this particular design, it can be seen that for every six subjects, each product appears equally often (twice) in each serving position and precedes and follows every other product equally

Table 4.9 Balanced-Block Design for a Three-Product Test[a]

	Product Serving Order				Product Serving Order		
Subject	First	Second	Third	Subject	First	Second	Third
1	1	2	3	19	1	3	2
2	2	3	1	20	2	1	3
3	3	1	2	21	3	2	1
4	1	3	2	22	1	2	3
5	2	1	3	23	2	3	1
6	3	2	1	24	3	1	2
7	2	3	1	25	2	1	3
8	3	1	2	26	3	2	1
9	1	3	2	27	1	2	3
10	2	1	3	28	2	3	1
11	3	2	1	29	3	1	2
12	1	2	3	30	1	3	2
13	3	1	2	31	3	2	1
14	1	3	2	32	1	2	3
15	2	1	3	33	2	3	1
16	3	2	1	34	3	1	2
17	1	2	3	35	1	3	2
18	2	3	1	36	2	1	3

[a]*See text for an explanation of the entries.*

often. We refer to this as "unfolding" in the sense that the aforementioned balance is maintained as the test progresses. Thus, any effects that are time dependent will be more easily isolated, and any question regarding serving order can also be addressed. With 36 subjects, each product appears in each position 12 times, as do the pairings (1–2, 2–1, and so on). This particular design should not be construed as a recommendation for only doing three product tests with 36 subjects; the only concern is that each product serving order is not compromised to an extreme.

A balanced-block design for a four-product test is presented in Table 4.10. There are 24 orders of presentation for the four products, so the number of subjects should be in multiples of 24 (6, 12, and so on). Of course, one can use fewer than 24 subjects and add replication as a means of achieving the desired balance. In this latter situation, one could use 12 subjects and have a single replicate; this would mean serving the 13th through 24th orders, shown in Table 4.10, as the replicate. There will also be test plans in which the number of subjects is not divisible by the serving orders. In Table 4.11, for example, is a design for 10 subjects and four sessions, such that there is a reasonable balance. Each product appears equally often in each position—first, second, third, and fourth—but not in each session. All four sessions are necessary for this to be realized, and a similar case is made for adjacent pairings. Each of the adjacent pairs occurs on 10 occasions across all four

Table 4.10 Balanced-Block Design for a Four-Product Test[a]

Subject	Serving Order			
	First	**Second**	**Third**	**Fourth**
1	4	3	1	2
2	2	1	3	4
3	1	2	4	3
4	3	4	2	1
5	4	1	2	3
6	3	2	1	4
7	4	2	3	1
8	1	3	2	4
9	2	3	4	1
10	1	4	3	2
11	2	4	1	3
12	3	1	4	2
13	3	2	4	1
14	1	4	2	3
15	2	3	1	4
16	4	1	3	2
17	1	3	4	2
18	2	4	3	1
19	2	1	4	3
20	3	4	1	2
21	3	1	2	4
22	4	2	1	3
23	1	2	3	4
24	4	3	2	1

[a]See text for an explanation of the entries.

sessions and the 10 subjects. When the balance is not possible within a reasonable number of sessions, some compromise must be made to minimize the imbalance that occurs. This particular issue is addressed in a design offered for a five-product test in which the number of serving orders is 120 (5! = 120). Table 4.12 lists the serving orders for 20 subjects evaluating each product twice or 40 subjects once, but with the serving order and the adjacent pairs kept in balance. However, because only 40 of the possible 120 serving orders were used to construct the design and with larger panels, one can select serving orders that maintain the balance.

In the situation in which the subjects are unable to evaluate all products in a single session but the desire is to stay with the balanced block, the serving orders can be prepared in the same way, as is shown in Table 4.11, and the products can be evaluated in separate sessions, as shown in Table 4.13. With 12 products, three

Table 4.11 Balanced-Block Design for a Four-Product Test with Replication[a]

Subject	Serving Orders Within and Across the Sessions			
	First Session	**Second Session**	**Third Session**	**Fourth Session**
1	1 2 3 4	3 2 4 1	2 3 1 4	4 2 1 3
2	2 4 1 3	2 1 3 4	3 4 2 1	1 4 3 2
3	3 1 4 2	4 3 1 2	1 2 4 3	3 4 1 2
4	4 3 2 1	1 4 2 3	4 1 3 2	4 2 3 1
5	2 3 4 1	3 2 1 4	1 2 3 4	1 3 2 4
6	3 1 2 4	2 4 3 1	2 4 1 3	2 1 4 3
7	4 2 1 3	1 3 4 2	3 1 4 2	4 1 2 3
8	1 4 3 2	4 1 2 3	4 3 2 1	1 3 4 2
9	3 4 1 2	2 1 4 3	2 3 4 1	2 4 3 1
10	4 2 3 1	1 3 2 4	3 1 2 4	3 2 1 4

[a]See text for an explanation of the entries.

Table 4.12 Balanced-Block Design for a Five-Product Test[a]

Subject	Serving Order				
	First	**Second**	**Third**	**Fourth**	**Fifth**
1	4	5	1	2	3
2	5	2	4	3	1
3	1	4	3	5	2
4	2	3	5	1	4
5	3	1	2	4	5
6	3	2	1	5	4
7	1	3	4	2	5
8	2	5	3	4	1
9	4	1	5	3	2
10	5	4	2	1	3
11	1	3	5	4	2
12	5	1	2	3	4
13	2	5	4	1	3
14	3	4	1	2	5
15	4	2	3	5	1
16	2	4	5	3	1
17	4	3	2	1	5
18	3	1	4	5	2
19	5	2	1	4	3
20	1	5	3	2	4
21	1	2	3	4	5

Table 4.12 Balanced-Block Design for a Five-Product Test[a] (*Continued*)

Subject	Serving Order				
	First	**Second**	**Third**	**Fourth**	**Fifth**
22	2	4	1	5	3
23	3	1	5	2	4
24	4	5	2	3	1
25	5	3	4	1	2
26	5	4	3	2	1
27	3	5	1	4	2
28	4	2	5	1	3
29	1	3	2	5	4
30	2	1	4	3	5
31	2	3	4	5	1
32	3	5	2	1	4
33	4	2	1	3	5
34	5	1	3	4	2
35	1	4	5	2	3
36	1	5	4	3	2
37	4	1	2	5	3
38	5	3	1	2	4
39	2	4	3	1	5
40	3	2	5	4	1

[a]See text for an explanation of the entries.

Table 4.13 Example from a Balanced-Block Design for a Twelve-Product Test[a]

Subject	Orders of Presentation Within and Across Sessions[a]		
	First Session	**Second Session**	**Third Session**
1	7 8 9 10	11 12 1 2	3 4 5 6
2	8 10 12 2	4 6 7 9	11 1 3 5
3	9 12 3 6	8 11 2 5	7 10 1 4
4	10 2 6 9	1 5 8 12	4 7 11 3
5	11 4 8 1	6 10 3 7	12 5 9 2
6	12 6 11 5	10 4 9 3	8 2 7 1
7	1 7 2 8	3 9 4 10	5 11 6 12
8	2 9 5 12	7 3 10 6	1 8 4 11
9	3 11 7 4	12 8 5 1	9 6 2 10
10	4 1 10 7	5 2 11 8	6 3 12 9
11	5 3 1 11	9 7 6 4	2 12 10 8
12	6 5 4 3	2 1 12 11	10 9 8 7

[a]See text for an explanation of the entries.

sessions will be required for each subject to evaluate each product once. If replication is necessary, then the design will reflect this requirement and the sensory professional will verify the serving orders for individual products and adjacent pairs.

In the case of balanced-incomplete-block designs, the rules remain the same insofar as individual products and pairings are concerned. Incomplete-block designs require more subjects, and in situations involving products with very pronounced sensory interaction (chili, cigars, etc.), these designs may be helpful. The number of permutations in an incomplete design is determined from the following formula:

$$P_{n,r} = \frac{n!}{(n-r)!}$$

when n is the number of products, and r is the number of products in a set. Thus, for a balanced-incomplete 4-of-5 design, there will be 120 orders; for a 3-of-5 design, there will be 60 orders. If all orders are not served, then every effort is made to maintain the balance of individual and adjacent pairs. Table 4.14 is a 4-of-5 design for 40 subjects that meets these criteria.

Table 4.14 Example of a Balanced-Incomplete-Block Design for a Five-Product Test[a]

Subject	Serving Order			
	First	Second	Third	Fourth
1	1	2	4	3
2	2	3	5	4
3	3	4	1	5
4	4	5	2	1
5	5	1	3	2
6	1	2	5	4
7	2	3	1	5
8	3	4	2	1
9	4	5	3	2
10	5	1	4	3
11	1	3	4	2
12	2	4	5	3
13	3	5	1	4
14	4	1	2	5
15	5	2	3	1
16	1	3	2	5
17	2	4	3	1
18	3	5	4	2

Table 4.14 Example of a Balanced-Incomplete-Block Design for a Five-Product Test[a] (*Continued*)

Subject	Serving Order			
	First	Second	Third	Fourth
19	4	1	5	3
20	5	2	1	4
21	1	4	5	2
22	2	5	1	3
23	3	1	2	4
24	4	2	3	5
25	5	3	4	1
26	1	4	3	5
27	2	5	4	1
28	3	1	5	2
29	4	2	1	3
30	5	3	2	4
31	1	5	3	4
32	2	1	4	5
33	3	2	5	1
34	4	3	1	2
35	5	4	2	3
36	1	5	2	3
37	2	1	3	4
38	3	2	4	5
39	4	3	5	1
40	5	4	1	2

[a]*See text for an explanation of the entries.*

A special consideration of the aforementioned designs is the method of serving order blocking. The objective of blocking within a serving design is to separate the sources of error (or experimental units) into groups such that the effect of the unit can be measured through statistical analysis. The variation within a block is expected to have less variability than the variation between the blocks. Although this method is similar to the multiple-factor designs described in Section 4.3, the blocking we address here emphasizes a slightly different design component, namely order of product presentation. An example of the results for a typical blocked design is illustrated in Figure 4.5 and demonstrates how an array of products that are within the same general category can be blocked by a unique character type (flavor). By creating a serving design such that all products within a flavor type

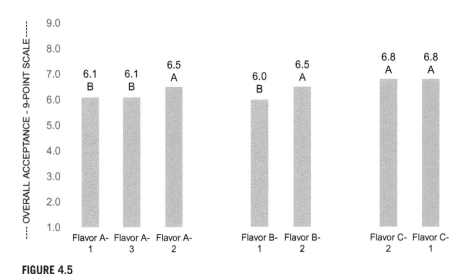

FIGURE 4.5

Blocking by serving order. Consumer acceptance results are summarized for products blocked by flavor type. Emphasis is placed on the differences among products within a flavor type and not across all flavors.

are evaluated prior to all other flavor types, the effect of flavor type is separated from the design. The blocking method is especially beneficial when the blocking variable is expected to have a large influence on perception and/or preference. As noted previously, the serving order presentation is helpful in reducing the potential psychological influences of contrast and convergence.

In addition to these designs, there will be situations in which the emphasis is solely on a comparison of one product (a reference) to a series of other products (competitive, experimental, etc.) and there is no concern about how the competitive products compare with one another. The initial idea might be a design in which the subjects evaluate the products in pairs—for example, 1 versus 2, 1 versus 3, and 1 versus 4. However, this design is inefficient in the use of the subjects because of the need to include the reference as one of each pair. There will be more responses for the reference product resulting in an unequal number of judgments in one cell, and there is a risk that the subjects may recognize the reference, resulting in some atypical response patterns. This particular approach is often used in the mistaken belief that including a reference with every evaluation increases test precision. Unfortunately, this approach also adds more variability to the testing process inasmuch as the reference product is variable and there is also an increase in adaptation—sensory fatigue. One solution to this problem is to not change the test design but use a specific ANOVA followed by Dunnett's test (intended for comparisons involving one product vs. all the others).

This particular issue warrants further consideration because it can be troublesome for sensory evaluation. Although the use of multiple paired comparisons is an acceptable design and there are appropriate statistical methods for analysis, the more basic issue is whether this is the most efficient use of the subjects' skills and the implication of always including a reference product in a test. From a strict sensory evaluation perspective, the use of multiple pairs is inefficient in the sense that for every test product response there will be a reference product response, which means that the database for the latter product will double that for the other products. This is not necessarily an undesirable situation; however, it will require an analysis that takes into account the unequal cell entries (if the data are pooled). The number of subjects required for a test is increased substantially, or if the same subjects are used, then the potential of product sensory interaction is concomitantly increased.

Products requiring extensive preparation may introduce additional variability, and the potential exists that the subjects recognize that one product in every pair is the same, and this could alter their responses. Subjects might make comparative judgments; that is, the response would be a combination of response to the product coupled with the difference between the products. This is not to imply that a paired test should never be considered and implemented but, rather, that alternatives should be thoroughly explored with the requestor before a final decision is reached. Finally, the requestor should be aware that the multiple-paired test plan makes it very difficult for post-test comparisons between the experimental products. Although there are procedures by which results from such a test can be converted to scaled values and thus enable such comparisons to be made, they must be planned in advance. In Chapter 3, we discussed paired versus scaled procedures, and we will have more to say about the topic in Chapter 7.

The sensory professional is confronted with a vast array of options in terms of how best to design a test. There is no single best test method or ideal subject, just as there is no single test design or method of analysis that is applicable in every situation. Each request must be considered with care to ensure that the test objective is consistent with the requestor's intent and that there is a clear understanding of how the results are expected to be used. The objective and the product will determine the specific method, the experimental design, and the method of analysis. Each element must be kept in perspective, and only through a careful and logical process can the best plan be formulated for that particular problem.

Discrimination Testing

CHAPTER OUTLINE

H. Stone, R. Bleibaum, H. A. Thomas: Sensory Evaluation Practices, fourth edition.
DOI: http://dx.doi.org/10.1016/B978-0-12-382086-0.00005-4

5.1 Introduction

Discrimination testing as a class of tests represents one of the two most useful analytical tools available to the sensory professional. It is on the basis of a perceived difference between two products that one can justify proceeding to a descriptive test in order to identify the basis for the difference, or the converse, products are not perceived as different, and appropriate action is taken; for example, the alternative ingredient can be used.

Within this general class of discrimination methods are a variety of specific methods; some are well-known, such as paired-comparison, triangle, duo–trio, and directional difference tests, whereas others are relatively unknown, such as the dual-standard test method. However, all the methods are intended to answer a seemingly simple question: "Are these products perceived as different?" Obviously, the response to this question can have major consequences. If the two products are perceived as different (at a previously established level of risk) and the development objective is to not be different, then the objective has not been achieved and the planned sensory testing sequence will likely be changed. Knowledge of perceived product differences enables management to minimize testing where it is either unwarranted or premature and to anticipate response to the products in the event of additional testing (e.g. products found to be different in a discrimination test will be described as different in a descriptive test). Reaching decisions about product differences based on carefully controlled tests, conducted in a manner that is well-defined and unambiguous, leaves little room for error in the decision-making process. To attain this level of sophistication, researchers must understand the behavioral aspects of the discrimination process and have an understanding of test products. This information should be combined with knowledge about the various test methods as part of the overall sensory testing program. As with any other scientific research, nothing can be done in isolation. In this chapter, we explore these issues in considerable detail in an effort to clarify what is possible along with the benefits and risks when using discrimination testing.

In a very early publication on discrimination testing, Peryam and Swartz (1950), in discussing flavor evaluation, observed that there existed a "tendency toward oversimplification in a complex field, this lack of appreciation of the true difficulties involved." This lack of appreciation for the complexities of the discrimination test may stem from the ease and simplicity with which the test is described and implemented. A typical instruction may be "Here are two products; which one is more salty?" or "Here are three products; which one is different from the other two?" The subject's task initially appears to be relatively simple; however, this apparent simplicity has led many researchers to create test variations without fully appreciating their consequences. In some instances, these variations were intended to increase the amount of information derived from the test; because the subject is there, why not get more information. In other instances, alternatives were proposed with the goal of providing a more precise measure of the difference or, at the very least, insight into the subject's decision-making process. As a result, the literature provides the researcher with an array of options in test methods and analyses, many of which create problems and may not enhance the decision-making process.

In discrimination testing, issues of oversimplification are due primarily to a lack of understanding of the basis for the discrimination test, lack of appreciation for subject qualification criteria, and considerable confusion regarding when the test is used, the discrimination method most appropriate for the research objective, and the interpretation of results (Lawless and Heymann, 2010; Roessler *et al.*, 1978; Stone and Sidel, 1978). The researcher must also be keenly aware of the limitations of the discrimination model. The subject's response is a discrete judgment, and as such, there are limitations as to what can be done with the result. As a consequence, misguided alternatives have been created, such as combining difference with a preference question or with a series of descriptive attributes, measuring the magnitude of the difference, or misinterpretation of results—for example, equating the level of statistical significance with the magnitude of difference. Although these efforts appear to have a research focus and/or an effort to extend or enhance the methodology, the sensory professional's failure to recognize the real likelihood for making a wrong decision as a consequence of the variations can have a far more significant and usually negative impact on test results and on the credibility of the sensory department within the company.

This problem of misinterpretation is exemplified in two modifications that involve the use of consumers in a discrimination test. It has been suggested that the difference test be used for screening consumers as a prelude to a preference test (Buchanan and Henderson, 1992; Johnson, 1973). The objective of this procedure was to eliminate individuals who were nondiscriminatory and hence could not have a preference and to obtain a truer measure of product preference. Unfortunately, this approach failed to take into account several basic issues, including the learning processes that all respondents go through when presented with a novel situation such as the discrimination test. This learning process requires many test trials, particularly when the evaluation involves products that stimulate the chemical senses. In addition, it would be impossible to determine after a single trial whether an individual could reliably discriminate differences because of the novelty of the task, a failure to understand the task, or an inability to perceive the difference. One last point is that consumers who participate in discrimination screening tests are no longer able to provide unbiased preference judgments. There is a response bias toward the different product—a factor of some importance if a preference question is asked after the discrimination task. This proposed prescreening of consumers had some early advocates; however, it appears to have been discarded based on the reasons listed here.

A second example of the misuse of the discrimination test, also involving consumers, is the use of large numbers of consumers. This approach is typically proposed as providing results representative of the population. With 100 consumers, 59 (59%) correct matches would be significant at the 0.05 probability level; with 200 consumers, 112 (56%) correct matches would be needed for significance. There are fundamental issues associated with this approach, including that it is easier to obtain a statistically significant but not practical result. This approach is not limited to discrimination but extends to any test in which large numbers of responses are obtained in the belief that "there is safety in numbers." Although the results are statistically correct, the practical implications could be quite different (Roessler *et al.*, 1978; see also the discussion in Chapter 4 on Type 1 and Type 2 errors). Should one be prepared to state that two products are different in a test involving large numbers of unqualified consumers? A related issue is the potential confounding of the variable being tested with typical product variation. The more product required for testing, the greater the likelihood of production variation impacting the variable being tested. There is no doubt that deciding on the number of responses needed to have confidence in that decision is difficult. In this particular example, the problem is also complicated by the use of consumers with unknown sensory acuity. Consumers are not equally skilled, even at the most simplest of sensory tasks. It is not a question of the consumer's lack of ability to complete the task, but much more important, there is no way to know whether the consumer understood the task and actually perceived a difference. We discuss this and other examples of misuse of the discrimination test later.

The sensory professional must recognize the inherent weakness and the risk associated with any approach based on faulty assumptions (whether these are subject related, method related, or product related). Failure to address these issues can only result in lack of credibility for a sensory program.

If the conclusions and recommended next steps from a discrimination test are to be accepted by management as reliable, valid, and believable, then it is important that each test be conducted with proper consideration for all aspects of the test, from design to product preparation and handling and implementation, to data analysis and interpretation. Failure to appreciate all the subtleties of a test increases the risk of data misinterpretation. From a business viewpoint, the ultimate value of sensory evaluation in general and the discrimination test in particular derives from the reliability and validity of test results and in the manner in which decisions are made, based on the results. Finally, credibility is enhanced when recommended changes are made and subsequent consumer results confirm the decision.

The difference test, a special case of an ordinal scale, is a type of threshold test in which the task is to select the product that is different (Dember, 1963; Laming, 1986; ASTM E-679-04 and ASTM E1432-04). In the classic threshold test, a series of products, some of which contain no stimulus, are presented to the subject one at a time. The subject's task is to indicate whether or not the stimulus was detected. The subject is not required to identify the stimulus, only to indicate that perception has occurred. When identification is required, this may be referred to as a recognition threshold test. Some types of directional difference tests (e.g. *n*-AFC) involve recognition or at least differentiation according to a specific characteristic.

In product tests (food, beverage, tobacco, cosmetic, etc.), the problem for the subject is especially difficult because the variable being tested rarely has a one-to-one relationship with the output. That is, a change in the amount of an ingredient such as salt cannot be expected to only affect perceived saltiness, and in fact, a change in saltiness may not be perceived at all. In addition, all the other sensory characteristics of the products are being perceived, and the subject must sort all these sensory inputs to determine which product is different. These latter two issues are extremely important. The failure to appreciate the sensory complexities introduced by a product has caused considerable difficulty for sensory evaluation and for users of the service. For example, perception of salt differences in water would not be representative of the complexities entailed in evaluating salt differences in a prepared food. Despite this, one continues to encounter use of salt thresholds as a basis for identifying individuals expected to perform above the average in evaluating products containing salt. The salt in a food will have numerous other flavor effects, and hence the direct extrapolation to model systems will be misleading. Finally, the method will contribute to task complexity. The methods are not equivalent; some methods involve only two products, whereas others involve three but require as many as three comparisons for a single decision. This latter issue will have a direct impact on the sensitivity of a subject. In the next section, we describe the most common test methods in preparation for addressing the more complex issues of test selection, data collection, and interpretation.

The subjects and their responses represent a third and equally important component. Difference testing involves choice behavior. The subject must either select one product as different from another or, according to specific criteria, select the one that has more of a specific characteristic (e.g. sweetness and saltiness). From the subject's viewpoint, the situation involves problem-solving behavior, a challenge in which there are correct and incorrect decisions; to be successful, one must make the correct choice using all of the information and all of the clues that are available.

To a degree, choice behavior has been taken into account in the statistical analysis of the responses (Bradley, 1953; Gridgeman, 1955a; Roessler *et al.*, 1953). However, on closer examination, it appears that even here the situation is not totally clear. That is, the mathematical model on which the analyses are based does not completely satisfy all test possibilities such as replication. We will have more to say about this issue later in the chapter. Investigators have focused increasing attention on the power analysis and decision criteria in an effort to devise mathematical models of the discrimination process and the appropriate sample size for sensory and consumer science (Bi and Ennis, 2001a,b; Ennis, 1993; Ennis and Jesionka, 2011; McClure and Lawless, 2010; Rousseau and Ennis, 2001). Power analysis has long been recognized (Baker *et al.*, 1954; Radkins, 1957; Stone and Sidel, 1995). These investigators use an approach proposed by Thurstone (1927) and developed further in signal detection theory (Green and Swets, 1966). It is evident that the discrimination process continues to present investigators with many challenges, not the least of which is a reaffirmation of the complexity of the process.

In view of the diversity of applications and the confusion that appears to exist about the various methods and their applications, it is useful to first consider the methods in detail, followed by their applications. Only in this way can we hope to develop realistic guidelines for testing and for reaching meaningful decisions about product similarities and differences. Although the early literature can provide us with insight into the development of specific methods as well as their modifications

Paired-comparison test

Name _____ Code _____ Date _____

In front of you are two samples; starting with the sample on the left, evaluate each and circle the sample which is most sweet. You must make a choice, even it if is only a guess. You may retaste as often as you wish. Thank you.

847   566

FIGURE 5.1

An example of a scorecard for use with the directional paired-comparison test. The specific characteristic is stated on the scorecard, as shown here for illustrative purposes.

(Bradley, 1963; Gridgeman, 1959; Peryam and Swartz, 1950; Schlosberg *et al.*, 1954), it remains for us to apply this information in the contemporary business environment without sacrificing quality, reliability, and validity. In so doing, we discuss the discrimination tests used by sensory professionals to make product decisions, without digressing into a discussion about competing psychological theories or models attempting to explain the discrimination process.

5.2 Methods

Basically, the three types of discrimination tests that are used most often are paired comparison, duo–trio, and triangle. A number of other test types have been developed, but because of their limited application, no more than a cursory examination is justified.

5.2.1 Paired-comparison test

The paired-comparison test is a two-product test, and the subject's task is to indicate, by circling or by some similar means, the one product that has more of a designated characteristic such as sweetness, tenderness, or shininess, with the designated characteristic having been identified before the test and stated on the scorecard. This method is also identified as a directional paired-comparison test, with the "directional" component alerting the subject to a specific type of paired test. An example of the scorecard for the directional paired test is shown in Figure 5.1, with the direction stated. Examples of test items provided to a subject are shown in Figure 5.2.

Note that the instructions require the subject to make a decision; that is, the test is forced choice and the decision must be either one or the other of the two products. Forced-choice tests are discussed later in this chapter in connection with statistical considerations (see also Bi, 2011; Frijters, 1980; Gridgeman, 1959).

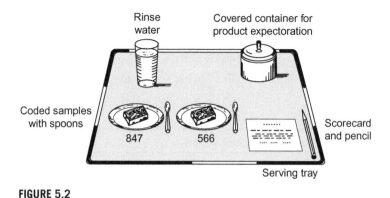

Rinse water

Covered container for product expectoration

Coded samples with spoons

847   566

Scorecard and pencil

Serving tray

FIGURE 5.2

Display of samples, scorecard, and other supplies provided to the subject.

The paired-comparison test is relatively easy to organize and to implement; the two coded products are served simultaneously, and the subject, after sampling both products, makes a decision. However, it is often difficult to specify the difference or be confident that the subjects will understand or recognize that difference. By "understand," we mean that the subjects are able to perceive that characteristic in the product. The effect of a single variable such as sweetness or saltiness cannot be totally specified; that is, a single ingredient modification may affect many other product characteristics, and assuming the subjects have responded only to that one is risky. If we must take the time to qualify subjects and train them to recognize a specific characteristic, then descriptive analysis is a more appropriate test. Despite this limitation of the directional paired test, it still has many applications. For example, by changing the instructions and requiring the subject to indicate only whether the products are the same or different by circling the appropriate word on the scorecard, one can eliminate the issues of directionality and semantics. Most subjects have an expectation about product differences; that is, they expect the products in any test to be different. If they are not informed of this change in instructions, their expectations will remain the same, and their responses could be atypical. The two orders of presentation in the directional paired test are AB and BA; however, the "same/different" form has four orders—AB, BA, AA, and BB. The latter two orders are necessary; otherwise, the subjects can state, and some will, that all of the product pairs are different and achieve 100% correct decision. This situation can be monitored if one tabulates responses according to the four orders of presentation. In Table 5.1 are two sets of responses: The first set of responses is consistent with expectation for the AA and BB series, whereas the latter set of responses is inconsistent and suggestive of misunderstood directions (or some similar problem).

Surprisingly little use is made of this form of the paired test, but it should have considerable application where product availability is limited and/or product characteristics (a strong-flavored product, spice mixes, tobacco, and so forth) suggest limiting subject exposure to the products.

Table 5.1 Number of Correct Decisions for the Nondirectional Paired Test According to Serving Order

Test	Serving Order				Totals (Correct/ Trials)
	AB	**BA**	**AA**	**BB**	
1	5[a]	5	4	5	19/40
2	7	7	0	1	15/40

[a]*Entries are typical and used here for illustrative purposes.*

Another version of the paired test is the A-not-A procedure. The subject is presented with and evaluates a single product that is then removed and is followed by a second product. The subject then makes a decision as to whether the products are the same (or different). This particular test procedure has considerable merit in situations in which non-test variables such as a color difference may influence results. Sensory evaluation professionals must recognize that a test method can be modified if the problem warrants it; however, any modifications should not change the basic integrity of the task or the statistical probability.

The paired-comparison procedure is the earliest example of the application of discrimination testing to food and beverage evaluation. In 1936, Cover described its use in the evaluation of meat (the paired-eating method). Additional attention was given to the discrimination procedure through the evaluation of beverages, as reported in a series of papers by the staff at the Carlsberg Breweries in Copenhagen by Helm and Trolle (1946) and as developed at the Seagram and Sons Quality Research Laboratory (Peryam and Swartz, 1950), and further utilized at the U.S. Army Quartermaster Food and Container Institute. Statistical issues were also addressed by these workers; however, a more thorough discussion of these statistical issues can be found in the work of Bradley (1953, 1963), Harrison and Elder (1950), and Radkins (1957); it is also discussed later in this chapter.

5.2.2 Duo–trio test

The duo–trio test, developed by Peryam and Swartz (1950) during their work with the Quartermaster Food and Container Institute, represented an alternative to the very popular triangle test that, for some, was a more complex test psychologically—that is, grouping three unknowns.

The duo–trio test was found to be useful for products that had relatively intense taste, odor, and/or kinesthetic effects such that sensitivity was significantly reduced. In the duo–trio test, the subject is presented with three products; the first is identified as the reference (or control) and the other two are coded. The subject's task is to indicate which product is most similar to the reference. This test can be viewed as a combination of two paired tests (R–A and R–B). A suggested, but not recommended, option is to remove the reference before serving the two coded products.

Duo–trio test

Name _____ Code _____ Date _____

In front of you are three samples, one marked **R** and the other two coded; evaluate the samples starting from left to right, first **R** and then the other two. Circle the code of the sample different from **R**.
You may retaste the samples. You must make a choice. Thank you.

      R   132   691

FIGURE 5.3

Example of scorecard used in the duo–trio test.

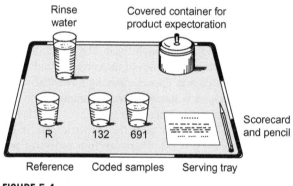

FIGURE 5.4

Example of items given to the subject for the duo–trio test.

Removal of the reference alters the subject's ability to refer back to the reference, if
necessary, before reaching a decision. The change may result in the test being more
a test of memory.

The original description of the method also included the possibility for presentation of a warm-up product to minimize first-sample effects. In many tests, product
availability may be limited or taste carry-over problems may be such that the use of
the warm-up technique would not be feasible. First-sample effects should be minimized through the use of qualified subjects and balanced designs.

The scorecard for the duo–trio test, as shown in Figure 5.3, is similar to that
used in the paired-comparison test, except that there is an "R" to the left of the two
coded products and instructions remind the subject to evaluate "R" before evaluating the other two products.

Examples of items provided to the subject are shown in Figure 5.4. Two design
options are available in the duo–trio test. The conventional approach is to balance the reference between the control and test products; however, in some situations, the reference may be kept constant, and the control is usually selected as

the reference. As Peryam and Swartz (1950) observed, if one product is considered to be more familiar to the subjects, then it would be selected as the reference (this is usually a control or product currently being manufactured). The use of this latter technique is encouraged where changes are contemplated in a product currently available in the marketplace. This approach, referred to as the constant-reference duo–trio test method, is intended to reinforce the sensory characteristics of the control and enhance detection of any product that is perceived as different. In situations in which formulation changes are necessary for a product currently available in the marketplace (because of ingredient availability, cost, etc.), it is important that the change be perceived by the fewest consumers possible. This objective stems from knowledge that the consumer, although not necessarily being aware, develops a sensory familiarity for a particular brand and would take a negative view of any perceived unadvertised changes. As a result, changes in a branded product are not intended to be perceived; that is, we do not wish to modify or change the product in any way that might be detected by the consumer. The reader should not assume that the intention is to fool the consumer but, rather, it is to ensure that the consumer's sensory reaction to the product remains unchanged. The use of the constant-reference design, by reinforcing the sensory properties of the current brand, increases the likelihood of perception of the difference, thus eliminating unwanted variables that might be perceived. This latter concern is given further consideration in the discussion on test strategy, risk, and the decision-making process.

The chance probability associated with the duo–trio test is identical to that of the other two product tests, $p = 1/2$. There are four orders of presentation with the balanced reference versus two orders with the constant reference. A series of experimental designs for this and other discrimination tests is presented later in this chapter.

5.2.3 Triangle test

The triangle test is the most well-known of the three methods. It has been used to a much greater extent because it was mistakenly believed to be more sensitive than other methods (i.e. based on the probability of 1/3). If this logic were extended, then a $p = 1/4$ would be even *more* sensitive; it is not. The method was often equated with sensory evaluation; that is, individuals claimed they had an active sensory program because they conducted triangle tests. This was never anticipated by its developers. As previously mentioned, the method was developed at the Carlsberg Breweries by Bengtsson and co-workers (Helm and Trolle, 1946) as part of an effort to use sensory tests for the evaluation of beer and to overcome difficulties associated with the directional paired method. The test was applied to a wide variety of products by other workers, most of whom found the method to be quite suitable.

The triangle test, as its name implies, is a three-product test in which all three products are coded and the subject's task is to determine which two are most similar or which one is most different from the other two. As shown in Figures 5.5 and 5.6, the subject is served three coded products and is required to indicate which

Triangle test

Name _____ Code _____ Date _____

In front of you are three coded samples, two are the same and one is different; starting from the left evaluate the samples and circle the code that is different from the other two. You may reevaluate the samples. You must make a choice. Thank you.

624   801   199

FIGURE 5.5

Example of scorecard for the triangle test.

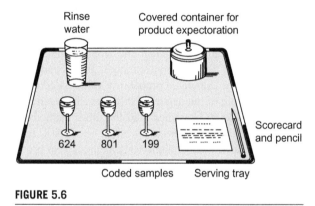

FIGURE 5.6

Example of items given to the subject for the triangle test.

one is most different. The chance probability associated with this test is only 0.33, which probably accounts for its claim of greater sensitivity. However, the fewer correct scores required for statistical significance should not be confused with the totally separate issue of sensitivity.

The triangle test is also a difficult test because the subject must recall the sensory characteristics of two products before evaluating a third and then making a decision. In fact, the test can be viewed as a combination of three paired tests (A–B, A–C, and B–C). These issues of test complexity and of sensory interaction should not be ignored by the experimenter (yet they often are) when organizing and fielding a triangle test.

After completing a series of olfaction studies comparing the triangle method with the three-alternative forced choice (3-AFC) signal-detection procedure, Frijters (1980) concluded that the triangle procedure is improved by modifying the test instruction to read which two of the three stimuli are "most similar?" Intuitively, this instruction makes sense because products from a single production lot can be different, thus presenting the subject with a real decision dilemma. This is not necessarily a new development as much as it is a reminder for the experimenter to be thorough in product screening before a test and to provide the subjects with clear test instructions.

5.2.4 **Other test methods**

Other methods for discrimination testing, such as the dual-standard method, have been developed, but few have ever been used routinely for product evaluation. The dual-standard method was proposed for use in quality control situations (Peryam and Swartz, 1950). The subject is served four products; two are identified as references A and B and two are coded. The subject's task is to match the reference product with the coded product. The designation of the two references could reflect quality control limits or current production and product outside the limit. Other multiproduct tests have been described and used by Mahoney *et al*. (1957) and Wiley *et al*. (1957). Both groups proposed comparing a series of treatments to a single reference; the hypothesis was that this would be relatively efficient and would enable the sensory professional to obtain more than just difference information. To this end, O'Mahony (1979, 1986) describes the R-index, a calculated probability value based on signal detection concepts, to provide a measure of the degree of difference between two products. The procedure requires more time and samples than the other methods and has not yet proven to be practical for routine business application. The 2-AFC (directional paired comparison) and 3-AFC (directional triangle test) methods also provide a degree of difference measure, d' (Bi *et al*., 1997; Frijters, 1979). The 4-AFC method has also been proposed (Bi *et al*., 2010). The more complicated analysis, issues related to use of directional difference tests, especially in complex systems, and relevance of using numbers of correct responses to imply degree of product difference have thus far resulted in limited used of these methods within the business environment.

Another method, labeled as the tetrad test (Delwich and O'Mahony, 1996; Ennis and Bi, 1998; Ennis and Jesionka, 2011; Masuoka *et al*., 1995; Pokorný *et al*., 1981), has been proposed. This is an alternative method (but similar to the dual-standard method) in which subjects are provided with four samples for evaluation—two samples from product A and two samples from product B. The subject can be instructed to select matching products based on an unspecified or a specified attribute and the probability of guessing is 1/3. The statistics demonstrate that the specified (directional) method (e.g. *n*-AFC and specified tetrad tests) is more powerful than the unspecified method. However, in most practical applications of sensory research, only measuring difference on one specified attribute can increase risks in decision making. Often, unspecified attributes may be more critical to the end user. The availability of other methods, especially descriptive analysis, can make specified discrimination methods less viable. This does not mean that these test methods should be ignored but, rather, that the experimenter should proceed with due consideration for the research objective, all elements of the test, subjects, products, and handling procedure before selecting a method. Sensory professionals should, at the least, be able to differentiate between basic research, statistical power, and potential applicability of new developments.

Some other types of methods amounted to nothing more than sorting tasks—for example, separating the nine samples into two groups of five and four (Sawyer *et al.*, 1962). The idea is that very few correct choices would be required to achieve significance. Unfortunately, most products possess a multitude of sensations, and these approaches, when combined with the human memory limitations, become much less feasible. Basically, one should be capable of handling most discrimination problems using the paired, duo–trio, triangle, or directional difference test methods. The multiple standards method is finding some application in quality control situations.

5.3 **Components of testing**
5.3.1 **Organization and test management**

Organization and management of discrimination test activities are essential to the success of any single test, as well as to the success of all testing independent of method selected. In this section, we discuss the various issues that occur prior to, during, and after a test. Some are philosophical, whereas others are easily identified as more practical in that they concern the number of subjects, the statistical significance, and so forth. For the sensory professional, however, these issues are interrelated and are discussed here as a practical guide.

Discrimination tests require thorough organization and planning to provide clarity on what research will be conducted and how the results can be used. Being organized for testing assumes that there exists a system by which requests are documented; products are screened; objectives are defined; subjects and specific test methods are selected, scheduled, and implemented; and results are reported in a way that is readily understood and actionable. Organization of the program is essential, and although it initially takes time to develop, it quickly yields benefits in terms of testing efficiencies, subject participation, and, perhaps most important, credibility of the results.

Organization of the sensory function should reflect realistic industry-relevant sensory practices. By realistic sensory practices by industry, we refer to rapid turnaround on test requests, complete record keeping, standard protocols for product preparation and handling, subject screening and qualification criteria, and so forth. For discrimination testing, once the necessary resources are organized, such records are relatively easy to maintain. Initially, this effort may result in an increase in database management activities; however, the benefits far outweigh any initial time spent establishing the library of information. Sensory departments that maintain sufficient records on past experiences and research will maximize their efforts to build a knowledge source within the organization. Companies frequently reintroduce technologies, ingredients, and/or products as a result of changes in the marketplace, changes in consumer behavior, changing economic conditions, or changes in management. Duplication of efforts can also occur simply through a change in sensory personnel who are not familiar with work conducted in prior years. The sensory professional must be able to quickly determine what has been done in the past and the extent to which the current request must be modified to reflect this past experience. For example, preparation procedures for a product may require special equipment or may preclude use of a particular method. A rapid search of existing information could save considerable time and resources.

Key elements of an organized approach to discrimination testing should include the following items:

1. A clear statement of the objective of discrimination testing (this should help prevent misuse of the method).
2. A brief description of each of the methods: paired, duo–trio, triangle, directional difference, and any other methods that might be used. Additional information should include standard scorecards for each method and a detailed description of testing protocols for each procedure, including suggested containers and serving procedures by product category, if necessary.
3. A brief description of the test request sequence, including the project briefing meeting with the requestor, if warranted; product list and description; test schedule; and report distribution.
4. Examples of the test request and report forms (paper and electronic).
5. A description of subject qualification and selection criteria, including screening procedures and performance monitoring.
6. Selected experimental designs with notes on their use.
7. Guidelines on test procedure, including product coding, amount of serving, and timing.
8. Methods for data analysis and their interpretation.
9. Examples of test reports for each discrimination method.
10. Subject recruitment and scheduling protocols.
11. Suggested motivational efforts for subjects.

A list of these items, along with examples of each, should be maintained within a sensory operations manual. As other industry-specific elements are identified and described, they should be added to the sensory operations manual. The list provided here is intended to serve only as a guide to those elements that have been found to be universally applicable. In the following sections, these elements are discussed in detail to enable the reader to develop a greater appreciation for the specific recommendations.

Before discussing these elements, however, it is useful to address sensory management approaches to make the system function to the greatest benefit of the organization. Sensory test management includes the determination as to whether products warrant a discrimination test, the specific test methodology to be used, the products to be served, the subjects to be selected, and the reporting of results. Sensory evaluation must ensure that each of these steps is considered to ensure integrity of the research function. A common weakness observed with sensory groups is to allow the requestor to stipulate the test method or to consistently use one discrimination method for all requests. Basically, the specific objective (see Section 5.2.3) and the characteristics of the product will determine which method is most appropriate. Failure to establish this qualification within a sensory operations manual will create some difficulties because requestors will recognize indecisiveness and/or willingness to use their requested method versus a method that may be more appropriate for a given research objective.

Sensory managers must also be able to address unpopular issues such as rejecting a request because the products may not warrant discrimination testing due to the differences being so obvious as to question the need for a test. Other issues include the frequency of tests; the frequency of subject participation (e.g. once per day or once per week); and the basis for qualifying, selecting, or rescreening individuals. Firm control of what sensory testing is being conducted must be demonstrated but without necessarily appearing to dictate the entire research process. The sensory professional's role as manager and educator is one of "controlled" flexibility. It is important to apply sensory test guidelines in a very precise way while at the same time adapting test procedures to best satisfy research objectives and researchers. The scientific component of sensory evaluation cannot be compromised to any great extent before the validity of the results is destroyed.

The following sections describe organizational issues and provide additional background information on the development of a sensory management approach. The topics selected include those most frequently encountered as critical issues in discrimination testing.

5.3.2 Test requests

The request for assistance (in this situation, the measurement of perceived differences) materializes in the form of a written test request. The use of a formal test request serves many purposes, as demonstrated in Table 5.2. The top part of the form is completed by the requestor, and the bottom part is completed by a sensory professional or designate. The top part includes name of requestor, date, priority, test objective, product identification (name, code, and how, when, and where manufactured), amount of product available, location, storage conditions, preparation specifications if known, individuals to be excluded from the panel, and report distribution. Depending on the sensory professional's experience with the product and the requestor, a meeting may be required for the sensory professional to fully understand the research objectives. Once completed, the form becomes an essential document for all subsequent test activities for that product, including detailed instructions to the experimenter as well as for report preparation. The information gathered on the test request should be specific and relatively easy to complete. Typically, a request form for discrimination testing should be no longer than a screen or two (or one page).

In the following sections, items for the test request form are discussed in detail, in addition to other elements of the discrimination test.

Table 5.2 Example of a Discrimination Test Request Form	
To be completed by the requestor	
Experimenter:	Date:
Test objective:	Priority:
Product:	Project number:
Sample location, description, and history (storage, etc.):	
If storage, withdrawal date:	
Sample amounts and availability:	
People to be excluded from testing:	
Report distribution:	
To be completed by sensory evaluation professional	
Receipt date:	Serving conditions:
Type of test method:	Sample quantity:
Suggested test date:	Sample temperature:
Design:	Carrier:
Number and type of subject:	Serving container:
Methods of sample presentation:	Lighting conditions:
Number of replications:	Other:
Experimenter comments:	

5.3.3 **Test objectives**

Identification and agreement on a specific test objective and, where appropriate, the overall project objective are essential for the success of a sensory test, both in terms of design and in terms of conclusions and recommendations. In practice, there seems to be some reluctance or, often, an inability to clearly state a test objective. Sensory professionals are often presented with requests in which the test objectives are stated as "do a triangle test between current production lemonade and test product code 42APB" or some similarly vague statement. This request is neither informative nor a correct statement of an objective. Nonetheless, the sensory professional must proceed to develop, in cooperation with each requestor, an objective that reflects the purposes for the test and how the results will be used. A more appropriate test objective might be to determine whether there is a perceived difference in lemonade containing an alternate supplier's lemon flavor versus the current flavor. The project objective may be to find a less costly flavor for lemonade that is not perceived as different from that of current production. This information is especially helpful because it not only identifies the specific test objective but also alerts sensory staff to the potential of additional testing and to the fact that current production is the target (and if the duo–trio test were selected, the constant reference rather than the balanced reference procedure would be used).

Alternatively, if the program objective were to screen new flavors, the test method might be the balanced reference duo–trio or the paired method followed by consumer acceptance testing. Some might choose to not conduct the discrimination test; however, the key here is having stated and documented an objective. Successful sensory programs depend on having well-defined project and test objectives.

In situations in which the request is not clear with regard to the basis for the test or how the information will be used, the sensory professional may find it necessary to communicate directly with others who might have more information—for example, consumer insights or the brand manager. In addition to the research objective statement, it is also important to determine action criteria and what decisions or actions will be made based on the test results; for example, if a difference is reported, will products be reformulated or will there be a test to determine its effect on preference.

5.3.4 **Test procedures**

Once the test objective is established, the sensory professional must turn its attention to the other elements of the test. Some of these elements might be considered as cooperative aspects, in that they require assistance from the requestor, whereas others are strictly a sensory decision (e.g. test method selection).

5.3.4.1 *Product screening*

After a request has been issued and before any testing is conducted, the sensory professional should benchtop screen the products submitted for testing. This should be conducted prior to any decision that the test is warranted, and it is especially important for requestors who lack familiarity with sensory testing procedures. Typically, most issues arise during the initial (organization) stages of a sensory evaluation program. However, sample screening prior to testing is recommended in all cases to prevent unnecessary testing of products that may have differences unrelated to the research objectives. These may occur for a number of reasons, such as during production, storage, handling, or due to unforeseen issues that may impact the product's sensory properties.

Requestors, especially those who are inexperienced, often submit products that do not warrant testing because the difference is obvious; that is, the likelihood of obtaining 90–100% correct decisions is high and thus the test represents an inefficient use of sensory resources, including facilities, experimenter, panel time, and effort. When meeting with the test requester, direct examination of the products should enable that decision to be more easily reached. During prescreening, it is often helpful to evaluate samples blind, arranged in a typical discrimination test format. In some situations, the sensory professional may be required to test products that have easily perceived differences. This effort can be considered in the context of an educational process of the researchers, an opportunity for the sensory staff to demonstrate its professional judgment, for the requestor to gain experience in recognizing "easily perceived differences" and their impact on product formulation, and for both to benefit from pretest screening. Product screening may also reveal inherent variability in manufacturing that results in larger differences within a treatment variable than between treatment variables, making the discrimination test untenable but leading to an alternative test method (e.g. a descriptive analysis).

Pretest screening will also enable the sensory professional to clarify any special requirements related to preparation and serving and to identify any additional background material required as it relates to the product or the test.

Occasionally, the sensory evaluation department may receive a test request but the product will not be available for preview until just before testing is scheduled to commence. This is obviously risky, and the sensory staff will have to devise some means for dealing with this kind of situation. If previous work with the product or product class and with the developer has been successful, then one could proceed; however, there should be at least a last-minute check (e.g. an ingredient was inadvertently left out of one sample). It is not so much a matter of mistrust but, rather, an issue of minimizing risk. Once data are collected, they cannot be disregarded (no matter what the requestor wishes), and the fact that the products were inappropriate may be forgotten. Product screening does not guarantee elimination of judgment errors; however, it will minimize these kinds of risk.

5.3.4.2 *Test selection*

Selection of the specific test method can be challenging for the sensory professional for several reasons. Requestors and sensory professionals often become attached to a particular method and, in the latter's situation, have organized their testing in such a way that it is most convenient to use just one method. Unfortunately, this situation tends to feed on itself. Continued reliance on a single method causes one to develop a rationale for its continued use, and requestors also become "comfortable" with the method and express discomfort when it is not used. To avoid disagreements with requestors, sensory evaluation will continue to use the method and will receive reinforcement for this decision from test requestors, usually in the form of no disagreement.

The literature provides a somewhat conflicting selection of conclusions regarding the sensitivity of the various test methods; some sensory professionals suggest the triangle is more sensitive than the duo–trio and the paired tests, whereas others have arrived at the opposite conclusion (Buchanan *et al.*, 1987; Byer and Abrams, 1953; Dawson and Dochterman, 1951; Gridgeman, 1955b; Hopkins, 1954, Hopkins and Gridgeman, 1955; Radkins, 1957, Schlosberg *et al.*, 1954). Both Rousseau *et al.* (1998) reported that the same–different method (overall) and Ennis *et al.* (2011) reported that the AFC methods (directional) are more powerful than the triangle and duo–trio. However, the AFC methods have limitations in that the attribute of difference must be specified *a priori*, which is often not possible. In addition, results can be misleading if the product differs on more than one attribute.

A careful examination of this literature provides useful insight concerning sensory evaluation methods research. Test method sensitivity appears to be more a matter of product, how the discrimination question is phrased, and the manner in which the investigator selected subjects than it is a matter of method (Buchanan et al., 1987). The issue of subject qualification and screening for sensory acuity has been given less consideration. This important task is discussed in detail in the next section. In addition, the type of product is recognized as having an influence on sensitivity. Although some statistical literature does suggest support for the triangle method (overall difference) and the AFC (directional difference) methods, this conclusion must be considered solely within the context of the products tested, whether the subjects were qualified, etc.

The reader is reminded that there are also conflicting findings with regard to this statistical conclusion (Bradley, 1963; Gridgeman, 1955b). As stated at the outset of this discussion, on a practical application basis, we find no discrimination test method to be any more or less sensitive than another. All discrimination methods have the same basic goal, and all are essentially different forms of the two-sample choice situation. Differences between methods can and do arise as a result of the characteristics of the products and the use of qualified subjects. For example, products that have intense flavors and aromas, that are spicy and/or are difficult to remove from the palate, or that have physiological effects (cigarettes and distilled beverages) usually preclude the use of the triangle test. Subjects must smell and/or taste three coded products and make six comparisons—A versus B, A versus C, B versus C—before reaching a decision. Although it might be argued that subjects do not always make the six comparisons, there is no question that there are more comparisons to be made in the triangle test than in the duo–trio; the fewest comparisons are made in the paired test. The point is that sensory evaluation must first consider product and objective before reaching a decision about the most appropriate method for that problem.

Whenever products are being compared with a current franchise (i.e. product now being manufactured), the duo–trio, constant-reference test method is most appropriate (Peryam and Swartz, 1950). As mentioned previously, this approach provides maximum sensory protection for the current franchise. Because there is one less comparison (R versus A, R versus B), there is a reduction in possible sensory interactions and there is maximum reinforcement of the sensory properties of the current product (it is evaluated twice versus once for the experimental product). Signal detection theory of perception indicates that the more familiar and focused one is with a stimulus, the more likely one will detect that stimulus which is different. Other factors could also reduce the choices to the paired procedure. The paired-comparison method remains the most widely used for strongly flavored products, those that have differences other than the variable being tested, and for many personal care products that require home use.

When products are presented simultaneously, the masking of visual difference so as not to influence subjects is often treated in rather bizarre ways—for example, blindfolded subjects, darkened rooms, or a variety of colored lights. These extreme situations can arise when a researcher is inflexibly committed to one specific test method. In fact, visual differences can be more easily managed by using a paired method with a sequential serving procedure. This particular issue is discussed further in the section on lighting.

Thus, test method selection is the result of a series of probing questions and answers by the sensory professional. The focus is on test objective, type and availability of product, preparation procedures, product sensory characteristics, test method capabilities, availability of qualified subjects, and the use of the results. Such an approach leads to a much higher quality database, and it also provides more reliable and valid product information.

5.3.4.3 Subject selection

The importance of subject selection for a discrimination test has long been acknowledged (Giradot *et al.*, 1952; Helm and Trolle, 1946), but this has often been neglected in many sensory test operations. Current procedures vary widely, with the most common practice being the use of individuals who have previously participated and/or who are located close to the testing area. This approach appears to be based on the misconception that sensory skill and technical training/experience are equivalent— that one's knowledge about a project, such as direct participation in that project, automatically qualifies one to be a subject. It is true that the more knowledge a subject has about product differences, the better he or she can focus attention, and the person's performance can be expected to improve (Engen, 1960). In fact, what happens is that these individuals (who work close to the test facility) base their judgments on what they think the response should be and not on what they actually perceive. Another approach makes use of large numbers of unscreened consumers (100 or more); however, the basis for this practice is not clear (Buchanan *et al.*, 1987; Morrison, 1981; Rousseau *et al.*, 2002). It has been suggested that use of a large number (100+) of unscreened consumers provides a more typical result of what can be expected in the marketplace. This approach overlooks several factors, including the fact that approximately 30% of the consumer population cannot discriminate differences above chance and identifying these individuals does not occur with one or two test trials (see Chapter 2). Using unscreened consumers results in a substantial amount of variability that will mask a difference, leading to a wrong decision. Alternatively, a significant difference could be obtained entirely by chance. Using large numbers of consumers increases the risk of finding statistical but not practical differences. If it requires as many as 100 responses to achieve statistical significance, it can be assumed that the difference is an artifact of the numbers involved. We further discuss this issue later in the chapter. Finally, it should be kept in mind that the purpose for the test (in most instances) is to minimize the risk of concluding there is no difference when, in fact, there is. Minimizing risk is achieved by using a homogeneous population of subjects qualified to participate based on their sensory skill. The issue of how one becomes qualified to participate is not a new or novel concept. For many years, use of threshold tests has been advocated, and they continue to be used, especially for some descriptive tests, as a means of selecting subjects. Unfortunately, the evidence indicated that such threshold measures had no relationship to subsequent performance with more complex stimulus systems (Giradot *et al.*, 1952). Mackey and Jones (1954) reported that the threshold measures did not serve as useful predictors of performance in product tests. Schlosberg *et al.* (1954), as part of a larger study on panel selection procedures, discovered only a weak relationship between selected life history questions and results from a series of discrimination tests, or even between one test trial and another, although the latter trials were sufficiently promising to warrant additional study. As a footnote, the authors reported that "performance during the first 2 days of testing had a fair predictive value during the following 20-day period."

In a related study, Sawyer *et al.* (1962) compared correct judgments from session to session, as well as performance within a session, and observed that it was possible to "predict the proportion of judges whose sensitivity can satisfy established specifications."

The literature clearly indicated that the selection process was critical to the quality (reliability and validity) of the results. Boggs and Hansen (1949), Peryam *et al.* (1954), Bennet *et al.* (1956), and Dawson *et al.* (1963a,b) all recommended that the subjects participating in a discrimination test be selected based on their skill. Later publications provided a reaffirmation of the positive impact of training on performance (Frijters *et al.*, 1982; McBride and Laing, 1979; O'Mahony *et al.*, 1988). The former researchers emphasized experience, subject attitude, and past performance as factors that should be taken into consideration in selecting subjects. From a practical standpoint, these requirements seemed reasonable and are consistent with our own experiences. It was clear from reviewing results from thousands of tests that screening individuals and excluding those who could not differentiate differences at greater than chance yielded more consistent conclusions, results exhibited greater face and external validity, and, from a practical standpoint, there were no decision reversals based on other test results. This empirical evidence led us to develop a set of guidelines for discrimination testing that have proven to be extremely useful:

1. The recruited subjects must be "likers and users" of the product category or must be "willing to try" the product as determined from the product attitude survey (PAS). Because the subjects will be asked to participate on numerous occasions, product liking is important. Greater familiarity with product subtleties is another potential benefit of product likers and users. Dislike for a product eventually manifests itself in tardiness, greater variability in performance, and general disruption of the testing process.

2. All subjects must be qualified (based on sensory skill) and must have participated in not less than two tests in the previous month. In general, the sensory forgetting curve is quite steep, and an individual must maintain the skill through regular participation. For different types of products, the frequency of participation is empirically derived; however, it will probably not be less than once per month. This should not be construed to mean that practice tests should be created if no regular requests have been received. It is rare for a sensory program to not have tests scheduled every week. Similarly, one can overuse an individual subject—for example, testing several times per day for months at a time. This always results in poor motivation and poor performance. The major concerns here, especially if employees are used, are absence of the subjects from their regular work and complaints from their management, along with test fatigue (i.e. too much testing).

3. Each subject must have a composite performance record that is greater than chance probability (separate $p = 1/3$ from $p = 1/2$). No subject should be selected if his or her performance is poorer than chance based on at least the previous three or four tests where the panel result has declared that a difference is present. Including subjects that are nondiscriminators increases the risk of incorrect decisions about product differences.

4. No subject should have specific technical knowledge about the products, test objective, and so on. A subject who has prior knowledge about the variables is likely to respond based on what the result is supposed to be rather than responding perceptually.

5. If using employees, no more than one or two subjects from the same specific work area should be scheduled at the same time. Although it may be highly desirable to complete a panel by scheduling everyone from a single location, this approach can have negative long-term consequences. Sensory evaluation should never empty a laboratory or work area, no matter how brief the test time; because participation is voluntary, it will alienate laboratory managers by causing undue disruption of their work.

6. Employee subjects should be rewarded (not monetarily) for their participation after each test and at other convenient times (e.g. holidays). Rewards for participation should never be taken for granted, and subjects must be made welcome and appreciated. Food, snacks, and beverage rewards and acknowledging past performance all are examples of this appreciation.

7. There is an upper limit to the frequency of testing for each subject. An individual could participate in several discrimination tests per week; however, situations may arise in which there is need for two or three tests per day for several days. Such a situation should be agreed to in advance. From a skills standpoint, testing more than once per day should lead to greater precision. From a practical standpoint, this approach will be quite disruptive of that individual's regular work routine and will lead to conflict. Consider the manager who finds that his staff of 10 is spending the equivalent of 1 or 2 days per week as test subjects.

These guidelines necessitate the development and maintenance of complete records for each subject. The importance of these records cannot be underestimated, for without them it would be difficult to make reasonable subject decisions. Once the decision is made to establish performance records, it then becomes a matter of determining how to organize the record such that it is accessible for both subject selection (monitoring performance) and adding new information (current results). The ideal system has the database stored electronically and it is accessible and searchable within the sensory facility. Performance records coupled with other subject records, including the PAS, will be of great value as a program develops.

One example of a discrimination performance record is given in Table 5.3. In this example, the first column is used for identification of product; all subsequent columns

Table 5.3 Example of a Performance Record for an Individual Subject[a]

Name				
Location				

		Test Type		
	Date	Paired Comparison	Duo-Trio	Triangle
Dry Cereal	Nov. 25	1/2	2/2	
Dry Cereal	Nov. 28		0/2	1/2
Vanilla Ice Cream	Dec. 4	2/2		
Vanilla Ice Cream	Dec. 5		1/2	

[a]Entries are the ratio of correct to total decisions. See text for additional explanation.

list the test date, test type, and rates of correct total decisions. Note that the $p = 1/2$ and $p = 1/3$ tests are separated to prevent errors in computation of performance. The actual layout of the record does not have to conform to Table 5.3 (i.e. the rows and columns can be switched). The important point is that there be a record that is understandable to the sensory professional. The performance record is used with particular emphasis on the most recent 60 days, which will be a realistic basis for selection. Earlier records provide a historical "perspective" on individual subjects and could be used for selection if an insufficient number of subjects were available for a test. This particular decision would be judgmental on the part of sensory evaluation.

In this discussion on subject selection, emphasis has been placed on the immediate past performance (e.g. the last 8–10 tests) of the subjects as a basis for selection. The expectation is that the individual subjects will maintain their proficiency; however, there should not be major concern if that does not occur in a single test. For example, if a subject is operating at 65% correct decisions, performance in a single test might diverge substantially, for example 0 or 100% correct, without cause for concern. Obviously, criteria for not accepting a subject will need to be developed. One such criterion might be 50% or less correct decisions for four consecutive tests (on tests in which the panel found a statistically significant difference and percent correct at $\geq 65\%$).

Another aspect of subject performance-selection criteria is the change in performance over time. Considerable emphasis has been given to identifying and using subjects who are discriminators—that is, subjects whose performance is consistent and greater than chance. Following screening, the subjects participate in tests. Cumulative performance records are maintained and used as the primary basis for continued selection in testing. However, over time it may appear that the level of performance is decreasing, as if the subjects have "lost" some of their sensory skills. In fact, this movement toward chance performance should be viewed in more positive terms; the product differences are smaller (hence more difficult to perceive) because the formulation/development specialist is better able to prepare products as a result of information derived from previous tests. This is a dynamic system

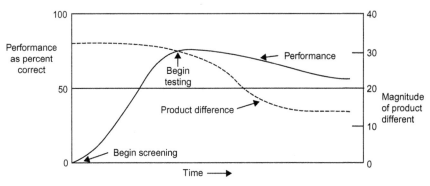

FIGURE 5.7

Visual representation of the changes that occur in the skill of the subjects over time as product formulation efforts yield products with smaller differences.

in which sensory skills that are improving with testing are being challenged by increasingly smaller product differences. Figure 5.7 provides a visual display of the relationship between performance and product formulation.

There is usually a keen desire on the part of subjects to assess their own performance in comparison with that of their peers, and this is particularly true for new subjects. Performance records permit graphing individual percent correct responses compared with the panel. The graphs can represent tests done on a chronological basis or a grouping of tests by product class. Each subject is given his or her own graph of correct matches or percentage compared with the panel. Confidentiality is easily maintained by sending individual graphs electronically to each subject. Subjects may also compare results with one another if they choose to do so. For some individuals, participation represents a mark of special significance relative to fellow workers, and to be more correct than other panel members can enhance this "special" identification. Annual awards recognizing best performances would further enhance their being special. Obviously, such an attitude is desirable for motivational purposes; however, it should not overwhelm the contribution of all other participants. From a sensory viewpoint, correctness of decision is important; but this must be placed within the context of the test because all responses are important, not just those that are correct.

Quite independent of these empirical approaches to subject selection, there are also statistical approaches, the most notable being the method described as sequential analysis. The method was adapted from the work of Wald and Rao (cited in Bradley, 1953), and a detailed description of the method can be found in reports by Bradley and others (Amerine and Roessler, 1983; Radkins, 1958; Steiner, 1966). Basically, the experimenter sets limits for the proportions of correct and incorrect decisions as well as the associated α and β risks. This information permits computation of two parallel lines, the distance between representing the area within which the subjects "remain" until their performance enables them to be accepted or rejected for a test. Depending on the levels of risk (and limits of sensitivity), these lines can be shifted. Radkins proposed using the technique as a basis for evaluating product differences, without depending on a fixed number of subjects. According to Radkins, this procedure has the "inherent advantage of smaller sample sizes," although the risks of such an approach compared with a fixed panel size were also recognized. However, all issues associated with subject participation must be kept in balance and assessed before any procedure that is used.

5.3.4.4 Product preparation and serving

The need for detailed attention to product preparation and serving conditions is often overlooked or understated. Previously in this discussion, emphasis was placed on product screening to ensure that the product and the test were suitably matched. The issue of product preparation was also considered. That particular problem should be resolved in advance so that preparation does not introduce another confounding variable. It must be clearly understood that all aspects of the test must be kept "constant" to ensure that the responses reflect only the variable being evaluated. After all, the subjects are well aware of the task and use all their senses to make a decision. In situations in which product differences are small, any cues, conscious or subconscious, will influence the decision. Such minor issues as serving temperature, amount of serving, or the nature of the serving can have a substantial impact on some subjects. For example, in food evaluation, some samples of an ice cream containing chocolate chunks could easily have more pieces of chocolate than others, thus incorrectly cueing or confusing some subjects.

Product serving is especially important in the discrimination test because of the usual simultaneous serving, which provides for direct comparison. Because the subject's task is only to indicate whether there is a difference, any cue is used. It may be necessary to use a sequential serving procedure when product differences, other than the test variable, cannot be eliminated, such as a visual difference. This particular problem is also discussed in the section on lighting.

The serving procedure could be changed for some products. For example, evaluation of fragrance samples can be accomplished by having subjects move from station to station and then make their decisions. The distance between stations can be kept to a minimum, approximately 3–6 ft apart. The fact that the subjects move rather than the products does not affect the results. It is important for the sensory professional to make every effort to utilize existing methodology and to consider all the options.

Finally, we would be remiss if we did not comment on certain practices involving discrimination tests in which attempts were made to circumvent non-test variables by means of subject instructions. The most common practice is to tell subjects to ignore a known difference (e.g. visual or textural) and evaluate the flavor or feel, as if the perceptual and decision-making processes were independent. No matter how well trained, subjects will almost automatically focus on the variable that they are supposed to ignore. Such an approach can only lead to confusion and to questionable testing practices for sensory evaluation. The product's characteristics must be carefully considered to be sure that it can be presented in a way that does not mask differences or provide cues. The responsibility for these decisions rests with the sensory staff.

5.3.4.5 Lighting

In discrimination testing, lighting will become an issue when visual differences occurred during product formulation that were unrelated to the variable being studied. Usually, the requestor cannot correct the difference without great inconvenience, so the test must be done despite the difference. In the discussion on booth lighting in Chapter 2, recommendations were made regarding the amount of light (controlled and up to 100 ft-candles illumination) and the type of light (incandescent for ease of handling). A common problem encountered with lighting is its overuse in an attempt to mask visual differences. There is widely reported usage of an array of colored lights (red, blue, yellow, or green lights), especially red lights, to mask color differences (usually in foods, beverages, and wines). Although it is assumed that the light will mask the color difference, there is little evidence to support this assumption, and the use of colored lights for some products may even enhance a visual difference. Some facilities using colored lights are designed such that light from the preparation area comes through the booth pass-through doors and shines on the products during serving, allowing subjects to easily observe the differences. Although it is possible to design a facility so as to provide complete control of the lights, the costs associated with such effort rarely warrant the investment. In other procedures, subjects are blindfolded or rooms are so darkened that vision is significantly impaired.

From a behavioral standpoint, these procedures introduce another level of risk in the decision-making process. Colored lights introduce additional variables in the sensory testing process and are rarely recommended and typically only at the basic research level. First, any conclusion about product differences (or lack thereof) must be stated in terms of the experimental condition—for example, a difference when served under red light (but is the difference due to the test variable, the effect of the lighting, or the combination of the two?). An aspect of this practice is the potentially negative impact on management. When managers review results with products present, considerable doubt may develop concerning the test conclusion. For more discussion on this topic, the reader is referred to Chapter 2. Test situations that are inherently weak or subject to misinterpretation must be avoided. Where differences unrelated to the test variable exist, sensory evaluation must make note in the record and then consider whether the discrimination test is appropriate. If so, the recommended procedure would be to use a monadic, sequential serving order; that is, serve the products one at a time, remove the first one before serving the second, and then request a decision (this is the nondirectional paired comparison). This procedure minimizes the visual difference and provides a more realistic test of the variable.

It can be argued that this latter serving procedure does not overcome a manager's concern about obvious visual differences; however, the approach is more realistic in terms of the test conditions and more representative of the manner in which the product would typically be evaluated or is typically consumed.

Another approach is to use colored containers for the product. However, as with any other experimental modification, the sensory professional should ensure that such an approach does not contribute cues or serve as a confounding variable. It should be apparent that sensory evaluation must evaluate each situation with care because often there are no easy or simple solutions.

Despite all these arguments, one continues to encounter the use of colored lights and other atypical means for masking unintended product differences. It is unfortunate because it casts the science in a rather unscientific "light."

5.3.4.6 Carriers

The use of carriers in discrimination testing is complex, especially when working in food and beverage evaluations. Carriers are defined as materials that accompany the product being tested. In food evaluation, a carrier could be milk served with a ready-to-eat breakfast cereal, cake with frosting, butter on bread, lettuce with salad dressing, and so forth. For other products, the use of carriers is generally less of a problem than the form of the product. For example, in evaluating different fabric treatments, the fabrics could be tested as swatches or as finished garments. On the other hand, evaluating differences in wine blends may be best accomplished as a test of the finished products.

In general, carriers tend to dilute and/or mask differences and are not recommended in discrimination testing. Nonetheless, they continue to be used and the issue must be examined in more detail. Probably the most persuasive argument advanced by proponents of the use of the carrier is "that is how the product is consumed." Although this observation is correct, it should be kept in mind that not all of a product is consumed with a carrier. In addition, the use of a carrier should be viewed within the context of the discrimination test and its relationship to product consumption in the home. The discrimination test is not intended to duplicate product use in the home. It is a unique test situation carried out in a controlled environment using a selected and homogeneous group of subjects who are screened and selected for their sensory acuity. The sole purpose of the test is to determine whether there is a perceived difference between two products, all other factors kept constant, and to determine the level of significance associated with that decision.

In most projects, the objective is to make a change (ingredient, process, etc.) that either is not detected or is detected by the fewest number of people. If the screened, selected, and experienced subject cannot perceive the change under the most controlled of conditions, then we can be confident the difference will not be perceived by the consumer. Therefore, our recommended test design incorporates many features that minimize the risk of an incorrect decision—that is, using qualified subjects (experienced, likers/users of product, or known discriminators), using a balanced serving order, and taking care to standardize product preparation and serving conditions. All of these "conditions" are designed to minimize errors in testing and reduce the chances of the subject's decision being influenced by a non-test variable.

Carriers introduce the potential for problems in terms of testing procedure and in the interpretation of the result. Each carrier will have unique physical and chemical characteristics that will require standardization and that in turn may limit the extent to which results can be generalized. Use of a carrier does not allow us to state that the difference is perceived without a carrier.

Consider the following situation. The objective of the test is to determine whether there is a perceived difference between products A_1 and A_2 (the null hypothesis is that $A_1 = A_2$). If a statistically significant difference exists (without use of a carrier), variable A_2 is rejected; that is, the objective has not been met. Even if a second test with a carrier shows no significant difference, the sensory professional's response to the requestor must first emphasize that the variable does have an effect; that is, A_2 is perceived as different from A_1. It is especially risky to ignore a difference that can be masked by carriers; sensory evaluation is essentially reporting no difference when in fact there is a difference. Consumers do taste products without a carrier, as well as with a multitude of different carriers, so their use in testing will create additional problems.

This does not mean that carriers should be forbidden from discrimination testing. Rather, our intention is to call attention to one of the more perplexing aspects of testing. If there is reason to believe that a particular carrier is essential due to specific flavor interactions, it might be appropriate to prescreen products and, if some doubt remains, to implement discrimination tests both with and without a carrier. Alternatively, some compromises are desirable in certain situations. For example, in the evaluation of a pizza sauce mix, it was agreed that flavor interactions with crust components resulted in a sauce flavor that could not be achieved by heating the sauce alone. However, due to variability in pizza crust within a brand, it was determined that the pizza would be cooked and the sauce scraped from the crust and tested by itself. Thus, the chemical reactions were allowed to occur and the subjects' responses were not influenced by crust thickness or other non-test variables. There can be situations, however, in which the use of the discrimination model is not the best solution to a request.

The sensory professional must evaluate each request in terms of the objective as well as the background of the request and the ultimate use of the information. In the case of carriers, all the options must be thoroughly explored before proceeding.

5.3.4.7 Coding

As noted in Chapter 4, products are coded using three-digit (or four-digit) codes, and no two products should be assigned the same code in a test. It is naive to assume that subjects will not communicate with one another after a test, for example, when passing one another in the corridor. Perhaps more troublesome is the subject who assumes that a certain coding is always used. Although one can never expect to eliminate this notion, adherence to three- or four-digit coding, different for each product, will minimize these effects and reduce bias for a particular number or code.

5.3.4.8 Experimental design

The design of discrimination tests is reasonably well-defined, with minimal opportunity for improvisation. The primary concern is attention given to the balanced order of presentation and sufficient responses for confidence in the conclusions (Gridgeman, 1958; Harries, 1956).

The designs described herein are based on the use of replicate responses by each subject and a pooled database of approximately 40 responses. The rationale for this approach is discussed in Section 5.3.5 (see also Roessler *et al.*, 1978).

For the paired comparison, the two orders of presentation are AB and BA; for the "same-or-different" form, the four orders of presentation are AA, AB, BA, and BB. Thus, one could easily develop appropriate experimental designs for any number of subjects. However, our desire to obtain replicate responses adds a degree of complexity—one that we will demonstrate is well worth the effort. Table 5.4 contains the serving orders for 20 subjects for the paired-comparison test, with direction specified (total N is 40 for statistical analysis).

Several characteristics of this design warrant comment. First, some subjects are served the same order twice. This is recommended to minimize subjects' theorizing that if the first sample of the first pair is the "stronger" one, then it will have to be

Table 5.4 A Serving Order for the Directional Paired-Comparison Test[a]

Subject	Serving Order	
	First Set	**Second Set**
1	AB	BA
2	BA	BA
3	BA	AB
4	AB	AB
5	BA	AB
6	AB	BA
7	AB	AB
8	BA	BA
9	BA	AB
10	BA	BA
11	AB	AB
12	AB	BA
13	AB	BA
14	BA	AB
15	AB	AB
16	BA	BA
17	AB	AB
18	BA	BA
19	BA	AB
20	AB	BA

[a]*One replication per subject.*

Table 5.5 A Serving Order for the Paired Same-or-Different Test[a]

Subject	Serving Order	
	First Set	**Second Set**
1	AB	AA
2	BB	BA
3	BA	AB
4	AB	BA
5	AA	BB
6	BB	AA
7	AB	BB
8	AA	AB
9	BB	AB
10	AA	BA
11	BA	AA
12	BA	BB
13	AA	BB
14	AB	BB
15	BB	AB
16	BA	AA
17	BB	BA
18	AB	BA
19	BB	AA
20	AA	AB
21	AA	BA
22	BA	BB
23	AB	AA
24	BA	AB

[a]One replication per subject.

the second sample of the second pair. Remember that the experienced subject, used on a regular basis, will become especially aware of the entire testing process (consciously or subconsciously). The design in Table 5.4 results in four orders for the two pairs, presented one (pair) at a time, and in our experience minimizes subjects' belief that they "know" the answers (or have broken the code). Second, this particular serving order can and should be varied so that the first subject does not always receive the same order, and so on.

For the nondirectional paired test, same-or-different, the 12 orders of two pairs are presented one (pair) at a time as indicated in Table 5.5. Thus, the level of statistical significance can be computed on a total of 48 responses. As noted previously, these serving orders can be changed; however, the more critical issue is to remind

Table 5.6 A Serving Order for the Duo–Trio Test, Balanced Reference[a]

Subject	Serving Order	
	First Set	**Second Set**
1	R_A AB	R_B AB
2	R_B BA	R_A BA
3	R_A BA	R_B BA
4	R_B BA	R_A AB
5	R_A BA	R_B BA
6	R_B AB	R_B BA
7	R_A BA	R_B BA
8	R_A AB	R_B AB
9	R_A BA	R_B AB
10	R_A BA	R_B BA
11	R_A BA	R_B BA
12	R_A AB	R_B AB
13	R_A BA	R_A AB
14	R_B BA	R_B BA
15	R_A AB	R_B BA
16	R_A BA	R_B AB

[a]One replication per subject.

subjects that a particular serving order is not related to previous orders and that their responses should reflect what they perceive in products they are evaluating at that moment.

For the duo–trio tests, balanced and constant-reference, the orders of presentation with replication are similarly organized. The layout for the balanced reference design for 16 subjects is presented in Table 5.6. For the constant reference, the design layout for 20 subjects is presented in Table 5.7; there are two serving orders for each subject, presented one set at a time, yielding a total of 40 responses.

The six orders of presentation for the triangle test, as in the previous design layouts, offer additional orders of presentation with replication within the session. Table 5.8 provides the design layout for 18 orders of presentation for two triads served one triad at a time, yielding a total $N = 36$.

In reviewing this layout, the reader should note that other orders of presentation are possible and that the different product is equally balanced between the A and B for each subject. One might also consider a triangle test design in which the different product in both triads could be the same for some subjects (e.g. AAB and BAA or ABB and BAB). This latter approach would be appropriate when the triangle test is used on a regular basis (i.e. daily) and when concern exists about the subjects' guesses as to which product is not the different one based on the first

Table 5.7 A Serving Order for the Duo–Trio Test, Constant Reference[a]

Subject	Serving Order	
	First Set	**Second Set**
1	R AB[b]	R BA
2	R BA	R BA
3	R AB	R AB
4	R BA	R AB
5	R AB	R AB
6	R BA	R AB
7	R AB	R BA
8	R BA	R BA
9	R AB	R AB
10	R BA	R AB
11	R AB	R BA
12	R BA	R BA
13	R AB	R AB
14	R BA	R AB
15	R AB	R BA
16	R BA	R BA
17	R AB	R BA
18	R AB	R AB
19	R BA	R BA
20	R BA	R AB

[a]One replication per subject.
[b]In this series, R is always the same, A or B.

test. The subject might conclude that if the first product was the different one in the first set, it will not be the different one in the second set. The extent to which this reasoning might apply is not known; however, one could speculate that if it were widely applicable, more significant differences could be expected to occur in the latter tests of a series (of tests). Nonetheless, it is an issue that the sensory professional must consider when monitoring test results on a long-term basis.

There is no question that additional design layouts should be developed for the various types of discrimination tests. Ideally, a series of these designs with explanatory notes should be available so that the sensory staff does not have to re-create them before each test. Some of the more advanced sensory software programs offer serving designs that are completely balanced for each discrimination method and allow the experimenter control over individual test designs.

Table 5.8 A Serving Order for the Triangle Test,
Balanced Order[a]

Subject	Serving Order	
	First Set	**Second Set**
1	ABB	ABA
2	BAB	AAB
3	BBA	BAA
4	AAB	BAB
5	BBA	ABA
6	ABB	AAB
7	BAA	BAB
8	ABA	BBA
9	AAB	ABB
10	BAA	BBA
11	ABA	ABB
12	BAB	BAA
13	AAB	BBA
14	BBA	AAB
15	BAA	ABB
16	ABB	BAA
17	ABA	BAB
18	BAB	ABA

[a]One replication per subject.

5.3.5 **Data analysis and interpretation**

Analysis of traditional paired, duo–trio, and triangle discrimination test responses is reasonably well-defined, based on the binomial distribution or approximations of the binomial distribution derived from other types of distributions (e.g. chi-square (χ^2)), the normal probability curve (Alder and Roessler, 1977; Bengtsson and Helm, 1946; Bradley, 1963; Ennis, 2011; Lockhart, 1951; McNemar, 1969; Roessler et al., 1953, 1978). In this discussion, attention is focused on the determination of statistical significance, the interpretation of the statistical value, and some of the associated issues, such as one-tailed and two-tailed tests and the importance of replication. In a subsequent section (and in Chapter 4), we discuss Type 1 and Type 2 errors and some of the options that are possible in terms of statistical power in the decision-making process. This discussion also covers the related topics of number of subjects per test and number of responses per subject per test.

Computing statistical significance was described by Roessler and co-workers in a series of publications (Baker et al., 1954; Roessler et al., 1948, 1953, 1956), such that one could refer to tables that would indicate whether a particular number of correct decisions was statistically significant at the 0.05, 0.01, and 0.001 significance levels. Formulae were also provided for computation of significance for values not included in the tables. The tables were well received because they were easy to use; there were no computations, and a conclusion could be reached within minutes of obtaining the last response. However, this simplicity had a disadvantage that could not have been foreseen at the time. The popularity of the tables led to some misuse and to misunderstanding relative to the derivation of the tabular entries. It was believed that the basis for the tables should be clarified and additional tables prepared that would provide exact probability values for measuring statistical significance (Roessler *et al.*, 1978; Stone and Sidel, 1978). These tables are reproduced here (Tables 5.9–5.14), and one can quickly see their value in establishing whether a particular result is significant. Now, most of these calculations are

Table 5.9 Minimum Numbers of Correct Judgments to Establish Significance at Various Probability Levels for Paired-Difference and Duo–Trio Tests (One-Tailed, $p = 1/2$)[a]

No. of Trials (n)	Probability Level						
	0.05	0.04	0.03	0.02	0.01	0.005	0.001
7	7	7	7	7	7		
8	7	7	8	8	8	8	
9	8	8	8	8	9	9	
10	9	9	9	9	10	10	10
11	9	9	10	10	10	11	11
12	10	10	10	10	11	11	12
13	10	11	11	11	12	12	13
14	11	11	11	12	12	13	13
15	12	12	12	12	13	13	14
16	12	12	13	13	14	14	15
17	13	13	13	14	14	15	16
18	13	14	14	14	15	15	16
19	14	14	15	15	15	16	17
20	15	15	15	16	16	17	18
21	15	15	16	16	17	17	18
22	16	16	16	17	17	18	19
23	16	17	17	17	18	19	20
24	17	17	18	18	19	19	20
25	18	18	18	19	19	20	21
26	18	18	19	19	20	20	22
27	19	19	19	20	20	21	22
28	19	20	20	20	21	22	23
29	20	20	21	21	22	22	24

Table 5.9 Minimum Numbers of Correct Judgments to Establish Significance at Various Probability Levels for Paired-Difference and Duo–Trio Tests (One-Tailed, $p = 1/2$)[a] (Continued)

No. of Trials (n)	Probability Level						
	0.05	0.04	0.03	0.02	0.01	0.005	0.001
30	20	21	21	22	22	23	24
31	21	21	22	22	23	24	25
32	22	22	22	23	24	24	26
33	22	23	23	23	24	25	26
34	23	23	23	24	25	25	27
35	23	24	24	25	25	26	27
36	24	24	25	25	26	27	28
37	24	25	25	26	26	27	29
38	25	25	26	26	27	28	29
39	26	26	26	27	28	28	30
40	26	27	27	27	28	29	30
41	27	27	27	28	29	30	31
42	27	28	28	29	29	30	32
43	28	28	29	29	30	31	32
44	28	29	29	30	31	31	33
45	29	29	30	30	31	32	34
46	30	30	30	31	32	33	34
47	30	30	31	31	32	33	35
48	31	31	31	32	33	34	36
49	31	32	32	33	34	34	36
50	32	32	33	33	34	35	37
60	37	38	38	39	40	41	43
70	43	43	44	45	46	47	49
80	48	49	49	50	51	52	55
90	54	54	55	56	57	58	61
100	59	60	60	61	63	64	66

[a]Values (X) not appearing in table may be derived from $X = (z\sqrt{n} + n + 1)/2$. See text.
Source: Reprinted from J. Food Sci. **43**, pp. 940–947, 1978. Copyright © by Institute of Food Technologists.

embedded into data analysis programs for rapid results. However, it is always wise for the sensory professional to fully understand how to calculate exact probabilities using the formula and existing tables.

The exact probability tables are especially useful because they enable the sensory professional to assess risk in terms more precise than would be possible using Tables 5.9 or 5.10, for example. In practice, most sensory decisions operate from statistical significance at the 0.05 probability level (also referred to as the α level);

Table 5.10 Minimum Numbers of Correct Judgments to Establish Significance at Various Probability Levels for the Triangle Test (One-Tailed, p = 1/3)[a]

No. of Trials (n)	Probability Level						
	0.05	0.04	0.03	0.02	0.01	0.005	0.001
5	4	5	5	5	5	5	
6	5	5	5	5	6	6	
7	5	6	6	6	6	7	7
8	6	6	6	6	7	7	8
9	6	7	7	7	7	8	8
10	7	7	7	7	8	8	9
11	7	7	8	8	8	9	10
12	8	8	8	8	9	9	10
13	8	8	9	9	9	10	11
14	9	9	9	9	10	10	11
15	9	9	10	10	10	11	12
16	9	10	10	10	11	11	12
17	10	10	10	11	11	12	13
18	10	11	11	11	12	12	13
19	11	11	11	12	12	13	14
20	11	11	12	12	13	13	14
21	12	12	12	13	13	14	15
22	12	12	13	13	14	14	15
23	12	13	13	13	14	15	16
24	13	13	13	14	15	15	16
25	13	14	14	14	15	16	17
26	14	14	14	15	15	16	17
27	14	14	15	15	16	17	18
28	15	15	15	16	16	17	18
29	15	15	16	16	17	17	19
30	15	16	16	16	17	18	19
31	16	16	16	17	18	18	20
32	16	16	17	17	18	19	20
33	17	17	17	18	18	19	21
34	17	17	18	18	19	20	21
35	17	18	18	19	19	20	22
36	18	18	18	19	20	20	22
37	18	18	19	19	20	21	22
38	19	19	19	20	21	21	23
39	19	19	20	20	21	22	23
40	19	20	20	21	21	22	24
41	20	20	20	21	22	23	24
42	20	20	21	21	22	23	25

Table 5.10 Minimum Numbers of Correct Judgments to Establish Significance at Various Probability Levels for the Triangle Test (One-Tailed, p = 1/3)[a] (*Continued*)

No. of Trials (n)	Probability Level						
	0.05	0.04	0.03	0.02	0.01	0.005	0.001
43	20	21	21	22	23	24	25
44	21	21	22	22	23	24	26
45	21	22	22	23	24	24	26
46	22	22	22	23	24	25	27
47	22	22	23	23	24	25	27
48	22	23	23	24	25	26	27
49	23	23	24	24	25	26	28
50	23	24	24	25	26	26	28
60	27	27	28	29	30	31	33
70	31	31	32	33	34	35	37
80	35	35	36	36	38	39	41
90	38	39	40	40	42	43	45
100	42	43	43	44	45	47	49

[a]Values (X) not appearing in table may be derived from $X = 0.4714z\sqrt{n} + [(2n + 3)/6]$. See text.
Source: Reprinted from J. Food Sci. **43**, pp. 940–947, 1978. Copyright © by Institute of Food Technologists.

Table 5.11 Minimum Numbers of Agreeing Judgments Necessary to Establish Significance at Various Probability Levels for the Paired-Preference Test (Two-Tailed, $p = 1/2$)[a]

No. of Trials (n)	Probability Level						
	0.05	0.04	0.03	0.02	0.01	0.005	0.001
7	7	7	7	7			
8	8	8	8	8	8		
9	8	8	9	9	9	9	
10	9	9	9	10	10	10	
11	10	10	10	10	11	11	11
12	10	10	11	11	11	12	12
13	11	11	11	12	12	12	13
14	12	12	12	12	13	13	14
15	12	12	13	13	13	14	14
16	13	13	13	14	14	14	15
17	13	14	14	14	15	15	16
18	14	14	15	15	15	16	17
19	15	15	15	15	16	16	17
20	15	16	16	16	17	17	18

Table 5.11 Minimum Numbers of Agreeing Judgments Necessary to Establish Significance at Various Probability Levels for the Paired-Preference Test (Two-Tailed, $p = 1/2$)[a] (*Continued*)

No. of Trials (n)	Probability Level						
	0.05	0.04	0.03	0.02	0.01	0.005	0.001
21	16	16	16	17	17	18	19
22	17	17	17	17	18	18	19
23	17	17	18	18	19	19	20
24	18	18	18	19	19	20	21
25	18	19	19	19	20	20	21
26	19	19	19	20	20	21	22
27	20	20	20	20	21	22	23
28	20	20	21	21	22	22	23
29	21	21	21	22	22	23	24
30	21	22	22	22	23	24	25
31	22	22	22	23	24	24	25
32	23	23	23	23	24	25	26
33	23	23	24	24	25	25	27
34	24	24	24	25	25	26	27
35	24	25	25	25	26	27	28
36	25	25	25	26	27	27	29
37	25	26	26	26	27	28	29
38	26	26	27	27	28	29	30
39	27	27	27	28	28	29	31
40	27	27	28	28	29	30	31
41	28	28	28	29	30	30	32
42	28	29	29	29	30	31	32
43	29	29	30	30	31	32	33
44	29	30	30	30	31	32	34
45	30	30	31	31	32	33	34
46	31	31	31	32	33	33	35
47	31	31	32	32	33	34	36
48	32	32	32	33	34	35	36
49	32	33	33	34	34	35	37
50	33	33	34	34	35	36	37
60	39	39	39	40	41	42	44
70	44	45	45	46	47	48	50
80	50	50	51	51	52	53	56
90	55	56	56	57	58	59	61
100	61	61	62	63	64	65	67

[a]*Values (X) not appearing in table may be derived from* $X = (z\sqrt{n} + n + 1)/2$. *See text.*
Source: Reprinted from J. Food Sci. **43**, *pp. 940–947, 1978. Copyright © by Institute of Food Technologists.*

Table 5.12 Probability of X or More Correct Judgments in n Trials (One-Tailed, p = 1/2)

n/x	0	1	2	3	4	5	6	7	8	9	10	11	12	13	14	15	16	17
5		0.969	0.812	0.500	0.188	0.031												
6		0.984	0.891	0.656	0.344	0.109	0.016											
7		0.992	0.938	0.773	0.500	0.227	0.062	0.008										
8		0.996	0.965	0.855	0.637	0.363	0.145	0.035	0.004									
9		0.998	0.980	0.910	0.746	0.500	0.254	0.090	0.020	0.002								
10		0.999	0.989	0.945	0.828	0.623	0.377	0.172	0.055	0.011	0.001							
11			0.994	0.967	0.887	0.726	0.500	0.274	0.113	0.033	0.006							
12			0.997	0.981	0.927	0.806	0.613	0.387	0.194	0.073	0.019	0.003						
13			0.998	0.989	0.954	0.867	0.709	0.500	0.291	0.133	0.046	0.011	0.002					
14			0.999	0.994	0.971	0.910	0.788	0.605	0.395	0.212	0.090	0.029	0.006	0.001				
15				0.996	0.982	0.941	0.849	0.696	0.500	0.304	0.151	0.059	0.018	0.004				
16				0.998	0.989	0.962	0.895	0.773	0.598	0.402	0.227	0.105	0.038	0.011	0.002			
17				0.999	0.994	0.975	0.928	0.834	0.685	0.500	0.315	0.166	0.072	0.025	0.006	0.001		
18				0.999	0.996	0.985	0.952	0.881	0.760	0.593	0.407	0.240	0.119	0.048	0.015	0.004	0.001	
19					0.998	0.990	0.968	0.916	0.820	0.676	0.500	0.324	0.180	0.084	0.032	0.010	0.002	
20					0.999	0.994	0.979	0.942	0.868	0.748	0.588	0.412	0.252	0.132	0.058	0.021	0.006	0.001
21					0.999	0.996	0.987	0.961	0.905	0.808	0.668	0.500	0.332	0.192	0.095	0.039	0.013	0.004
22						0.998	0.992	0.974	0.933	0.857	0.738	0.584	0.416	0.262	0.143	0.067	0.026	0.008
23						0.999	0.995	0.983	0.953	0.895	0.798	0.661	0.500	0.339	0.202	0.105	0.047	0.017
24						0.999	0.997	0.989	0.968	0.924	0.846	0.729	0.581	0.419	0.271	0.154	0.076	0.032
25							0.998	0.993	0.978	0.946	0.885	0.788	0.655	0.500	0.345	0.212	0.115	0.054
26							0.999	0.995	0.986	0.962	0.916	0.837	0.721	0.577	0.423	0.279	0.163	0.084
27							0.999	0.997	0.990	0.974	0.939	0.876	0.779	0.649	0.500	0.351	0.221	0.124
28								0.998	0.994	0.982	0.956	0.908	0.828	0.714	0.575	0.425	0.286	0.172
29								0.999	0.996	0.988	0.969	0.932	0.868	0.771	0.644	0.500	0.356	0.229
30								0.999	0.997	0.992	0.979	0.951	0.900	0.819	0.708	0.572	0.428	0.292
31									0.998	0.995	0.985	0.965	0.925	0.859	0.763	0.640	0.500	0.360
32									0.999	0.997	0.990	0.975	0.945	0.892	0.811	0.702	0.570	0.430
33									0.999	0.998	0.993	0.982	0.960	0.919	0.852	0.757	0.636	0.500
34										0.999	0.995	0.988	0.971	0.939	0.885	0.804	0.696	0.568

Table 5.12 Probability of X or More Correct Judgments in n Trials (One-Tailed, $p = 1/2$) (Continued)

n/x	0	1	2	3	4	5	6	7	8	9	10	11	12	13	14	15	16	17
35										0.999	0.997	0.992	0.980	0.955	0.912	0.845	0.750	0.632
36										0.999	0.998	0.994	0.986	0.967	0.934	0.879	0.797	0.691
37											0.999	0.996	0.990	0.976	0.951	0.906	0.838	0.744
38											0.999	0.997	0.993	0.983	0.964	0.928	0.872	0.791
39											0.999	0.998	0.995	0.988	0.973	0.946	0.900	0.832
40												0.999	0.997	0.992	0.981	0.960	0.923	0.866
41												0.999	0.998	0.994	0.986	0.970	0.941	0.894
42													0.999	0.996	0.990	0.978	0.956	0.918
43													0.999	0.997	0.993	0.984	0.967	0.937
44													0.999	0.998	0.995	0.989	0.976	0.952
45														0.999	0.997	0.992	0.982	0.964
46														0.999	0.998	0.994	0.987	0.973
47														0.999	0.988	0.996	0.991	0.980
48															0.999	0.996	0.993	0.985
49															0.999	0.998	0.995	0.989
50																0.999	0.997	0.992

18	19	20	21	22	23	24	25	26	27	28	29	30	31	32	33	34	35	36

0.001																		
0.002	0.001																	
0.005	0.003	0.001																
0.011	0.007	0.003	0.001															
0.022	0.014	0.006	0.002	0.001														
0.038	0.026	0.012	0.005	0.002	0.001													
0.061	0.044	0.021	0.010	0.004	0.002	0.001												
0.092	0.068	0.035	0.018	0.008	0.004	0.002	0.001											
0.132	0.100	0.055	0.031	0.015	0.007	0.004	0.002	0.001										
0.181	0.141	0.081	0.049	0.025	0.012	0.007	0.003	0.002	0.001									
0.237	0.189	0.115	0.075	0.040	0.020	0.012	0.006	0.004	0.002	0.001								
0.298	0.243	0.155	0.108	0.061	0.033	0.020	0.010	0.007	0.003	0.002	0.001							
0.364	0.304	0.203	0.148	0.088	0.049	0.033	0.017	0.012	0.006	0.004	0.002	0.001						
0.432	0.368	0.256	0.196	0.121	0.072	0.049	0.027	0.019	0.010	0.007	0.003	0.002	0.001					
0.500	0.434	0.314	0.250	0.162	0.100	0.072	0.040	0.030	0.016	0.012	0.005	0.003	0.002	0.001				
0.566	0.500	0.375	0.309	0.209	0.134	0.100	0.059	0.044	0.024	0.019	0.008	0.006	0.003	0.002	0.001			
0.629	0.564	0.437	0.371	0.261	0.174	0.134	0.082	0.063	0.036	0.030	0.014	0.010	0.006	0.003	0.002	0.001		
0.686	0.625	0.500	0.436	0.318	0.220	0.174	0.111	0.087	0.052	0.044	0.022	0.016	0.010	0.005	0.003	0.001		
0.739	0.682	0.561	0.500	0.378	0.271	0.220	0.146	0.116	0.072	0.063	0.033	0.024	0.016	0.008	0.006	0.002	0.001	
0.785	0.734	0.620	0.563	0.439	0.326	0.271	0.186	0.151	0.097	0.087	0.048	0.036	0.024	0.013	0.009	0.004	0.001	
0.826	0.780	0.674	0.620	0.500	0.383	0.326	0.231	0.191	0.126	0.116	0.068	0.052	0.036	0.020	0.015	0.007	0.003	0.001
0.860	0.820	0.724	0.674	0.557	0.440	0.383	0.280	0.235	0.156	0.151	0.092	0.072	0.052	0.030	0.022	0.012	0.005	0.002
0.889	0.854	0.769	0.724	0.612	0.500	0.442	0.333	0.284	0.196	0.191	0.121	0.097	0.072	0.043	0.032	0.016	0.008	0.003
0.913	0.884	0.809	0.765	0.667	0.557	0.500	0.388	0.333	0.235	0.235	0.156	0.126	0.097	0.056	0.043	0.022	0.012	0.005
0.932	0.908	0.844	0.804	0.716	0.612	0.558	0.443	0.385	0.280	0.284	0.196	0.161	0.126	0.076	0.059	0.030	0.016	0.008
0.948	0.928	0.874	0.839	0.760	0.664	0.615	0.500	0.443	0.329	0.336	0.240	0.196	0.156	0.097	0.076	0.043	0.022	0.016
0.961	0.944	0.903	0.874	0.804	0.716	0.667	0.557	0.500	0.385	0.388	0.284	0.235	0.191	0.126	0.101	0.059	0.032	
0.970	0.957	0.924	0.899	0.839	0.760	0.716	0.612	0.557	0.442	0.444	0.333	0.284	0.235	0.156				
0.978	0.957	0.924	0.874	0.804	0.716	0.612	0.500	0.388	0.284	0.196	0.126	0.076	0.043	0.022	0.012	0.005	0.002	0.001
0.984	0.968	0.941	0.899	0.839	0.760	0.664	0.556	0.444	0.336	0.240	0.161	0.101	0.059	0.032	0.016	0.008	0.003	0.001

Table 5.13 Probability of X or More Correct Judgments in n Trials (One-Tailed, $p = 1/3$)

n/x	0	1	2	3	4	5	6	7	8	9	10	11	12	13
5		0.868	0.539	0.210	0.045	0.004								
6		0.912	0.649	0.320	0.100	0.018	0.001							
7		0.941	0.737	0.429	0.173	0.045	0.007							
8		0.961	0.805	0.532	0.259	0.088	0.020	0.003						
9		0.974	0.857	0.623	0.350	0.145	0.042	0.008	0.001					
10		0.983	0.896	0.701	0.441	0.213	0.077	0.020	0.003					
11		0.988	0.925	0.766	0.527	0.289	0.122	0.039	0.009	0.001				
12		0.992	0.946	0.819	0.607	0.368	0.178	0.066	0.019	0.004	0.001			
13		0.995	0.961	0.861	0.678	0.448	0.241	0.104	0.035	0.009	0.002			
14		0.997	0.973	0.895	0.739	0.524	0.310	0.149	0.058	0.017	0.004	0.001		
15		0.998	0.981	0.921	0.791	0.596	0.382	0.203	0.088	0.031	0.008	0.002		
16		0.998	0.986	0.941	0.834	0.661	0.453	0.263	0.126	0.050	0.016	0.004	0.001	
17		0.999	0.990	0.956	0.870	0.719	0.522	0.326	0.172	0.075	0.027	0.008	0.002	
18		0.999	0.993	0.967	0.898	0.769	0.588	0.391	0.223	0.108	0.043	0.014	0.004	0.001
19		0.999	0.995	0.976	0.921	0.812	0.648	0.457	0.279	0.146	0.065	0.024	0.007	0.002
20			0.997	0.982	0.940	0.848	0.703	0.521	0.339	0.191	0.092	0.038	0.013	0.004
21			0.998	0.987	0.954	0.879	0.751	0.581	0.400	0.240	0.125	0.056	0.021	0.007
22			0.998	0.991	0.965	0.904	0.794	0.638	0.460	0.293	0.163	0.079	0.033	0.012
23			0.999	0.993	0.974	0.924	0.831	0.690	0.519	0.349	0.206	0.107	0.048	0.019
24			0.999	0.995	0.980	0.941	0.862	0.737	0.576	0.406	0.254	0.140	0.068	0.028
25			0.999	0.996	0.985	0.954	0.888	0.778	0.630	0.462	0.304	0.178	0.092	0.042
26				0.997	0.989	0.964	0.910	0.815	0.679	0.518	0.357	0.220	0.121	0.058
27				0.998	0.992	0.972	0.928	0.847	0.725	0.572	0.411	0.266	0.154	0.079
28				0.999	0.994	0.979	0.943	0.874	0.765	0.623	0.464	0.314	0.191	0.104
29				0.999	0.996	0.984	0.955	0.897	0.801	0.670	0.517	0.364	0.232	0.133
30				0.999	0.997	0.988	0.965	0.916	0.833	0.714	0.568	0.415	0.276	0.166
31					0.998	0.991	0.972	0.932	0.861	0.754	0.617	0.466	0.322	0.203
32					0.998	0.993	0.978	0.946	0.885	0.789	0.662	0.516	0.370	0.243
33					0.999	0.995	0.983	0.957	0.905	0.821	0.705	0.565	0.419	0.285
34					0.999	0.996	0.987	0.965	0.922	0.849	0.744	0.612	0.468	0.330
35					0.999	0.997	0.990	0.973	0.937	0.873	0.779	0.656	0.516	0.376

	14	15	16	17	18	19	20	21	22	23	24	25	26	27	28
36							0.998	0.992	0.978	0.949	0.895	0.810	0.697	0.562	0.422
37							0.998	0.994	0.983	0.959	0.913	0.838	0.735	0.607	0.469
38							0.999	0.996	0.987	0.967	0.928	0.863	0.769	0.650	0.515
39							0.999	0.997	0.990	0.973	0.941	0.885	0.800	0.689	0.560
40							0.999	0.997	0.992	0.979	0.952	0.903	0.829	0.726	0.603
41								0.998	0.994	0.983	0.961	0.920	0.854	0.761	0.644
42								0.999	0.995	0.987	0.968	0.933	0.876	0.791	0.683
43								0.999	0.996	0.990	0.974	0.945	0.895	0.820	0.719
44								0.999	0.997	0.992	0.980	0.955	0.912	0.845	0.753
45								0.999	0.998	0.994	0.984	0.963	0.926	0.867	0.783
46									0.998	0.995	0.987	0.970	0.938	0.887	0.811
47									0.999	0.996	0.990	0.976	0.949	0.904	0.836
48									0.999	0.997	0.992	0.980	0.958	0.919	0.859
49									0.999	0.998	0.994	0.984	0.965	0.932	0.879
50									0.999	0.998	0.995	0.987	0.972	0.943	0.896

Table 5.13 Probability of X or More Correct Judgments in n Trials (One-Tailed, $p = 1/3$) (Continued)

n/x	0	1	2	3	4	5	6	7	8	9	10	11	12	13
0.001														
0.002														
0.003	0.001													
0.006	0.002													
0.010	0.003	0.001												
0.016	0.006	0.002												
0.025	0.009	0.003	0.001											
0.036	0.014	0.005	0.002											
0.050	0.022	0.008	0.003	0.001										
0.068	0.031	0.013	0.005	0.001										
0.090	0.043	0.019	0.007	0.002	0.001									
0.115	0.059	0.027	0.011	0.004	0.001									
0.144	0.078	0.038	0.016	0.006	0.002	0.001								
0.177	0.100	0.051	0.023	0.010	0.004	0.001								
0.213	0.126	0.067	0.033	0.014	0.006	0.002	0.001							
0.252	0.155	0.087	0.044	0.020	0.009	0.003	0.001							
0.293	0.187	0.109	0.058	0.028	0.012	0.005	0.002	0.001						
0.336	0.223	0.135	0.075	0.038	0.018	0.007	0.003	0.001						
0.381	0.261	0.164	0.095	0.051	0.025	0.011	0.004	0.002	0.001					
0.425	0.301	0.196	0.118	0.066	0.033	0.016	0.007	0.003	0.001					
0.470	0.342	0.231	0.144	0.083	0.044	0.021	0.010	0.004	0.001					
0.515	0.385	0.268	0.173	0.104	0.057	0.029	0.014	0.006	0.002	0.001				
0.558	0.428	0.307	0.205	0.127	0.073	0.038	0.019	0.008	0.003	0.001				
0.600	0.471	0.347	0.239	0.153	0.091	0.050	0.025	0.012	0.005	0.002	0.001			
0.639	0.514	0.389	0.275	0.182	0.111	0.063	0.033	0.016	0.007	0.003	0.001			
0.677	0.556	0.430	0.313	0.213	0.135	0.079	0.043	0.022	0.010	0.004	0.002	0.001		
0.713	0.596	0.472	0.352	0.246	0.161	0.098	0.055	0.029	0.014	0.006	0.003	0.001		
0.745	0.635	0.514	0.392	0.282	0.189	0.119	0.070	0.038	0.019	0.009	0.004	0.002	0.001	
0.776	0.672	0.554	0.433	0.318	0.220	0.142	0.086	0.048	0.025	0.012	0.006	0.002	0.001	
0.803	0.706	0.593	0.473	0.356	0.253	0.168	0.105	0.061	0.033	0.017	0.008	0.003	0.001	
0.829	0.739	0.631	0.513	0.395	0.287	0.196	0.126	0.076	0.042	0.022	0.011	0.005	0.001	0.001

Source: Reprinted from J. Food Sci. **43**, pp. 940–947, 1978. Copyright © by Institute of Food Technologists.

Table 5.14 Probability of X or More Agreeing Judgments in n Trials (Two-Tailed, $p = 1/2$)

n\x	3	4	5	6	7	8	9	10	11	12	13	14	15	16	17	18	19
5	0.625	0.312	0.062														
6		0.688	0.219	0.031													
7			0.453	0.125	0.016												
8			0.727	0.289	0.070	0.008											
9				0.508	0.180	0.039	0.004										
10				0.754	0.344	0.109	0.021	0.002									
11					0.549	0.227	0.065	0.011	0.001								
12					0.774	0.388	0.146	0.039	0.006								
13						0.581	0.267	0.092	0.022	0.003							
14						0.791	0.424	0.180	0.057	0.013	0.002						
15							0.607	0.302	0.118	0.035	0.007	0.001					
16							0.804	0.454	0.210	0.077	0.021	0.004	0.001				
17								0.629	0.332	0.143	0.049	0.013	0.002				
18								0.815	0.481	0.238	0.096	0.031	0.008	0.001			
19									0.648	0.359	0.167	0.064	0.019	0.004	0.001		
20									0.824	0.503	0.263	0.115	0.041	0.012	0.003		
21										0.664	0.383	0.189	0.078	0.027	0.007	0.001	
22										0.832	0.523	0.286	0.134	0.052	0.017	0.004	0.001
23											0.678	0.405	0.210	0.093	0.035	0.011	0.003
24											0.839	0.541	0.307	0.152	0.064	0.023	0.007
25												0.690	0.424	0.230	0.108	0.043	0.015
26												0.854	0.557	0.327	0.169	0.076	0.029
27													0.701	0.442	0.248	0.122	0.052
28													0.851	0.572	0.345	0.185	0.087
29														0.711	0.458	0.265	0.136
30														0.856	0.585	0.362	0.200
31															0.720	0.473	0.281
32															0.860	0.597	0.377
33																0.728	0.487
34																0.864	0.608
35																	0.739

Table 5.14 Probability of X or More Agreeing Judgments in n Trials (Two-Tailed, $p = 1/2$)[a] (Continued)

n^x	3	4	5	6	7	8	9	10	11	12	13	14	15	16	17	18	19
36																	0.868
37																	
38																	
39																	
40																	
41																	
42																	
43																	
44																	
45																	
46																	
47																	
48																	
49																	
50																	

20	21	22	23	24	25	26	27	28	29	30	31	32	33	34	35	36	37

					0.001	0.001										
		0.001	0.001	0.001	0.001	0.002										
		0.001	0.001	0.002	0.003	0.004										
		0.002	0.003	0.005	0.006	0.008										
0.002	0.002	0.004	0.007	0.009	0.011	0.014	0.001									
0.004	0.006	0.008	0.014	0.017	0.020	0.024	0.003									
0.009	0.013	0.016	0.024	0.029	0.034	0.038	0.005	0.001								
0.019	0.024	0.030	0.041	0.047	0.053	0.060	0.009	0.002	0.001							
0.036	0.043	0.050	0.065	0.073	0.081	0.088	0.017	0.006	0.002	0.001						
0.061	0.071	0.080	0.099	0.108	0.117	0.126	0.028	0.012	0.004	0.001	0.001					
0.099	0.100	0.121	0.143	0.154	0.164	0.174	0.044	0.020	0.008	0.003	0.001					
0.150	0.163	0.175	0.200	0.211	0.222	0.233	0.066	0.032	0.014	0.005	0.002	0.001				
0.215	0.229	0.243	0.268	0.280	0.291	0.302	0.096	0.049	0.023	0.010	0.004	0.001				
0.296	0.310	0.324	0.349	0.360	0.371	0.382	0.135	0.072	0.036	0.016	0.007	0.002	0.001			
0.392	0.405	0.418	0.441	0.451	0.461	0.471	0.184	0.104	0.054	0.026	0.011	0.005	0.002	0.001		
0.500	0.511	0.522	0.542	0.551	0.560	0.568	0.243	0.144	0.079	0.040	0.019	0.008	0.003	0.001		
0.618	0.627	0.636	0.652	0.659	0.665	0.672	0.312	0.193	0.111	0.059	0.029	0.013	0.006	0.002	0.001	
0.743	0.749	0.755	0.761	0.771	0.775	0.775	0.392	0.253	0.152	0.085	0.044	0.021	0.009	0.004	0.001	
0.871	0.875	0.878	0.880	0.883	0.885	0.888	0.480	0.322	0.203	0.119	0.065	0.033	0.015	0.007	0.003	0.001

Source: Reprinted from J. Food Sci. **43**, pp. 940–947, 1978. Copyright © by Institute of Food Technologists.

however, the final decision can involve other issues. The tables do not lead to any decision regarding the appropriate level of statistical significance, appropriate number of responses, or impact of replication on the final decision. The sensory professional must consider all of these issues before the test and then proceed with the interpretation.

Most discrimination tests are one tailed; the tail refers to a segment of the distribution curve (i.e. the tail end). Duo–trio, triangle, A-not-A, and same-or-different tests are one tailed because there is a single correct outcome. The directional paired-comparison test may be either one tailed or two tailed depending on the number of outcomes of interest to the experimenter. Because the discrimination test is a perceptual task rather than an instrument measure, it is not always correct to assume that the result must be in the same direction as the physical or chemical difference between two products. A product having less sweetener may be perceived as sweeter than, less sweet than, or equally as sweet as another product depending on the remaining sensory components of those products. If all possible outcomes in the directional paired comparison are of interest, the two-tailed table is recommended. More correct decisions are necessary to achieve statistical significance, which is a more conservative approach. For additional discussion about one-tailed versus two-tailed tests, see McNemar (1969).

Discrimination tests are forced-choice tests; the subject must make a decision and the issue should not be in doubt (Gridgeman, 1959), but one periodically encounters a sensory test operation where "no-decision" responses by subjects are permitted. This practice is not recommended because it is inconsistent with a basic principle of the discrimination test and the fact that guessing is accounted for in the design. From a practical standpoint, when "no-decision" responses are permitted, the total database is reduced. As Gridgeman noted, the decision regarding what to do with these "no-decision" responses poses a dilemma. One cannot ignore them, but they are not included in the methods for analysis and they cannot legitimately be divided equally between correct and incorrect decisions. Finally, there is the problem of subjects using this option to avoid more difficult decisions; it becomes an "easy way out." This particular problem can be avoided by simply requiring subjects to make decisions and by not providing the "no-decision" option.

At the outset of any testing, sensory evaluation must be firm about certain aspects of the test—for example, promptness, following test instructions, and making a decision. Subjects will generally adhere to these requirements, and individuals who either cannot comply or believe they are exempt from rules should be excused from testing. The fact that differences are relatively small is no reason for not making a decision; the subject must remember that the test is designed to account for small differences. It should also be kept in mind that in most operations, the differences are relatively small (remember that easily perceived differences are not worth testing), and if the "no-decision" option is included, more subjects will take that option. Finally, the developers of the methods did not recommend the "no-decision" option.

An attempt to work around this problem has come from developments in signal detection theory, although the authors did not have that purpose in mind at the outset. Nonetheless, signal detection theory has provided additional insight into the decision-making process. This application of signal detection has also led to some proposed practical applications that warrant some attention. O'Mahony (1979) and O'Mahony *et al.* (1979, 1980) devised a multiple-difference procedure that yielded measures of difference based on signal detection principles; it was suggested that the method would be especially useful as a means for assessing very small differences, however "very small" was not defined. A subject evaluates a set of products, one at a time, and indicates one of four response alternatives: "yes," "maybe yes," "maybe no," or "no." Based on the number of products in the test, one then has a response matrix from which R values or $P(A)$ values (indices of difference) can be computed. The primary advantages one might gain from the use of the procedure is that it takes into account the subject's uncertainty in a decision and that it could be used to evaluate many samples within a brief time period. Whether knowledge of a subject's uncertainty results in a more meaningful decision about product differences remains to be demonstrated.

In industrial operations, the sensory professional is working with experienced subjects and it is possible that uncertainty about a decision may provide no advantage and could reduce sensitivity by providing this option where none existed. Other factors that need to be taken into account in comparing this procedure and those used in traditional discrimination testing include the amount of product required for a test, number of evaluations per subject, and time required for a decision. O'Mahony (1979) noted that the method has potential for flavor measurement; however, further study is required before the technique can be used effectively within the industrial environment. Developments such as these warrant consideration; however, they must be evaluated relative to current activities and in their entirety, including product preparation, subject test time, analysis, and interpretation.

Decisions regarding number of subjects for a test and number of responses per subject are issues that are not well-defined. A perusal of the current literature would lead one to assume that the decision is arbitrary, and that any number of subjects from 10 to 50 (or more) would be adequate or that statistical power calculations dictate number of observations (Ennis, 1993; Schlich, 1993). Until recently (Bi and Ennis, 2001a,b; Ennis and Bi, 1998), less consideration appears to have been given to replication. By replication in this situation, we mean a subject evaluating more than one set of the test products in a single session. The actual procedure would be for one set of products and scorecard to be served, a decision made, followed by an appropriate interval of 1–3 minutes, and the procedure would be repeated (this would result in two responses from each subject within a single session).

Generally, most companies have not utilized replication effectively in test design. If subjects are allowed to consume the product rather than "sip and spit," it is easy to understand why replication is not successful. Sensory fatigue (decrease in sensitivity) causes the percentage of correct responses to decrease in the second trial. This problem is easily corrected, and the information obtained from the replicate responses substantially increases the power of the test and the confidence that the sensory professional has with the results. Response patterns can be extremely informative, as shown in the following table:

Test	Serving Order		Pooled Results
	First	Second	
A[a]	10/25[b]	18/25	28/50
B	19/25	9/25	28/50
C	14/25	14/25	28/50

[a]Each test was either paired comparison or duo–trio: $p = 1/2$.
[b]Entries are the ratios of correct to total decisions.

The three test examples—A, B, and C—show the same total number of correct decisions; however, the response patterns are different. In test A, the number of correct decisions was almost twice as great for the second serving order (18/25), suggesting that some form of learning may have occurred. After completing the first set, the subjects were able to perceive the difference (more easily). It can be argued that such an occurrence could represent a lack of skill; however, a new variable could introduce flavor changes not familiar to the subjects and thus could account for the observed response pattern. This problem could be minimized through the use of "warm-up" samples; however, sensory fatigue, as described next, might occur.

In test B, the situation is reversed and performance has declined from the first set (19/25) to the second set (9/25). This pattern suggests that some type of fatigue may have occurred; the interval between sets may not have been sufficient and/or the subjects did not rinse sufficiently to recover. Finally, in test C, the number of correct decisions is similar in both orders, indicating none of the problems encountered in tests A and B. All three tests yielded a total of 28 correct decisions with a probability value of 0.240, which is not sufficient for rejecting the hypothesis that the products are perceived as similar.

However, there is a question regarding whether these results are consistent. For tests A and B, we may be in error if we conclude that there is no significant difference between the products. Although it is true that the total correct decision is not sufficient for statistical significance, sensory evaluation would be remiss if the results by serving order were not discussed. In fact, both patterns suggest the likelihood of a statistically significant difference, and depending on product and subject availability, the test might be repeated. Practically, repeating a test usually is not possible, most often because of product supply, and a decision must be made using available information; hence the conclusion that there is a perceived difference. In the nondirectional test, the number of correct decisions can be monitored by order of presentation, as well as by the specific pair of products. Here, the four possible serving orders—AA, BB, AB, and BA—are listed with their respective results:

Test	Same		Different		Pooled
	AA	**BB**	**AB**	**BA**	**Pooled**
D	2/10	2/10	9/10	9/10	18/30
E	5/10	4/10	9/10	7/10	18/30

Assuming subjects understood the task, the serving orders of AA and BB should yield approximately 50% correct decision. In test D, only 4 of the 20 decisions are correct, which suggests some subjects did not understand the task and circled "different" on every scorecard. Thus, the 18 correct decisions for the AB and BA orders might be considered to be the result of a combination of true perception of a difference as well as a misunderstanding of the task. Examination on a subject-by-subject basis may also provide further insight as to the source of the problem. For test E, the responses are more consistent with expectation. These examples serve as a reminder of the importance of examining response patterns on a subject basis as well as by serving order before preparing any conclusion and recommendation.

This discussion about replication has thus far focused on problems that are traced back to the experimenter and the subjects; however, it is also possible that the product can be at fault. Product changes can and do occur between test planning and the experiment, and the sensory professional must be vigilant to the prospect of this problem. Often, product variation due to production variation may be as large as the variable being tested. This could result in a nonsignificant conclusion, which would be incorrect. One added benefit of replication is the use of more product so that the product variations do not have as great an effect. Of course, where it is known that typical production variation cannot be avoided and it always results in significant differences, then it is necessary to consider using an alternative methodology such as descriptive analysis (discussed in Chapter 6). Finally, one should not mix all the control product so as to blend the variability. Such an approach will yield a result that in and of itself may be quite reliable but have absolutely no practical value (or external validity). Of course, major concerns about product are addressed before data collection; however, judgmental situations do arise, and the solution becomes apparent only after a test is performed. Regardless of the problem, replication enables the sensory professional to identify problems to avoid in future tests with those products and also to reach a more realistic and meaningful decision.

It can be argued that replication within a test session violates the statistical rule of independence of judgments and therefore the standard tables for determining statistical significance are not appropriate (Bi and Ennis, 2001a; Ennis and Bi, 1998). This particular issue warrants comment.

The χ^2 statistic can be used to determine if the proportions of correct decisions in the replicate are significantly different from those in the first trial, and so on. When the number of correct decisions is not significantly different, results from the repeated trials can be combined. The traditional tables (Roessler et al., 1978) used for determining statistical significance can be used.

Table 5.15 Number of Testers and Their Performance in a Duo–Trio Discrimination Test with One Replication

Second Trial	First Trial Correct	First Trial Incorrect	Total
Correct	5^a	$6_{(r_2)}{}^b$	11
Incorrect	$3_{(r_1)}{}^c$	1^d	4
Total	8	7	15

[a]Entry is number of subjects (five) who were correct on both trails.
[b]Entry is number of subjects (six) who were correct only on the second trial (the replicate).
[c]Entry is number of subjects (three) who were correct only on the first trial.
[d]Entry is number of subjects (one) who were incorrect on both trials.
Note: $m = r_1 + r_2$; see text for explanation.

A quick method for determining if there is a significant difference between proportions of responses from the same subjects has been proposed by Smith (1981). In Smith's example, the same panel participates in two different tests. The total number of correct responses for the first test (r_1) is added to the total number of correct responses for the second test (r_2) to produce a value m (i.e. $r_1 + r_2 = m$). Using Tables 1, 2, or 3 prepared by Roessler et $al.$ (1978), Smith used m to represent the number of trials (n) and r_1 and r_2 are significant.

Smith's (1981) procedure may be expanded to the situation in which the same subjects evaluate products on two trials. A worked example for a duo–trio test is presented in Table 5.15; of the nine subjects who were correct only once, three were correct on the first trial and six were correct on the replication. Using Smith's notation, we determine that $r_1 = 3$, $r_2 = 6$, and $m = 9$. Consulting Table 1 in Roessler et $al.$ (1978), and substituting M for n, we note that neither r_1 nor r_2 is >8, which is the table value required for statistical significance with an n of 9. Therefore, there is no evidence that the proportions from these two replications are different, suggesting that it would then be reasonable to pool the responses.

Another statistical solution to the handling of replication would be to use only the data from subjects who were correct on both trials for analysis. The rationale is that these data may be considered as representative of discrimination. Exact probabilities may be obtained from the tables of the cumulative binomial probability distribution. When such tables are not available, a Z formula for computing a good approximation to the exact probability has been provided by Roessler et $al.$ (1953, 1978) and is modified here as follows:

$$Z = \frac{(X - np) - 0.5}{\sqrt{npq}}$$

where X is the number of subjects correct on both trials, n is the total number of subjects, p is the probability of two correct decisions $[(p_1)(p_2) = p]$ (in the triangle test, $p_1 = 1/3$; in the duo–trio and paired comparison tests, $p_1 = 1/2$), and $q = p - 1$. Tables 5.16 and 5.17 have been prepared to accommodate tests in which

Table 5.16 Probability of X or More Correct Judgments on Two Successive Trials [One-Tailed, $p = (1/2)(1/2) = 1/4$]

n	1	2	3	4	5	6	7	8	9	10	11	12
12	0.968	0.842	0.609	0.351	0.158	0.054	0.014	0.003	0.000			
13	0.976	0.873	0.667	0.415	0.206	0.080	0.024	0.006	0.001	0.000		
14	0.982	0.899	0.719	0.479	0.258	0.112	0.038	0.010	0.002	0.000		
15	0.987	0.920	0.764	0.539	0.314	0.148	0.057	0.017	0.004	0.001	0.000	
16	0.990	0.937	0.803	0.595	0.370	0.190	0.080	0.027	0.007	0.002	0.000	
17	0.992	0.950	0.836	0.647	0.426	0.235	0.107	0.030	0.012	0.003	0.001	0.000
18	0.994	0.961	0.865	0.694	0.481	0.283	0.139	0.057	0.019	0.005	0.001	0.000

Table 5.17 Probability of X or More Correct Judgments on Two Successive Trials [One-Tailed, $p = (1/3)(1/3) = 1/9$]

n	1	2	3	4	5	6	7	8
12	0.756	0.391	0.140	0.036	0.007	0.001	0.000	
13	0.783	0.432	0.168	0.048	0.010	0.002	0.000	
14	0.807	0.470	0.198	0.061	0.014	0.003	0.000	
15	0.829	0.508	0.228	0.076	0.019	0.004	0.001	0.000
16	0.848	0.544	0.259	0.093	0.026	0.006	0.001	0.000
17	0.865	0.577	0.291	0.111	0.033	0.008	0.001	0.000
18	0.880	0.609	0.322	0.131	0.042	0.011	0.002	0.000

Table 5.18 Probabilities for Data in Table 5.15

Trials	Duo–Trio or Paired Comparison[a]	Triangle[b]
1	0.50	0.08
2	0.06	0.00
Combined	0.10	0.00

[a]Assumes results entered in Table 5.15 were derived from a duo–trio or paired-comparison test.
[b]Assumes results in Table 5.15 were derived from a triangle test.

12–18 subjects participate in a trial plus replicate session for the duo–trio, paired-comparison, or triangle test. When the number of subjects exceeds these table values, the previous Z formula is recommended.

Returning to the worked example in Table 5.15 and using Table 5.16 (if a duo–trio or paired test) and Table 5.17 (if a triangle test), we obtain a p of 0.31 and 0.02, respectively, with 5 of 15 subjects correct on both trials. Alternatively, analyzing the data in Table 5.15 for each trial separately and pooled, we obtain the results shown in Table 5.18. Although the analysis applied to the data in Table 5.15 allowed us to pool the responses, and the results from the discriminators (subjects who were correct on both trials) are in general agreement with the pooled data, observation of the different probabilities from the first and second trial reaffirm the value of separate trial-by-trial analysis prior to pooling or any other analysis. Equipped with this information, the sensory professional is in a better position to determine whether the products were found not to be different or a learning effect occurred.

A second issue related to replication concerns the number of subjects participating in the test. As noted previously in this section, the literature is not totally clear on this issue. The tables for estimating statistical significance do not differentiate between 30 responses derived from 1, 2, 10, or more subjects. However, as noted by Roessler et al. (1978), there can be a difference in the interpretation that would be applied to the results. Sensory evaluation might be reluctant to recommend, with confidence, a "no difference" (or a difference) based on responses from 1 subject tested 30 times. One reason for a panel of subjects is to move away from reliance on one or two individuals. Obviously, a panel size that is larger than 1 subject and less than 100 should be sufficient for reaching conclusions with confidence.

From an operational standpoint, the fewest subjects should be used for the least amount of time that will yield a result with minimal risk and maximal confidence. Originally, we observed that approximately 16–20 subjects per test with a replicate was satisfactory. However, concerns raised by brand and product technical managers necessitated recommending increasing panel size to 25. This did not change the guidelines listed here:

1. Subjects should not have to spend more than approximately 15 minutes away from their regular workstation, including travel time.

2. For most products, evaluating two sets is optimal relative to minimizing sensory interactions (for products with no chemoreceptor stimuli, the number of pairs could be increased).
3. A total of 50–60 responses is sufficient to minimize risk in the decision about product differences (quality control procedures will have a different set of conditions and fewer responses needed).
4. This panel size minimizes product usage, product preparation and scheduling time, and total test time.
5. From a design perspective, a total of approximately 25–30 subjects is a practical panel size (most serving orders are multiples of 2, 4, or 6).

Several practical issues associated with the number of subjects in a discrimination test have been identified, and we now approach the problem from a statistical standpoint. In a discrimination test (as in any sensory test), a hypothesis about the products is established *a priori* the test; for example, there is no difference between product A and product B (or products A and B are perceived as the same), which is expressed as

$$H_0 : A = B$$

and referred to as the null hypothesis. The alternative hypothesis is expressed as

$$H_a : A \neq B$$

A is not the same as B. To reach a decision as to whether to accept or reject the hypothesis selected, a particular probability (or confidence) and a level of risk are assigned. The decision regarding which probability level is most appropriate is quite arbitrary; for further discussion, see McNemar (1969), Hays (1973), or Winer (1971). The traditional values used in sensory evaluation are 0.05, 0.01, and 0.001 (Amerine et al., 1965). By probability values of 0.05, 0.01, and 0.001, we mean that the probability of the difference occurring by chance is 5 times in 100 tests, 1 time in 100 tests, and 1 time in 1000 tests, respectively (on all other occasions, this was due to a perceived difference). The number of correct responses sufficient to achieve the statistical significance is not a measure of the magnitude of difference between two variables.

Whenever significance is considered, there is always risk, albeit a small one, that our decision may be wrong. Obviously, this possibility of making a wrong decision is one that we would like to minimize, but the cost of achieving no risk may be very high (e.g. testing all consumers of a particular product). One option is to increase N to some very large number, but this is not practical. In addition, it should be noted that as N increases, the number of correct decisions for statistical significance becomes proportionally smaller, such that one could have statistical significance with just one or two correct decisions above chance. How practical is this result? Obviously, compromises are made that allow testing within the controlled environment with a relatively small number of subjects but with low risk and high external validity. This approach leads again to the problem of setting an acceptable level of risk. As noted previously (Chapter 4), when the probability is 0.05, there is a possibility of 1 error in 20 decisions when we say that the products are the same (H_0: A = B). This particular error is referred to as a Type 1 error, and the probability of it occurring is given by α (alpha). A second type of error that could occur when we say the products are the same, when in fact they are not, is a Type 2 error, and its probability is β (beta). These probabilities are also expressed as follows:

Rejecting the null hypothesis when it is *true* is called an error of the first kind (Type 1), and its probability is α (alpha).

Accepting the null hypothesis when it is *false* is called an error of the second kind (Type 2) and its probability is β (beta).

Note that the probability of α is the significance (or probability) value found in the aforementioned tables; that is, when we report that the number of correct decisions is statistically significant at the 0.05 level, we are stating the probability of a Type 1 error. These two errors are not of equal importance, and as we will show, the decision about their relative importance requires an appreciation for all components of the test before the decision is reached. Historically, the emphasis has been on α and minimizing this error; inadequate attention has been given to the Type 2 error and its effect on the conclusion (Cohen, 1977). There is a relationship between the two errors in that once α is specified, β is also determined. The two errors are also inversely proportional; the smaller we set α (e.g. $\alpha = 0.01$ or 0.001), the larger β will be (which can be offset to some degree by increasing N with replication and/or more qualified subjects). From a business perspective, it turns out that β is often more important than α. In most discrimination tests, the objective is to minimize detecting a difference by the fewest number of consumers (any perceived change is considered negative). This particular situation is very common; where the availability, cost, and quality of raw materials change frequently, it is important that the impact of these changes does not change preference.

As stated previously, the goal for most of these product modifications is to not have the change perceived by the consumer. To achieve this, we select our best subjects and possibly use the duo–trio test with a constant reference. We create a situation in which the likelihood of perceiving the difference is emphasized, and if the difference is not perceived at the appropriate level of significance, we are confident that the change can be made. There is minimal risk of detection by the consumer if our best discriminators under ideal conditions do not perceive the difference. We could also institute a statistical change: Let $\alpha = 0.10$; thus, fewer correct decisions will be necessary before there is statistical significance. This approach will result in β decreasing (all other factors being equal), and we are more likely (than if we operated at $\alpha = 0.05$) to reject some products that are not different. Because our objective is to protect the current franchise, we will "sacrifice" α for β. That is, we would rather reject some products that may not be different to ensure the acceptance of that "one" which we know is least likely to be perceived as different.

When we make this trade-off, we are reducing the risk of saying there is no difference when in fact there is. Of course, such a change in probability (from 0.05 to 0.10) may be cause for concern by the test requestor, who may wonder why the change is being made. In this case, it may be desirable to not change α.

It is interesting to note that the sensory literature contains very little information on computation of β risk. Two early publications (Blom, 1955; Radkins, 1957) and a subsequent discussion by Amerine *et al.* (1965, see especially pp. 439–440) provided procedures for the computation and additional insight into β risk. Using the equation provided in Amerine,

$$N = \left[\frac{z_\alpha \sqrt{P_0(1 - P_0)} + z_\beta \sqrt{P_a(1 - P_a)}}{P_a - P_0} \right]^2$$

where N is the number of judgments; z is the z value corresponding to α and β, respectively; P_0 is the probability of the null hypothesis; and P_a is the probability of the alternate hypothesis.

This formula enables one to determine β with a specific N or the N necessary for specific α and β values. However, the situation is not totally clear because the experimenter must also specify the P_a and P_0 values.

Whereas it may be relatively easy to establish α, usually at 0.05, β may not be so easy to specify. If the desire is to minimize risk and we set β small such as $\beta = 0.01$), the number of judgments required in the test may be quite large (e.g. >100). The problem is complicated by our inability to mathematically account for subject sensitivity and for replication in the statistical formula. In addition, the experimenter must decide on the values to be assigned to P_0 and P_a and how large or small a difference is reasonable between the null hypothesis, P_0, and the alternative hypothesis, P_a. This difference is based on the true difference between the samples; however, it is not possible to know the value (which will vary from product to product). We usually set $P_0 = 50$; that is, the likelihood of A = B is approximately 50%. But what probability can be assigned to P_a?

For discrimination testing, it may be necessary to approach this topic from our knowledge of the general sensory skills of the subject. That is, in most product situations, a qualified subject is expected to perceive product differences once the magnitude of the difference reaches approximately 20%. For some of the senses (vision and audition), the just-noticeable difference is less than 20%, whereas for others it will be greater (e.g. olfaction; see Geldard, 1972; Schutz and Pilgrim, 1957; Stone, 1963; Wenzel, 1949). Because products stimulate many of the senses, we could arbitrarily set $P_a = 0.70$; the difference between P_0 and P_a is set at 0.20 (or 20%). In other words, we are substituting product difference for the subject's estimated ability to perceive the difference. With that as a guide, we compute N for a variety of test conditions as well as for several levels of sensory skill (Table 5.19).

It is obvious that the magnitude of the difference between products, P_a and P_0, will cause an increase or decrease in N, the number of judgments, as does the

Table 5.19 Number of Judgments Required in a Discrimination Test at Specific α and β Values[a]

$P_a - P_o$	$\alpha = 0.05$, $\beta = 0.10$	$\alpha = 0.05$, $\beta = 0.05$	$\alpha = 0.05$, $\beta = 0.01$	$\alpha = 0.10$, $\beta = 0.05$
	N	*N*	*N*	*N*
10	209	263	383	208
15	91	114	165	90
20	49	62	89	48
25	30	37	53	29
30	20	25	34	19

[a]Entries are total number of judgments per test for selected α and β values. Computed from Amerine et al. (1965).

Table 5.20 Computed β Risk Values for Discrimination Tests as Determined from the Number of Judgments[a]

N	Level of α	
	0.05	0.10
44	0.1358	0.0672
40	0.1671	0.0866
38	0.1853	0.0981
36	0.2050	0.1111
30	0.2757	0.1606
20	0.4377	0.3900

[a]Entries are β values computed using the formula from Amerine et al. (1965).

change in α and β. These data demonstrate that the greatest changes occur as the magnitude of difference decreases: The smaller the difference, the more responses required before this difference will be perceived. However, the reader should reflect carefully on the practical implications. By increasing N, one can expect to find statistical significance between two products from a production line selected within 1 minute of each other and the true difference could be 0.001 units or less. One will find significance, but it will have no practical value.

Is it advantageous to shift the levels of α and β, which could mean using an unbalanced design or a large number of judgments? Before attempting to answer the question, we should establish the risk associated with the current procedure of limiting judgments to approximately 30–40. In Table 5.20, computed β values are reported when $\alpha = 0.05$ and 0.10, at a fixed level of sensitivity of 20% ($P_a - P_o$). It is apparent that β risk almost doubled when α was decreased from 0.10 to 0.05. For $N = 40$, the risk of committing a Type 2 error is approximately 17% (0.1671), approximately one time in six, which may not be that serious.

Associated with this risk is the related issue of the statistical power of the test. Cohen (1988) defines power as "the probability that it will lead to the rejection of the null hypothesis." In other words, power is equal to 1 minus the probability of a Type 2 error (Winer, 1971). Cohen (1977), McNemar (1969), and Winer (1971) all direct attention to the importance of power and the extent to which it is most often overlooked. In general, it can be stated that as α is reduced, for example, to 0.01 or smaller, the increase in β risk results in a decrease in statistical power. Furthermore, one can assess the relative importance (or seriousness) of the respective errors. Because power (P) is equal to $1 - \beta$ (and the greatest power is achieved when β is very small), then for $N = 40$ and $\alpha = 0.05$ and $\beta = 0.1671$, the statistical power is 0.8329.

We can also consider the situation in terms of its relative importance; that is, the ratio of Type 1 to Type 2 errors is 0.1671/0.05, or 3.4:1. We consider a Type 2 error as 3.4 times more likely to occur (or more serious) than a Type 1 error.

It should be kept in mind that these computations have not taken into account subject replication, a factor that further reduces β risk. A reasonably high degree of confidence in results can be obtained from using approximately 25 qualified subjects with a test involving one replication (a total of 50 judgments). By maintaining subject performance records and using the appropriate test method, one should expect to further reduce β risk from that computed by as much as 50%. Empirically, we observed that using the procedures described here has not ever resulted in a wrong recommendation. The relevant research to support these results remains to be done.

5.3.6 The just-noticeable difference

In Chapter 1, the importance of the just-noticeable difference (JND) was discussed in relation to the perceptual process and the discrimination model. There also are situations in which it is useful to estimate JND as an element in establishing product manufacturing specifications. This practice is a step removed from the more typical applications of the discrimination model as described in this chapter. In effect, systematic changes made in a particular variable are used as a basis for estimating how much change is needed before it is perceived. Rather than determining whether or not ingredient "x" can be substituted without being perceived, the objective is to determine how much change can be tolerated. Although this particular issue is also considered in the discussion on quality control in Chapter 8, it warrants some discussion here because of its unique discrimination methodology requirement.

To determine the JND for a particular variable and to express the relationship mathematically, the constant-stimulus method is most likely to be used (Guilford, 1954; Laming, 1986; Perfetti and Gordin, 1985; Schutz and Pilgrim, 1957; Stone, 1963). This procedure involves a series of paired comparisons, with the subject's task being to note which one of the pair is more intense. One of the pair is always the same (the constant), and the other is either weaker or stronger. A proper spacing of the latter products results in proportional judgments from which an equation for the line of best fit can be computed (Bock and Jones, 1968). An alternative single-stimulus method could also be used (Schutz and Pilgrim, 1957); however, the result is identical in all other aspects.

For the test to be successful, it is necessary that the particular variable be controlled so that a set of test products can be prepared that bracket the "norm." Although it is relatively easy to implement this control in model systems, it is often quite difficult in product manufacturing. Nonetheless, if a predictive equation is to be formulated with any degree of confidence, then it is essential that no evaluation be made unless the products are within the experimental limits. Obviously, if the products are too close together, the predictive equation will be based on too few points and be of less value.

However, once the JND is known, it can be used to gauge the impact of machine variability and the importance of such change on preference. The implications are far-reaching and have considerable economic importance.

5.4 **Special problems**

Mention was made earlier in this chapter about the importance of utilizing the test procedures as they were developed, or at the least to carefully consider modifications within the context of the sensory task. For example, a monadic sequential serving procedure is a reasonable alternative to the use of lighting changes (e.g. red or yellow), which can introduce other undesired variables into the test. Whereas such a modification is reasonable, many other modifications that have been described create considerably more problems than they propose to solve. Some modifications offer clearly demonstrated statistical advantages but entail considerable disadvantage from a behavioral standpoint and are often quite impractical in a typical business environment. In general, these modifications can be characterized as being of three main types; preference after difference, magnitude of difference, and description of difference (descriptive). In each case, it appears that the overall intent of the modification is to increase the amount of information obtained from the test. If one considers that a subject may take as much as 5–10 minutes to provide two responses (the usual number, but more responses are possible), it would seem to be a relatively small output compared with that of the descriptive test, for example, in which the same amount of time might yield 10 times the number of responses. Many of the modifications are supported by statistical computations (Bradley and Harmon, 1964; Ferdinandus *et al.*, 1970; Gridgeman, 1964; Radkins, 1958; Steiner, 1966); however, most modifications have limited behavioral support. In the discussion that follows, we take a behavioral approach to provide the sensory professional with a greater appreciation for the risks as well as the statistical advantages offered by each method.

5.4.1 **Is there preference after difference?**

This particular question is one that seems to have lessened throughout the years because there are many problems associated with it. However, this approach of discrimination followed by preference has been used with consumers, providing still one more example of the misapplication of sensory methodology.

In the laboratory situation, the subject makes a discrimination decision followed by the preference task. An alternative has been to remove the products and scorecard after the discrimination task and then to serve the two products again and request the preference. Either approach leads to a high-risk situation. First, odd sample bias exists in discrimination tests (Schutz, 1954), and one cannot design away the bias. By odd sample bias, we mean avoiding the product that is different from the other two in a three-product test and avoiding the product that is not "the same as …" in the paired situation. Some subjects will assume that the product that is different is not "as good."

A second weakness is inherent in the subjects themselves; they were selected based on discrimination skill, and their attitudes about the products may be completely skewed. The panel size is small (approximately 25 subjects) and the database is equally small (approximately 50 responses)—hardly sufficient for any meaningful preference information. Perhaps most important is the context of test tasks; there is no assurance that the subjects have made the psychological transition from perceiving a difference to stating a preference. The assignment may seem to be relatively simple, such as "tell me which product you prefer," and the subjects mark their scorecard; the fact is that two very different tasks are involved, and some subjects may not make the mental transition, thus resulting in a response that is confounded.

A third objection concerns the preference responses by subjects who did not make correct decisions—that is, subjects who did not match the same pair or a panel that had an insufficient number of correct decisions to achieve statistical significance. If no statistically significant difference exists, then no difference in preference for the two products can be expressed. Whereas this latter situation is clear, the preference results pose a problem. Does one only report preferences for the correct decisions, which means that one is selectively reporting results, or does one include results for subjects who do not perceive a difference? Although it is true that there are statistical procedures for analysis of the results (Ferdinandus *et al.*, 1970; Woodward and Shucany, 1977), the basic concerns we describe have been ignored.

The report by Ferdinandus and co-workers (1970) is especially troublesome for several reasons: the combining of difference with preference; the establishment of a new, combined probability of 1/6; and the concept of nonrecognition. In fact, the authors have chosen to discuss discrimination using the term "recognition." Behaviorally, one might be on rather tenuous ground with the term "recognition" versus "discrimination." A subject may be capable of perceiving a difference but may not recognize the sensation(s) responsible. The concept of nonrecognition is also very confusing, especially in relation to accepting or rejecting a particular hypothesis. It is difficult to define nonrecognition. Is it an inability to perceive a difference, a momentary loss of concentration, or a case of sensory fatigue? Could it be that the concentration difference was too small for that subject at that time? None of these options can be separated before, during, or after the test. Finally, what use is derived from knowledge that a subject did not "recognize" the difference on one particular occasion but did "recognize" the difference on another occasion? The problem with such approaches is the rationale; all products are "different," and given sufficient products and subjects, one can demonstrate that a statistically significant difference exists. Telling the requestor that a significant number of subjects did not recognize a difference might mislead that individual to assume that there was no perceived difference. Similarly, when there was a statistically significant difference, there could also be statistically significant nonrecognition.

The discrimination–preference combination procedure has also been used in marketing research with consumers (Buchanan *et al.*, 1987; Morrison, 1981). It is assumed that if consumers cannot perceive product differences, they cannot have a valid preference. By eliminating the nondiscriminators, one is left with individuals whose preferences are more precise. Unfortunately, the rationale, although possessing a certain appeal, is flawed. As discussed previously in this chapter, the idea that a single (or even two) discrimination test will identify a discriminator is inconsistent with the evidence. Screening subjects takes considerably more than a single trial; most individuals must first learn how to take a test. This does not mean that an individual cannot make a reliable judgment about which product is liked. As noted, the two tasks are quite different. Thus, one will have a group of consumers, some of whom are participating based on a chance decision but not based on actual perception.

The typical database used in these tests is derived from several hundred consumers. With such a large database, it is relatively easy to have statistical significance, leading the investigator to conclude that consumers can participate in discrimination tests, that one can observe differences not identified with laboratory panels, and, by implication, that the results are valid. The flaws in this approach were described previously in this discussion and elsewhere in this chapter. Sensory evaluation must address these issues directly and must delineate the characteristics of the discrimination test and ensure that it is not used as part of a consumer preference test.

5.4.2 **Magnitude or degree of difference**

It has been suggested that one might also be interested in determining the magnitude of the difference between products in discrimination testing. In this procedure, the subject first makes a discrimination decision followed by a measure of the perceived magnitude. The measurement is usually achieved with a category-type scale—weak, moderate, strong, and so on—which is converted to numerical values for subsequent analysis. A detailed discussion about the method, including a procedure for data analysis that incorporates all responses from subjects, was given by Bradley and Harmon (1964). Although their approach utilized the triangle test, the authors noted that the procedure could be extended to the duo–trio test. It is emphasized that if one intends to collect magnitude of difference responses, then the approach described herein is certainly recommended.

Remember that prior to any test, products are usually screened by the sensory professional, and where the differences are obvious, the request will be rejected.

Gradually, formulation efforts result in submission of products whose differences are small, usually at or near the JND, and therefore warrant testing. The responses will be clustered around the lower end of the scale but still it is a sensory magnitude. For sensory evaluation, the problem is considerably less complex in the sense that knowledge of JND provides general direction to the requestor. In the chemical senses, a JND of approximately 20% is typical (Schutz and Pilgrim, 1957; Stone and Bosley, 1965); for the other senses, a JND of approximately 15% is reasonable (Geldard, 1972). Basically, the JNDs for the individual senses are smaller and larger than these values; however, one must take into account the complexity of products and the extent to which ingredients interact with one another, hence the guideline of approximately 20%. This approach is obviously less formal; however, the problem of sensory magnitude affords one little solace. Sensory evaluation must still establish the perceptual meaning of that value—one JND, two, and so on— to provide a meaningful course of action for the requestor. From a practical standpoint, such effort provides no advantage for sensory evaluation, but it does require further analysis and interpretation.

A similar approach to magnitude of the difference is degree of difference (DOD), proposed by Aust *et al.* (1985) for heterogeneous products. Scaled difference ratings were used to contrast the difference between the control and the test samples to a baseline difference between control lots. The difference between the control and test was compared to within- and between-lot variability for controls to determine if the difference was significant. There have been some recent publications and updates on this DOD method (Pecore *et al.*, 2006; Young *et al.*, 2008).

A basic issue with these types of discrimination methods concerns the value of knowledge of the magnitude or degree of the difference and whether such knowledge will be useful in product formulation and quality control.

Assume the test objective is to replace ingredient A with B such that the samples are not perceived as different (i.e. H_o: A = B). If no statistically significant difference exists, then we accept H_o, and any magnitude of difference information is of no value. If a statistically significant difference does exist, we reject H_o, and we turn our attention to the value of the magnitude value. The magnitude value is a sensory magnitude, and the relationship between the sensory response and a specific amount of ingredient is not established. For example, knowledge that the difference is two scale units may not be very helpful to sensory evaluation or to the test requestor because these are sensory units. Sensory evaluation must still make a professional judgment in guiding the requestor.

5.4.3 **Equivalency and similarity testing**

At the other extreme of the spectrum, Meilgaard *et al.* (1999) described an equivalency procedure and then a unified approach to difference and similarity testing (Meilgaard *et al.*, 2006) to determine whether there is no perceivable difference between two products. Although the idea of similarity or equivalency testing is to minimize a Type 2 error (the chance of missing a true difference), these techniques have seen less use in comparison to traditional discrimination approaches (Bi, 2005, 2011; Castura, 2010).

Not finding a significant difference does not mean that products are the same. Such a conclusion is misleading because at some level (of significance) the products are perceived as different. It is just that at the significance level selected for the test, the rejection of the null hypothesis was not realized. There are many reasons that could account for not finding a significant difference when it exists, including lack of best practices for sample preparation and serving, naive consumers with unknown discrimination abilities, inadequate test instructions, poor environmental conditions, and lack of replication (Lawless and Heymann, 2010). It must also be kept in mind that all products are different from all other products. The fact that results did not yield a statistically significant difference at the 5% probability level does not mean no one detected the difference and therefore the products are the same. It may be significant at the 15% probability level; it depends on the decision criteria.

5.4.4 Description of difference

Another modification to the discrimination test requires the subject to indicate the basis for each decision; that is, it was the color, the sour taste, etc. This task is facilitated by providing the subjects with a checklist. Examples of this type of scorecard are common in many companies and appear to have been adapted from quality control sources (see especially Amerine *et al.*, 1965, p. 341). The problem with these kinds of improvements is that they create their own problems, of which the main one is what to do with the incorrect matches (a common feature of all test add-ons). One is now using selective reporting, a practice that is not correct. Second, assuming that the words have the same meaning to all subjects also has no basis. As we know from descriptive analysis, it takes several hours for subjects to develop a common basis for use of specific words. Using a checklist will cue subjects to focus on attributes that may not be there. What appears to be a simple addition to a simple test adds far more problems than it solves. If requestors believe that the basis for any differences is more important than just knowing if products are perceived as different, then a descriptive test is more appropriate. The failure to appreciate the behavioral–product relationships eventually impacts the credibility of sensory information and the program itself.

These particular issues have been addressed previously (see Chapter 2) and are mentioned here to allow the reader to refer to the arguments associated with these issues. Often, the requestor will misuse the results and the blame will go back to the sensory evaluation department. One should use the discrimination test strictly as it was designed—to determine whether the products are perceived as different and to establish the level of significance for that difference.

5.5 Summary

The discrimination test is a powerful sensory evaluation method in terms of its sensitivity and providing reliable and valid results. The test method has evolved in response to a greater appreciation for the perceptual processes involved in the discrimination task. In this exposition of discrimination testing, considerable emphasis has been placed on the use of methods according to strict guidelines. Of paramount importance is the strict adherence to detection of a perceived difference. Although the literature includes numerous references to modifications of the methods, most, if not all, modifications introduce problems, especially with regard to subject selection, data analysis, and interpretation. To a considerable degree, many of these problems arise from a lack of appreciation for the use of the method in relation to other test methods and from a failure to follow accepted testing guidelines. It is important that the sensory professional recognize the following innate qualities of discrimination tests:

1. In practice, all discrimination tests are generally equally sensitive.
2. The discrimination test is forced choice; subjects must make a choice.
3. The discrimination test usually precedes other testing (e.g. descriptive and preference).
4. Products not perceived as different cannot be assigned different preferences.
5. Products perceived as different can be equally liked for different reasons.
6. Not all products warrant testing.

To minimize problems and to develop a resource that is accepted, the following guidelines are recommended:

1. Use a discrimination test only to measure whether there is a perceived difference.
2. Select a particular method based on the test objective and the product's characteristics.
3. Subjects should be selected on the basis of specific performance criteria, including sensory acuity, and experience within the product category; a performance better than chance for at least the previous three or four tests.
4. Use "sip-and-spit" procedure for foods and beverages; however, there can be situations in which swallowing is required. Subjects must be consistent in what they are doing.
5. Replication is required.
6. Use a balanced serving order.
7. Do not use discrimination testing as part of a consumer acceptance test.
8. Design the test environment to minimize all non-product variables, and do not attempt to simulate any other environment.
9. Avoid or minimize product carriers where possible.
10. Express results from a discrimination test in probabilistic terms.

Careful planning and attention to details are required to develop a capability that management can accept as reliable and valid. The results of discrimination testing have proven effective and have saved companies substantial amounts of time, money, and effort for a relatively modest investment.

Descriptive Analysis

6.1 Introduction

Descriptive analysis is the most sophisticated of the methods available to the sensory professional. A descriptive test provides word descriptions of products, a basis for comparing product similarities and differences, and a basis for determining those sensory attributes that impact preferences. The results enable one to relate specific ingredient or process variables to specific changes in some (or all) of the sensory attributes of a product (i.e. establish causality). These applications are discussed later in this chapter and in Chapter 8. From a product development

H. Stone, R. Bleibaum, H. A. Thomas: Sensory Evaluation Practices, fourth edition.
DOI: http://dx.doi.org/10.1016/B978-0-12-382086-0.00006-6

viewpoint, descriptive data are essential in focusing efforts on those variables that are perceived to have an effect relative to a target and thus facilitate the development effort. For these and other reasons, considerable attention is focused on the methodology, as evidenced by conversations at conferences and by examining the current sensory literature. Some of this attention is focused on developing methods, and some is focused on other stages in the process, such as language development, measuring intensity, subject performance, and data analysis. This research reflects different philosophies about the descriptive analysis process and provides opportunities for users to assess the state of the field.

Before describing specific methods, it is useful to define descriptive analysis as discussed here as well as to provide a brief historical perspective to the development of the methodology.

> *Descriptive analysis is a sensory methodology that provides quantitative descriptions of products, obtained from the perceptions of a group of qualified subjects. It is a complete sensory description, taking into account all sensations that are perceived—visual, auditory, olfactory, kinesthetic, etc.—when the product is evaluated. The word "product" is used here in the figurative sense; the product may be an idea or concept, an ingredient, a protocept, prototype, or a finished product as purchased and used by the consumer. The evaluation can be total, for example, as in the evaluation of a shaving cream before, during, and after use or the evaluation of a food from package opening through consumption. Alternatively, the evaluation can focus on only one aspect, such as during use. The evaluation is defined in part by the product characteristics as determined by the subjects and in part by the nature of the problem.*

Considering the information obtained from a descriptive test, the high level of interest in the use of results is understandable. The value of descriptive analysis was appreciated by its earliest practitioners, brewmasters, perfumers, flavorists, chefs, and others. Wine and food writers and reviewers also practice a form of descriptive analysis; however, these are purely qualitative expositions, and one should treat that information accordingly. The food industry's use of chefs for developing new and innovative flavors is an indirect example, particularly when their descriptions become a part of the product message. Brewmasters, flavorists, and other specialists described products and made recommendations about the purchase of specific raw materials; they evaluated the effect of process variables on product quality (as they determined product quality), especially the determination that a particular product met their criteria (and that of their company) for manufacture and sale to the consumer. These activities served as the basis for the foundation of sensory evaluation as a science, although at the time it was not considered within that context. It was possible for the expert to be reasonably successful as long as the marketplace was local or regional and less competitive. Limited choices were available to the consumer, as were ingredient and technology options.

All this changed with the scientific and technological developments in most consumer products industries and especially in the food and beverage industry. It was also one of the earliest of the packaged goods industries to recognize the importance of measuring consumer attitudes. The impact of technology, the use of more complex raw materials, and related developments made it increasingly difficult for experts to be as effective as in the past and certainly not effective in a national or global sense. The expert's functions were further challenged by the increasingly competitive and international nature of the marketplace, the rapid introduction and proliferation of new products, and the sophistication of the consumer's palate. In this same time period, sensory evaluation began its development, providing specialized test methods, scales for measurement, and statistical procedures for the analysis of data. Formal descriptive analysis and the separation of the individual expert from the science of sensory evaluation and use of sensory panels received its major impetus from development of the Flavor Profile method (Cairncross and Sjöström, 1950; Caul, 1957). These investigators demonstrated that it was possible to select and train a group of individuals to describe a product without dependence on the individual expert. The method was distinctive for two reasons: (1) No direct judgment was made concerning consumer acceptance or quality of the product, although acceptance was inferred based on the reporting process, and (2) no data analyses were needed because the collective judgments were summarized and reported as a consensus judgment, not unlike the expert. The method attracted considerable interest as well as controversy (discussed later in this chapter); however, there is no question as to its "historical" importance to the field. Since then, other methods have been described and their applications have grown substantially (Meilgaard *et al.*, 2006; Stone *et al.*, 1974; Williams and Langron, 1984).

Before describing specific test methods, it is useful to review the fundamental issues on which all descriptive methods are based. In particular, we refer to the subject selection process; the number of subjects; the duration and the type of training, including the development of the descriptive language (for the array of products being evaluated); the intensity judgments; the data analysis; subject performance; and associated issues. Many practitioners today use a method but appear to not understand the basis for what is done and, as a result, provide information that is not actionable or is incorrect. This discussion is also particularly important because there are numerous decisions made by the panel leader in the course of establishing a panel and evaluation of a set of products. These decisions and actions derive from that person's knowledge and understanding of the perceptual process in general and the descriptive process specifically. Unlike discrimination and acceptance tests, in which subjects exhibit choice behavior in a global sense—that is, all perceptions are taken into account to yield a single judgment—the descriptive test requires each subject to provide a judgment for all of a product's attributes on a repeated basis. Finally, it is mentioned here and elsewhere in this chapter that one always obtains responses regardless of what method is used. Obtaining responses does not, by itself, constitute a basis for assuming results are valid.

A descriptive test is a small panel procedure; it relies on the responses of as few as 6 to as many as 20, with 10–12 being typical. The lower number of 6 was the size of a Flavor Profile panel. As with all sensory tests, there is no specific number required; however, one can compute a number based on assumptions about the sensitivity of the subjects, the magnitude of the difference among the products, etc. In the end, however, most sensory professionals establish their own guidelines based on available evidence. Regardless of the criteria selected, the most important priority is having qualified subjects. After all, when data collection is complete and results are in hand, one must be confident that the specific differences obtained are reliable and valid and not the result of one or more atypical responses or representative of a unique perception. As noted previously, a Flavor Profile panel relied on six subjects, all of whom went through an extensive screening and training program that took approximately 3 months. Considering the limited number of subjects, the impact of one individual would be significant, each accounting for approximately 16% of the information that was provided. If any subject was not focused on the task, it would negatively impact the decision-making process; alternatively, a dominant personality will influence the results regardless of the responses of the other five panelists. Because, as originally developed, the method was qualitative, reaching agreement would be difficult. To address this potential problem, decisions were based on a consensus; that is, there were no data for analysis. This approach assumed that the screening and training would create a panel that had minimal differences, minimal variability, and therefore no need for analysis. It would seem that the long hours spent screening and training were designed to modify behavior to create subjects who function like experts. It is interesting to note that many of the procedures used to train subjects continue to be used today despite a lack of evidence that training actually yields better skilled subjects. In general, most sensory professionals agree that subjects should be qualified, but there is less agreement regarding what that actually means. For some sensory professionals, this continues to mean measuring absolute thresholds to various taste and odor stimuli; for others, it also includes measuring absolute thresholds as well as providing names for selected taste and odor stimuli. Developers of the QDA method chose, instead, to rely on demonstrated sensory discrimination ability using products within the category that will be tested. As emphasized in Chapter 2, the ability to discriminate differences at better than chance is essential to minimize the risk of decision errors and is also fundamental to use of any higher order measurement techniques. This empirical approach has enabled us to categorize volunteers relative to their ability to participate in the descriptive analysis testing. Approximately 30% of those who volunteer will not meet qualifying criteria. As a result, it is not surprising that one continues to encounter questions about descriptive information that does not yield expected differences or is inconsistent with other product information. It is discouraging to encounter continued use of threshold tests as a basis for qualifying subjects, as recommended by Meilgaard *et al.* (2006) and Powers (1988), especially in view of earlier evidence that it provided no indication of a subject's subsequent performance (Mackey and Jones, 1954). Although some sensory professionals may have the time to obtain threshold values, there is little to be gained from the exercise. Anecdotally, we have observed that subjects qualified based on sensitivity to basic tastes and/or odorants (proposed as standards) usually repeat these words when they develop a language; that is, there appears to be an assumption on their part that these stimuli are expected to be present in the products. We have also observed subjects who were qualified to participate on basic tastes unable to detect the same basic taste differences when evaluated in a complex food and beverage system. In addition, there is a clear lack of appreciation for the messages that are "communicated" to subjects without the experimenter appreciating this and its effects on their responses (for more on this topic, see Orne, 1981). Still another approach (called "free-choice profiling") proposed no subject screening; anyone could participate (Williams and Arnold, 1985; Williams and Langron, 1984), and a type of factor analysis was used to identify meaningful results. The use of multivariate analyses in these kinds of situations has increased in recent years, and most of it is inappropriate (May, 2004). Subsequent to its introduction, free-choice profiling was revised to include training that appears to require more hours than required by QDA, for example, despite initial claims that it was a faster method. Although mathematical approaches to describing behavior are a topic of considerable interest and there have been some successes, predicting how a subject will "perform" in a future test remains a challenging issue. Clearly, there are differences in the approaches used for qualifying subjects, and the panel leader must decide what procedure(s) will be used.

When developing the QDA methodology (Stone *et al.*, 1974), it was determined that the discrimination methodology was the most effective procedure for identifying subjects who could and could not perceive differences (after first determining that they are regular users of the specific product category being tested). It takes relatively little time to identify individuals who satisfy these criteria—as few as three sessions over 3 days. Using approximately 30–40 discrimination trials fielded over 2 or 3 days (with each session lasting approximately 90 minutes), one can select a pool of discriminators from among a group of totally naive individuals. For more information on this procedure, the reader should review Chapters 2 and 5. This fundamental screening procedure has several purposes: It eliminates the insensitive nondiscriminator; identifies the sensitive and reliable discriminator; familiarizes all subjects with the sensory properties of the products; teaches the subjects how to function in a sensory test (to read a scorecard and to follow instructions); and helps to identify and excuse the disruptive volunteer. In effect, one wants to start with a panel that is representative of the discriminating segment of the population because these are the consumers who are most important to the company; they are most likely to detect and respond to product differences. Such a group will also likely be more homogeneous in terms of their error variances. As with all screening methods, only after data collection can one determine empirically that the screening process was effective. When individuals who did not meet the screening qualifications were included, their responses were more variable, thus providing the basis for excluding them from further panel activities. As noted previously, the screening process is a dynamic system, adjusted based on test results from recently completed testing. The eventual result is a subset of 10–15 product pairs that can best predict the potential of that individual volunteer. For each product category for which there are available descriptive capabilities, sensory staff will have a set of guidelines with details about the screening and qualifying process.

Screening must be product category specific, as is the subsequent training effort. For example, if one were testing coffee, then screening is done with various coffee samples and not pizza or ice cream. In this way, one minimizes the risk of including subjects sensitive to pizza but not coffee. This does not mean that one is confronted with the task of having to repeat the qualifying process in its entirety every time a different set of products is being tested. As a general rule, most subjects are capable of meeting the qualifying criteria for many different categories of products, particularly after they have participated in several tests. These are issues that a sensory professional must consider whenever a test is planned and must reconsider after a test, based on examining subject performance records.

A frequently asked question is whether one can qualify subjects for different kinds of products in this screening procedure. This question seems to be based on the idea that this group could be used for descriptive analysis of different kinds of products and one can avoid having to repeat screening and save time. There are several reasons why this may not be a useful approach. First is the matter of performance; that is, a subject may demonstrate qualifying for one product and not the other. Second, it could require that the number of trials be increased to account for the second category, thus extending the screening time. Finally, once subjects are selected for a test, it will be 2 or 3 weeks before they can participate in a test of this second product category. As discussed later in this chapter, the QDA method incorporates subject performance measures, enabling the sensory staff to assess the empirical evidence regarding a subject's skill (sensitivity and reliability) for evaluating the other product category. It assumes that the subjects have the time and are not otherwise committed to testing products from the first category. The idea seems to be based on not wanting to spend more time screening the subjects; however, the actual screening might not be more than a single session. As the volume of testing increases, reliance on a limited number of subjects is problematic, regardless of whether they are employees or local residents. It is better to develop a larger pool of qualified subjects and thus reduce reliance on a small subset of people for all testing needs.

In summary, it is necessary that subjects for a descriptive test demonstrate their ability to perceive differences at better than chance among the products that they will be testing. This will take as many as 30–40 trials to demonstrate; approximately 30% of those who volunteer will fail to meet the chance probability requirement. Once subjects have participated in testing, the need for screening is removed or limited, and data analyses provide direct measures for each subject's performance on each test.

The training process, which follows screening, also has a number of alternative approaches. The purposes for training are focused on developing or familiarizing subjects with a descriptive language that is used as a basis for evaluating the products. The details of the training include grouping attributes by modality (i.e. appearance attributes, aroma attributes, etc.), listing them by occurrence within a modality, developing working definition for the attributes, identifying references for use during training, developing an evaluation protocol, and having the subjects practice using the scoring procedure. Depending on the test method, panel training can be quite different. If subjects are new to descriptive analysis (i.e. inexperienced and no existing language is available), then this process, as developed for Tragon QDA, should take approximately 7–10 hours. Typically, a panel will meet daily for 5 consecutive days for approximately 90 minutes each session. Inexperienced subjects presented with the task of learning to use/modify an existing language might need as much as 7 hours, whereas experienced subjects might need approximately 5 hours. These times are presented as a guide inasmuch as the products and the skill of the panel leader will have an impact on the training process. In all these situations, the subjects work individually and as a group to ensure that the attributes are fully understood and that all of the product's attributes have been fully accounted for. In QDA, all of this is done under the direction of a panel leader who does not participate in developing the language. The QDA methodology was the first and still is one of the few methods that exclude the panel leader from directly participating in the language development or data collection, contrary to the statement by Powers (1988). This is true whether one is working with an experienced group or one that has never participated before. The panel leader's primary responsibility is to facilitate communications among subjects. The panel leader who participates as a subject is a biased subject because this person is most likely aware of the product differences and the test objective. Not only does this action communicate the wrong message to the subjects (subjects will tend to defer to the panel leader, whom they assume has the correct answer) but also the end result is more likely to produce a group judgment rather than a group of judgments (for more on this issue, see Jones, 1958).

Developing a sensory language or using one that already exists is an interesting process and certainly one that is essential for the successful outcome of a test. For some, the language assumes almost mystical importance, such that a considerable body of literature has been developed in which lists of words, predesignated intensity values, and associated brand names of products are published (identified as lexicons) for specific types of products (see, for example, Johnson and Civille, 1986; Meilgaard *et al.*, 2006; Muñoz, 1986). Some measure of credibility was achieved as a result of publication of the list, with little scientifically relevant information on how the individual intensity values were established. Such lists are not very different from the efforts of product specialists or experts of 50 or more years ago when they developed quality scales and corresponding word descriptions for products as part of an effort to establish food quality standards. In addition to their interest in the evaluation of their respective company's products, their technical and trade associations often formed committees for the express purpose of developing a common language for describing the flavor (or odor) of the particular category of products. For purposes of this discussion, we chose the publication by Clapperton *et al.* (1975) on beer flavor terminology as an example of the efforts (of the association of brewers) to develop an internationally agreed terminology. Another example of this approach to standardizing the sensory language is described by Noble *et al.* (1987) for wine. The former authors stated the purpose as "to allow flavor impressions to be described objectively and in precise terms." Various literature and professional sources were screened for words (or descriptors) describing the flavor of beer, and after reviews, the committee arrived at an agreed-on list. As the authors noted, the issue of chemical names compared with general descriptive terms was resolved with the inclusion of both, which seems to be an odd decision.

From a technical standpoint, the use of chemical names was appealing because it was believed they could be related to specific chemicals in the beer. An example is the term diacetyl, which could be ascribed to a specific chemical, compared to buttery, a less precise term that might be ascribed to several chemicals. By including both, it was stated that the terminology would be of value to the flavor specialist (the expert) and the layman (the consumer). Although the concept of an agreed, in advance, terminology should have appeal, careful consideration of this approach reveals significant limitations, especially in relation to its application in sensory evaluation. It is risky to decide *a priori* what words subjects should use to describe a particular sensation or group of sensations. After all, the language is context driven based on the set of products being evaluated and the word choices of those subjects. The fact that product changes—for example, a formulation or a process change—rarely change just one attribute emphasizes the complex nature of the process. To restrict subjects to specific words assumes that language is unchanging in terms of these words and the meanings assigned to them. Although the chemical terminology should have specific meaning to experts, it is unlikely to have the same meaning to the consumer or to trained subjects (see Chapter 8, Section 8.3). It is interesting to note that these relationships among experts are themselves subject to disagreement. From a behavioral standpoint, this approach must therefore be viewed as not directly applicable to descriptive analysis. Regardless of the source, a language that does not provide for subject review and, if necessary, change is not likely to yield uncomplicated sensory responses. Subjects are influenced by the information given to them and are much less likely to question it because of its source. Although it can be argued that such approaches are merely intended to help panel leaders or the subjects, the temptation is very strong to use this approach rather than allowing subjects to use their own terminology. There is a difference between subjects developing their own language and telling subjects what to perceive.

To the student of the history of psychology, descriptive analysis can be considered as a type of introspection, a methodology used by the school of psychology known as structuralism in its study of the human experience (for a lengthy discussion on this topic, see Boring, 1950; Harper, 1972; Marx and Hillix, 1963). Structuralism required the use of highly trained observers in a controlled situation verbalizing their conscious experience (the method of introspection). Of particular interest to us is the use of the method of introspection as an integral part of the descriptive analysis process and specifically the language development component.

However, the obvious difference is that products are included (in descriptive analysis) so that the subject's responses are perceptual and not conceptual. In addition, the subjects are encouraged to use any words they want, provided they use words that are a common everyday language, and that they define the meaning of each word-sensation experience, if for no other reason than to ensure that they will be able to score the products using a terminology with which they are familiar. Whereas each subject begins with his or her own set of words, all subjects work as a group, under the direction of a panel leader, to come to agreement as to the meaning of those words—that is, the definitions or explanations for each word-sensation experience and also when they (the sensations) occur. All of these activities require time; in the QDA methodology, there can be as many as four or five sessions, each lasting approximately 90 minutes, and they are essential if this sensory language is to be understood and the subjects are to be capable of using it (and the scale) to differentiate the products. These training sessions help to identify attributes that could be misunderstood (and will be by some subjects) and also enable the subjects to practice scoring products and discussing results, on an attribute-by-attribute basis, under direction of the panel leader. Of course, all this effort cannot make subjects equally sensitive to all attributes. In fact, when evaluating products, subjects rarely, if ever, achieve equal sensitivity or complete agreement for all attributes, nor is this ever expected (if this did occur one could rely on the N of 1!). These are observations that have proven throughout the years to be correct; to wit, all subjects are different from each other, hence the need to have more attributes than one might want and the need for replication to account for individual differences in sensitivity. The effectiveness of this effort can be determined only after a test has been completed; that is, each subject has scored the products on a repeated trial basis, the appropriate analyses have been completed, and decisions can be reached that are consistent with what is known about the product variables and with the results of the analyses. The idea that there should be a specific number of attributes is questionable, as is the issue of whether one has the correct or right attributes. How does one know that all the attributes have been developed or that they are the right ones? In fact, such questions as these are beside the point because they are implying that there are correct answers (determined by the panel leader or possibly some higher authority). The answer to the former is empirical (and in part answered in much the same way as one determines the number of angels that can be fit on the head of a pin). The answer to the latter is also empirical; that is, given a set of product variables, do some sensory attributes exhibit systematic changes as a function of those variables? One of the relatively newer methods, free-choice profiling, requires no subject screening or training and allows subjects to use any words they want relying on statistics (e.g. Procrustes analyses) to decide which attributes are related (statistically) to each other (Williams and Langron, 1984). However, in some publications in which the method was used, approximately 5+ hours of training was reported as necessary to obtain useful results (Marshall and Kirby, 1988). This method appears to be suggesting that questions as to the number of attributes or their correctness are not at all relevant; that, in fact, it makes no difference what words subjects use; and that the appropriate statistical analysis will yield the answer(s). This method is discussed further in the next section.

Words used to represent sensations are nothing more than labels developed by the subjects after frequent samplings and discussions; they provide a common basis for their scoring an array of products. There is no reason to assume that these words represent anything beyond that—for example, that each represents a unique sensation (for comments on this issue, see Stone, 1999). It has been suggested by

O'Mahony and co-workers (Ishii and O'Mahony, 1990, 1991; O'Mahony *et al.*, 1990) and by Civille and Lawless (1986) that the words represent concepts, and that for a descriptive panel to be effective and the results useful, concepts must be aligned; that is, subjects must agree on all the sensations (or attributes that represent those sensations) to be included in a concept. Ishii and O'Mahony (1991) stated that "to understand fully the mechanisms involved in formation of the ad hoc descriptive language it becomes necessary to consider the mechanisms of concept formulation." However, one must be careful to not confuse the issues associated with the search for these underlying mechanisms as having anything other than a very indirect effect, at best, on the descriptive process. If the language is not consistent with the proposed mechanism of concept formulation, then the mechanism needs to be changed. The descriptive analysis of thousands of products for a variety of applications serves as ample evidence of its value to companies worldwide.

Sensory evaluation, as treated in this book, is an applied science that deals with measuring responses to selected products or product attributes. Those measures are used, directly or indirectly, to describe or predict consumer behavior. There is little impetus or value to deal with cognitive mechanisms that may or may not explain that behavior. Using unobservable (and usually unmeasurable) variables, "underlying structures" and "hypothetical constructs," does little to clarify sensory issues and assist in the decision-making process. The search for those mechanisms that explain a behavior after it has happened is an academic activity, in which practical value and usefulness need not be measures for success. No evidence has been presented that would support the ideas of these authors about concept formation and alignment improving the reliability or the validity of a descriptive test. If anything, the converse would be true; that is, any theory about concept formation and alignment must fit the empirical evidence. One could hypothesize that the process by which subjects discuss their judgments for each attribute and the definitions for those attributes could be considered as concept formation and alignment. This is an integral part of the QDA training process (Stone *et al.*, 1974); however, attaching a label to this activity seems quite unnecessary other than for academic purposes. It also introduces circular thinking in that the construct is used first to describe an observed behavior and then to explain that behavior, much the same way as Freudians use the construct of id, ego, and superego to first describe and then explain behavior. It is surprising that so much attention and effort is given to the terminology, as if there is a true, correct language that, when identified, will somehow make results magically correct and respected by everyone. As stated at the beginning of this discussion, the terminology is simply a set of labels that panel members have agreed enables them to fully communicate their description of the sensory properties of the products being evaluated. The label likely represents a complex perception integrating many qualitative and quantitative aspects of the product. For example, increasing the amount of sucrose in some products can influence perceptions for attributes in addition to perceived sweetness. To assume some innate correctness or relationship between a label and an ingredient or process oversimplifies the issue and ultimately misleads the researcher.

It should be remembered that subjects will not agree completely on all the sensations that are perceived—that the sensations are themselves interactive, leading to multiple words to represent them. In addition, the individuality of each subject (in terms of sensitivity, genetics, motivation, focus, and personality) further adds to the complexity of the process. During the first few training sessions, it is not unusual for a panel to provide as many as 80–100 words. During the subsequent training sessions, this list will be reduced by as much as 50% as the subjects evaluate and discuss their responses and realize that many of the words have a common sensory basis. However, the number of attributes will always be considerably more than is necessary from a strictly mathematical standpoint; for example, scorecards for many foods will have 30 or more attributes. This should not be surprising if one considers that there will be attributes for each modality—that is, for appearance, aroma, taste, texture, and aftertaste. The fact that there are more attributes than are needed (based on mathematical computations) should not be unexpected or of concern because not all subjects will be sensitive to all attributes. Rather, one should be more concerned about having too few attributes, such that subjects are not able to differentiate product differences. It is better to have too many attributes and rely on the analysis to establish which ones were used to differentiate the products.

In addition to a descriptive language and definitions, it is useful but not essential to have references for use in training or retraining subjects (Muñoz, 1986; Rainey, 1986; Stone and Sidel, 1998; Stone et al., 1974). Here, too, one finds different opinions as to the types of references and how they are to be used. For example, Muñoz presented a comprehensive list of references and how they are to be used (in texture analysis), including their respective intensity scores for scale extremes. However, these references are based on commercially available products, all of which are variable in the normal course of production, additional to the intended changes based on changing technologies, ingredients, and/or market considerations. The author then advised the reader that these references and their intensities may change for the aforementioned reasons, in which case the original references and their intensities will have to be re-established. So what then is the value of such references? References have a potential value by providing a sensory context, helping subjects relate to a particular sensation (or group of sensations) that is not easily detected or not easily described. In this way, references are referred to as qualitative references and are designed to provide a common sensory experience. However, references should not, by themselves, be a major source of variability, introduce additional sensory interaction/fatigue, or significantly increase training time. In most training (or retraining) situations, the most helpful references are usually a product's raw materials. Of course, there will be situations in which totally unrelated material will prove helpful to some subjects and it is the panel leader's responsibility to obtain such materials. There will be situations in which no reference can be found within a reasonable time period; however, asking the subjects for assistance often proves to be effective. Training should not be delayed just because a reference cannot be found. Although most professionals agree that references are helpful, there is no evidence that without them a panel cannot function or those results are unreliable and/or invalid. We have observed that so-called expert languages usually include numerous references, have prespecified intensity values, and subjects take considerably longer to learn such a language and the associated scale value (if they ever do) than they do a language developed by themselves and using everyday language (this is evident in the 14+ weeks of training required in the Spectrum method as in Flavor Profile). This should not be surprising. After all, references are themselves a source of variability; the prespecified intensity value may not provide a sensation relative to the distance on the scale, they will introduce other attributes unrelated to their purpose, and they will increase sensory interactions. The more technical or broad ranging they are, the more difficult it is to relate to an actual product. Also, pretest training with simple references leads subjects to expect both qualitative and quantitative perceptions in the products they will evaluate. Where those simple references do not translate well in the complex product, and the subjects embark on a search for that perception, the result is often a "phantom" (i.e. it is not there) product attribute. Therefore, the panel leader must consider the use of references with appreciation for their value as well as for their limitations, and he or she must decide when and what references will be used. In our experience, they are of limited value; they may be used in the language development/training but not in the initial language sessions because they could mislead the subjects into thinking that there is a correct language and the panel leader will provide it. In retraining or when adding new subjects to a panel, references can be helpful in enabling these individuals to experience what the other subjects are talking about and possibly to add their comments to the sensory language.

Some comments also are warranted about data analysis, although information was provided in previous chapters and more specific information is provided later in this chapter. A descriptive test yields a large sensory database, as much as 10,000 or more data points (in comparison with discrimination or acceptance tests), enabling a wide range of statistical analyses. One of the main features of the QDA methodology (Stone *et al.*, 1974) was the use of statistical analysis of the data, which represented a significant development for sensory evaluation at that time. With the availability of statistical packages and of PCs, panel leaders have unlimited and low-cost resources, providing an online capability for obtaining means, variance measures, ranks, and pairwise correlations and for analysis of variance, including multivariate analyses such as factor analysis, multiple regression, cluster analysis, and discriminant analysis.

As with any readily available resource, statistics are often misunderstood and/ or are misused, particularly when responses are highly variable or when the panel leader uses multivariate procedures without determination of the quality of the basic data. In other instances, factor analysis and/or clustering techniques are used during training as a basis for excluding subjects who are not in agreement with the panel or to eliminate attributes that the panel leader believes are not being used to differentiate products. Lyon (1987) and Powers (1988) described such an approach during training as a means of reducing the number of attributes on the scorecard. One must be careful about using procedures that, *a priori*, decide which subjects or attributes will be used. After all, the subjects are still in training and to have them score products and use the results as a basis for eliminating words and/or subjects makes little scientific or practical sense. One of several objectives of training is to enable subjects to learn how to evaluate products using the words that they determined were helpful to them (at that stage of the training). To proceed from one session to another and substantially reduce the list of attributes is communicating a message that there is a correct list of words and with a little patience it will be provided. This approach could lead to a group of subjects who are more likely to agree with each other, but how does one know that those who agree are correct? Using factor analysis to directly reduce the number of attributes is troublesome, particularly during training, because it assumes a cause-and-effect relationship for those attributes correlated with a specific factor, and such assumptions are risky. After a test, one can use various statistical procedures not only to help explain results but also to identify specific subjects and attribute problems that can be addressed before the next test; however, there is much greater risk if one does this before training is completed.

In summary, the descriptive test is a dynamic system in which the panel leader must make numerous decisions when organizing a panel, whether one is recruiting and screening a new panel, reviewing performance records for the next panel, deciding on the appropriate design, deciding when and where products will be evaluated, or reporting results. Without sufficient knowledge about human behavior and the objectives for a test, inadequate or incorrect product decisions could be reached to the detriment of the science and the product category.

6.2 **Test methods**

In this section, we focus on those descriptive analysis methods described in the literature so as to enable the reader to assess their usefulness relative to what method they will use. The methods are qualitative or quantitative, although one could convert the latter to the former. As shown in Table 6.1, five methods are assigned specific names, one qualitative and four quantitative, reflecting a relatively wide range of approaches to descriptive analysis. A sixth method, designated as diagnostic descriptive analysis, is included because it represents a broad category of methods, all of which are quantitative. A number of different procedures fit within this designation. Experts are listed here as qualitative for illustrative purposes, although one might want to assign numerical values to the responses as in the case of wine evaluation as described by Amerine *et al.* (1959).

6.2.1 **Flavor profile**

The Flavor Profile method (Cairncross and Sjöström, 1950; Caul, 1957; Sjöström and Cairncross, 1954) is the only formal qualitative descriptive procedure and is probably the most well-known of sensory test methods. As previously noted, it was the first formal descriptive method and was an attempt to be less reliant on the expert.

Table 6.1 Classification of Descriptive Analysis Methods	
Qualitative	**Quantitative**
Flavor Profile[a]	Texture Profile[b]
Product experts (perfumer, flavorist, brewmaster, etc.)	QDA[c]
	Spectrum analysis[d]
	Free-choice profiling[e]
	Diagnostic descriptive analysis[f]

[a]*Cairncross and Sjöström (1950) and Caul (1957).*
[b]*Brandt et al. (1963) and Szczesniak et al. (1963).*
[c]*Stone et al. (1974, 1980).*
[d]*Meilgaard et al. (2006).*
[e]*Williams and Langron (1984).*
[f]*Cross et al. (1978), Larson-Powers and Pangborn (1978), and Lyon (1987).*

The Flavor Profile method utilizes a panel of six screened and qualified subjects who first examine and then discuss the product in an open session. Once agreement is reached on the description of the product, the panel leader summarizes the results in report form. One of its most important benefits to the product technologist is the speed with which results can be reported. Subjects are selected for training based on interviews and an extensive series of screening tests, including sensory acuity, interest in product evaluation, attitude, and availability. Some of these requirements are common to most descriptive methods, and in general one can find little disagreement with this approach to screening and subject selection. In practice, however, the sensory acuity tests are concerned only with basic taste and odor sensitivity, skills that appear to have minimal connection with product evaluation (Mackey and Jones, 1954). Nonetheless, they provide some form of differentiation of individuals based on acuity. For Flavor Profile, this database is combined with the attitudinal information, and six subjects are selected for further training, which includes instructional information about the senses as well as direct experience evaluating selected products.

For this method, the key individual is the panel leader, who coordinates the testing, participates in the discussions, samples products, and reports results. This individual assumes a leadership role, directing the conversation and providing a consensus conclusion based on the results. This role as panel leader can have significant consequences without some independent controls. Subjects could be led to a conclusion without being aware that this had occurred. In addition, the six subjects take turns serving as the panel leader, which could have a further impact on results. Nonetheless, as a sensory test, the method had considerable appeal because results could be obtained rapidly once the qualifying process was completed (~14 weeks). The subjects meet, as a group, for approximately 1 hour to evaluate a product, reach a consensus about its sensory properties, and provide the requestor with a result. The developers of the method emphasized that there would be confidence based on the collective professional judgment of the panel and this would obviate the need for statistics.

One modification to Flavor Profile that elicited interest was the dilution procedure recommended by Tilgner (1962). The purpose for the method was to be able to evaluate products in which flavor interactions were minimized by means of diluted samples, and comparisons with full-strength samples were not possible. The method aroused interest when first introduced, but this interest was not sustained. The major weakness appeared to be the failure of the method to demonstrate the value of the dilution results compared with those from finished products. In a later publication, Tilgner (1965) described results from a series of products using the method but not addressing the more practical issue of application. Despite this omission, the method represented an attempt to broaden the scope of Flavor Profile. Tilgner also instituted a number of other modifications, including the use of a 5-point intensity scale and statistical procedures for analysis of results. These changes were in a direction that was consistent with the more typical sensory test; however, the use of the dilution method appeared to be primarily applicable to research.

6.2.2 **Texture profile**

Chronologically, the next descriptive method of importance was the Texture Profile method developed at the General Foods Research Center (Brandt *et al.*, 1963; Szczesniak, 1963; Szczesniak *et al.*, 1963). This method represented advancement in descriptive analysis from a methodological perspective; however, there were concerns from a behavioral perspective. These concerns centered around such issues as the language, the scales, and the references/anchors for each scale that were linked to specific physical measures. The actual evaluation used the Flavor Profile method with the subjects discussing results in order to reach a conclusion.

Brandt and co-workers (1963) defined a texture profile as "the sensory analysis of the texture complex of a food in terms of its mechanical, geometrical, fat and moisture characteristics, the degree of each present and the order in which they appear from first bite through complete mastication." This definition could be applied to any descriptive analysis; in this situation, the focus is on texture, which implies independence of responses from other sensations. We discuss this issue later.

In developing the method, the objective was to eliminate problems of subject variability, allow direct comparison of results with known materials, and provide a relationship with instrument measures (Szczesniak *et al.*, 1963). The authors claimed these objectives were accomplished through the use of standard rating scales for each texture term and specific reference materials to represent each scale category for each of the terms. At the outset, texture terms and definitions were compiled, sorted, and categorized. This information included both scientific/technical and popular terms, the latter from laboratory or field situations in which common usage terms were used to describe textural sensations. This led to the development of a classification of terms that were believed to encompass texture sensations. As shown in Table 6.2, the information was organized into three main categories and subsets within each of these.

Table 6.2 Relationship between Textual Parameters and Popular Nomenclature		
Mechanical Characteristics		
Primary Parameters	**Secondary Parameters**	**Popular Terms**
Hardness		Soft, firm, hard
Cohesiveness	Brittleness	Crumbly, crunchy, brittle
	Chewiness	Tender, chewy, tough
	Gumminess	Short, mealy, pasty, gummy
Viscosity		Thin, viscous
Elasticity		Plastic, elastic
Adhesiveness		Sticky, tacky, gooey
Geometrical Characteristics		
Class	**Examples**	
Particle size and shape	Gritty, grainy, coarse, etc.	
Particle shape and orientation	Fibrous, cellular, crystalline, etc.	
Other Characteristics		
Primary Parameters	**Secondary Parameters**	**Popular Terms**
Moisture content		Dry, moist, wet, watery
Fat content	Oiliness	Oily
	Greasiness	Greasy

*Source: Reprinted from J. Food Sci., **28**(4), (1963), 388. Copyright © by Institute of Food Technologists.*

All of these terms were defined; for example, hardness was defined as the force necessary to attain a given deformation, and viscosity was defined as the rate of flow per unit force. The technical nature of these definitions should be meaningful to the chemist, but one might question their perceptual meaning to a subject in a sensory test. For each parameter, a scale of five, seven, eight, or nine categories (depending on the specific parameter) was developed, and for each category (of a scale) there was a specific product that represented that sensation. For the hardness scale, a panel rating of "1" was derived from a specific brand of cream cheese, "5" was a giant-size stuffed olive, etc. Except for the chewiness scale, the authors stated that the distances from point to point were equivalent; however, no data were presented in support of this claim.

Although some workers questioned definitions for some of the parameters (see the discussion by Sherman, 1969), the method required subjects to score products according to these aforementioned parameters. It was claimed that comparison of results with the specified references enabled product formulation to occur according to known physical and chemical parameters (Szczesniak *et al.*, 1963). Having a reference (high and low) for every scale may seem like an ideal means of focusing responses (and minimizing variability); however, it will not elim- inate variability, as noted in the introductory discussion of this chapter. Recall that the references will have their own variability. Having to sample three "products" and provide a judgment for each attribute would be expected to increase variabil- ity just from sensory fatigue. To date, no evidence has been presented to demon- strate the value of this reliance on references during actual data collection. Use of products as scale anchors also presents its own set of problems. These products are not invariant and change over time as a function of marketing and other considera- tions. Normal production variation will cause a subject's response to a product to be offset to some (greater or lesser) extent as a function of that variability. It is also reasonable to expect that a subject's liking (or dislike) for a reference will further impact response behavior. Obviously, the solution to these problems is to avoid use of such references during data collection. In effect, one is sending a message to the subjects that their judgments must be consistent with the references. Subjects will modify their perceptions to "adjust" for the references.

A second concern with this method is the *a priori* decision as to the attributes (chewiness, hardness, etc.) to be used. There are inherent risks in experimenter-assigned attributes; the subjects could ignore a perception or use one or more of the listed words to "represent" the particular perception. Although a survey may have elicited these aforementioned texture attribute (see Table 6.2), these are not necessarily the only ones or the most appropriate to reflect texture perceptions for a particular product. It is not a question of right and wrong attributes but, rather, an issue of risk. Training subjects to use a specific set of attributes and comparing these responses with instrument analyses yielded a high degree of agreement or repeat measure reliability. However, the question of appropriateness and validity has not yet been addressed from a sensory standpoint. Although the words are different and they can be related to specific physical measures, no evidence has been provided to show that they are perceptually independent. In fact, some data show quite the opposite; that is, there is considerable redundancy among the words and among the physical measures that were designated. Using different words for the scales does not mean that they are perceived as different.

A third concern associated with the method is the separation of texture from other sensory properties of a product, such as appearance/color, aroma, and taste. As a rule, perceptions are interdependent, and the exclusion of some attributes from a score- card does not eliminate their perceptions. The subject is likely to use other attributes to acknowledge these perceptions, and the visible manifestation is increased vari- ability and decreased sensitivity. In addition, these other perceptions will influence the responses to the textural perceptions and vice versa. The reader should take note that this particular problem, a perception with no place for it on the scorecard, is not unique to this method. It is a weakness of every sensory test that requires the subject to respond to a limited set of product descriptors for one modality to the exclusion of another—for example, texture but not color, taste, aroma, or aftertaste. By measuring responses to all perceptions, the experimenter can derive a more complete picture of the product's sensory properties. One can gain an appreciation for this by examining the results from a factor analysis output. The descriptors are not grouped by modality (i.e. by aroma, appearance, etc.) but, rather, there will most likely be a mingling of them. In other words, there is considerable overlap of the attributes, independent of how they were grouped on the scorecard. Without allowing the subjects the opportunity to add attributes and/or modify those available, the potential for missing information is dramatically increased.

It is important for the sensory professional to recognize these limitations to the method and to more properly assess the results in relation to other sensory information. This criticism does not mean that the method is invalid or is not useful. The sensory professional must be able to assess the weaknesses and strengths of each method relative to the particular problem under consideration. From a purely technical standpoint, there is considerable appeal to the direct link between specific instrumental measures of the rheological properties of a product and the responses of a panel to specific sensory attributes—for example, texturometer units and hardness sensory ratings (Szczesniak *et al.*, 1963). One would expect these relationships to be stronger if all the sensory properties of the products have been measured.

6.2.3 Quantitative descriptive analysis (the QDA method)

Development of the Flavor Profile and Texture Profile methods stimulated interest and research on new descriptive methods and especially methods that would overcome the weaknesses previously identified—reliance on qualitative information, use of product attributes established by the experimenter, reliance on a limited number of subjects, lack of comprehensive statistical analysis, and so forth. Further interest in descriptive methods developed as a result of the growth of new products and competition in the marketplace for products with unique sensory properties, as well as by advances in measurement and improved data processing systems. The QDA method (Stone and Sidel, 1998, 2003; Stone *et al.*, 1974) represented an opportunity for sensory evaluation to satisfy these needs; however, it was also a substantive departure from existing methods in the sense that the approach incorporated current knowledge about human behavior and measurement. Other features included requiring the subjects to develop and agree on the language, use of a scale to obtain measures of attribute (language) strength, replication for assessing subject and attribute sensitivity and identifying specific product differences, and defined statistical analyses. It was determined that the method would be more than a simple rephrasing of the test questions or the use of a particular scale. In effect, the method required a different approach (at that time) to the concept of descriptive analysis, beginning with the subject selection procedure and concluding with communication of results in an understandable and actionable manner. The development of the method evolved from a number of considerations, including the following:

- Identify and measure all the sensory properties of a product
- Reliance on a limited number of subjects for a test

- Subjects qualified before participation
- Able to evaluate multiple products
- Products evaluated individually
- Use a language development process free from leader influence
- Be quantitative and use a repeated trials design
- Have a useful data analysis system

These features of the methodology are further discussed here to enable readers to determine how the method can be applied to their particular needs.

6.2.3.1 Identify and measure all the sensory properties of a product

The method provides a complete word description for all of a product's sensory properties or attributes, and it is applicable for consumer products beyond food and beverage. By products, we mean that it can be an existing product (currently in the marketplace), an ingredient, an idea, or an entirely new product for which there is no existing competition. It can also include the opening of the package, preparation, and consumption. A complete description of all of a product's sensory properties is essential if product similarities and differences are to be fully documented. One cannot expect to train subjects, for example, to ignore the appearance/color of a set of products. Although the neural pathways for each of the senses are unique and well-defined, considerable interaction occurs at high centers of the brain, and these sensations culminate in the behavioral responses that are not one-dimensional in character. In addition, the previous experiences of the subjects will also have an influence on their responses. Thus, the response of a subject to the sweetness of a beverage will include the interactive effects of all sensory attributes such as color, aroma, and so forth, as well as past experiences. To ensure that product flavor (i.e. taste in the mouth), for example, is fully accounted for, it is necessary to also measure other attributes. If there is no opportunity to express a sensation, then most likely it will be embedded in another response, resulting in increased variability for that subject and less information for the requestor.

For these reasons, there were no limitations with regard to perceptions for a product, and subjects were encouraged to use as wide a range of words (i.e. the language) as possible. For example, subjects evaluating a fabric could include one or more aroma measures if those attributes were perceived. It would also include the feel of the fabric before and during wear. Of course, the test objective will impact the test plan as to how and where a product will be evaluated. The verification procedure for all perceptions is covered in more detail in a subsequent discussion. Implicit in this discussion is the applicability of the QDA method to any type of product—food, beverage, shaving cream, fragrance, fabric, furniture polish, paper towels, writing instruments, sporting goods, and so on.

When training is initiated, the subjects are instructed to use any words to describe a product; they are encouraged to use words that they understand and can explain to their fellow panel members, and no restriction is placed on the number of words except that preference or preference-related judgments, such as good, bad, etc., are discouraged. They are instructed to examine the product and write on a blank paper (or type on a screen or tablet) the words that represent their perceptions. When finished, the panel leader asks subjects to state what they wrote and then lists those words on the board for all to see. This sequence of activities is repeated a second time with the panel leader using check marks for those words already listed. This listing of words serves as a useful training tool, enabling subjects to see the degree of commonality in their perceptions as well as to see and hear the variety of words that are used. With the third sample, the panel leader may suggest the subjects think about what they notice first (e.g. appearance or aroma) and how they do this. Do they look at the sample while it is flat on the surface or do they tilt it, etc.? This process continues for the remainder of the 90-minute session. The panel leader collects the sheets of paper or downloads the material from the tablets to verify all written attributes have been identified. Throughout the years, it has been interesting to note that subjects throughout the world exhibit some common behaviors. That is, no one subject contributes all the language; some contribute more than others, and this does not necessarily translate into greater sensitivity. It takes the combined effort of a panel to provide a comprehensive attribute list to describe the products that will be tested. This is one reason for recommending a panel of approximately 12 participate in language development when a new scorecard is being developed. In a typical situation, as many as 80–100 words will be listed during the first language development session. By the second session, subjects are encouraged to group words by modality, and in the process they discuss use of the words and identify any possible duplication (i.e. using different words to represent the same sensory experience). It is a responsibility of the panel leader to encourage this kind of discussion while at the same time asking for any new language. After the third session, it is not usual to identify new language; there are some products that have a long ingredient list and subtle perceptions that take longer to be expressed in an agreed language. Throughout training, the subjects are encouraged to be sure that they have fully accounted for all the products' sensory properties. To ensure that the subjects respond to all the sensory properties, they must evaluate as many of the test products as possible during the language development sessions. Where possible, all products should be sampled during these sessions, but when the number of products is large (e.g. 15), it may not be possible and the panel leader will have to select the appropriate pairings for the sessions.

6.2.3.2 Able to evaluate multiple products

Evaluation of more than a single product in a test is intended to capitalize on the subjects' skill in making relative judgments with a high degree of precision. As is known, humans are very poor judges of absolutes but very good judges of relative differences. Thus, a test involving several products will capitalize on the subjects' skill in making precise judgments about relative differences. The multiproduct test is also very efficient, first by providing a much more complete context of how products differ one from the other, and second by providing information on more than a single product. In view of the effort required to qualify subjects and teach them how to verbalize their perceptions using an everyday language, it is good business practice to achieve optimal output from the investment. Of course, more products in a test provide more information to analyze and to apply additional types of analyses to further understand product relationships such as in an experimental design study. This emphasis on the benefits of multiple product testing should not be construed to mean that a two-product test could not be done. As mentioned previously, it depends on the program and test objective(s).

6.2.3.3 Subjects qualified before participation

Each subject must be qualified prior to participation following a protocol described in Chapter 2. At the outset of the QDA methodology, there are certain requirements for all subjects. They must be volunteers, available for protracted time periods for the language development sessions and then for data collection, and must be users or potential users of the product or of the product category that will be evaluated. This latter qualifying criterion is necessary because the individuals will have repeated exposure to the products, and those individuals who do not like or would never use the product tend to be less sensitive and more variable to product differences. Whether this is due to the individual's dislike for the product (and therefore lack of attention to differences) or represents a true difference in sensitivity is not known; however, the crucial point is to eliminate these individuals during screening. For QDA testing, subjects must also demonstrate their ability to perceive differences within that class of products. This is an empirical measure of a subject's skill. For new subjects, one expects a minimum of 65% correct across the trials (use a test with a $p = 1/2$ outcome). Selecting 65% correct as a minimum requirement for training is an arbitrary decision; one can raise or lower this, based on the number of subjects who meet this criterion. Empirically, we observe that subjects with poorer skills (e.g. 50–55%) tend to be more variable and less sensitive in subsequent product tests. Other qualifications include the ability to verbalize and to work as a group (both are important to the success of the language development process) and the ability to perform during actual testing. Of course, with experienced subjects for whom performance records exist, the need for requalification is minimized and their performance in actual testing becomes the primary criterion. In the process of screening and then language development sessions, the subjects develop a thorough familiarity with the products and with the differences among them. These activities, when combined with replication, enhance the likelihood of finding differences.

A basic strength of the QDA method is the ability to independently verify (after each test) that individual subjects perceive differences among products based on each attribute. This is directly measured with a one-way analysis of variance (ANOVA) for each subject for each attribute (a topic discussed later in this chapter). The need to monitor the performance of each subject in each test reflects an awareness of the sensory limitations of humans. There is no guarantee that a screened and qualified subject will always perceive differences among products on all attributes. By establishing a formal, data-based selection procedure, the system becomes less dependent on a limited number of subjects whose responses are atypical. It also recognizes that products are variable, and to some extent this variability will be embedded in a subject's response to some attributes. With use of a repeated measures design and ANOVA, one can partition (separate) the subject effect from the product effect and gain greater insight to product differences as well as the individual subject's ability to differentiate among the products on an attribute basis.

6.2.3.4 Use a limited number of subjects

All descriptive tests use 20 or fewer subjects. The QDA methodology was initially started with approximately 20 subjects, but based on empirical evidence the number of subjects was reduced to 12. This was achieved by examining results based on 20 versus 18 subjects, versus 16, etc., to determine if decisions were changed as a function of the number of subjects. The core question was whether having more subjects improved the product differentiation, and it did not. Business concerns were the availability of subjects and the cost of the test. Both were and still are reasonable considerations with regard to the panel size. This does not mean that one cannot use more than 12, and some companies use as many as 20 because there is concern that small differences may be missed. If a panel leader has technical knowledge about the expected differences (e.g. very small), then more subjects could be used or the replication could be increased. We recommend the latter and less so the former. In addition, when there are less than 10 subjects, the panel should be supplemented with more screened and qualified subjects to avoid trying to interpret results when there is an insufficient number of judgments. The problem is the same that was directed at Flavor Profile—having each subject contribute a significant amount of the total variability. The overall contribution of each subject to the total variability increases accordingly, such that too much dependence is placed on too few subjects. This dependence is not lessened with replication. Alternatively, increasing panel size to more than 12 (e.g. 20 or more) can present other challenges, including scheduling, the extent to which each subject has the opportunity to contribute to the discussion during language development sessions, and the potential need for parallel sessions. Social psychological studies have shown that having more than approximately 12 people in a group discussion requires an assistant to the panel leader to help maintain focus during the discussion. The panel leader must decide on the number of subjects based on past experience (the empirical evidence relative to the subjects and the products). Experience with many trained panels leads us to the conclusion that 12 is an optimal number of subjects for a panel for most situations. One final note of caution concerns reliance on the same subjects for all testing, which is also not recommended. Such a practice will cause a variety of problems to develop, including elitism, atypical response behavior, and test fatigue. Instead, one should develop a pool of qualified subjects (as discussed in Chapter 2).

6.2.3.5 Use a consensus language development process free from leader influence

The success of a descriptive test is, of course, also dependent on the sensory language developed for the products being evaluated. The subjects use the language as a basis for differentiating the products. To be useful, the language must be easily learned and meaningful in terms of its applications in product development, in relation to consumer preference measures, and so on. We concur with Andani *et al.* (2001) that a vocabulary derived from consumers is more likely to provide a better representation of product characteristics as perceived by those buying and consuming products. This language is also found to be easily correlated with formulation variables as well as with physical and chemical measures and, most important, with preferences and related imagery information. Developing a QDA language is an iterative process, and depending on the product, it could take as much as 8–10 hours (in 60- to 90-minute sessions) before agreement is reached on the attributes. In addition to developing attributes, they must be grouped by modality, and the subjects must define/explain each attribute and, with assistance, if needed, from the panel leader, develop an agreed upon evaluation procedure. The definitions serve as a guide for the subjects during testing to minimize confusion over the meaning of each attribute (a source of variability). Linking attributes with definitions enables the subjects to proceed very rapidly and misunderstandings are minimized. As a further aid in language development, reference materials may also be identified and are made available during the training. The reference material can be an ingredient, a product, or some other material that "represents" the particular sensation and the attribute (or attributes) used to describe that sensation. In most instances, ingredients are useful as references; however, not all sensations will have a reference and some references will provide more than a single sensation. The combination of definitions and references is important in helping to standardize the language development process, especially when training new subjects or retraining existing subjects who are experiencing difficulty differentiating products with a particular attribute.

There is no limit to the number of attributes that subjects can use provided there is a consensus among the subjects with regard to their meanings and their order of occurrence. Theoretically, this could result in 100 or more words being used to describe the products. In practice, however, the number of words initially is quite large but subsequently is reduced to some smaller subset. As part of the training, the subjects will score many, if not all, of the products, and in the ensuing discussion they identify those attributes with a common sensory basis, thus resulting in an approximately 50% reduction in the final number of attributes.

In the QDA language development process, the panel leader's function is almost entirely nonparticipatory to minimize influencing the subjects. The panel leader's responsibility is to decide which products will be evaluated in each session, to facilitate conversation between subjects, and to provide an operational framework for each session based on previous results and the project objective. The panel leader must also determine when group discussions are finished and the panel is ready to evaluate products—the data collection process. This language development process is therefore dependent on the skill of the panel leader. The subtleties of leader influence can never be totally eliminated; however, a key to minimizing a leader's influence is to keep this person as an impartial observer and to never convey the impression of knowing what attributes should be used or what are the expected product differences. To further minimize panel leader or subject-to-subject influences, actual product evaluations are usually done in a conventional sensory booth according to the procedures developed during the training sessions. Alternatively, the evaluations might be done in a typical use environment or at home (discussed later in this chapter).

If a language is already available, then it is necessary to familiarize the subjects with the attributes and the definitions, but always in concert with the products. This is a quicker process, particularly for the experienced subject, requiring as few as two or three 90-minute sessions. The key to success in this process is to be sure that the subjects have ample opportunity to sample products and to change or refine their language. As long as there is a consensus, the existing language can be changed and the original word or words are placed in the definition section. For inexperienced subjects, however, the time required to learn an existing language can require as many as four sessions. There is no rule as to the number of sessions required. A panel leader will have to determine when the subjects are ready for actual product evaluation. New perceptions or a request to change an attribute and/or a definition is the collective responsibility of the panel. If an individual subject indicates a sensation that is not accounted for, then the panel leader must ask the panel to reconvene for discussion. Subjects are reminded of this additional responsibility at the outset of each test. This particular aspect of descriptive analysis permits the communication process to remain current and provides for flexibility in the use of different words to connote the same sensation but, through the definition, to maintain a specific anchor regardless of the label. In the succeeding sections, the quantification process is discussed relative to verifying that the subjects individually and as a group perceive the sensations and differentiate between products with a measured degree of reliability and face validity.

In recent years, a number of alternative techniques have been proposed to facilitate the language development through use of feedback derived from multivariate analyses. We discuss these ideas later in this chapter; however, suffice to note that such procedures assume that there are correct responses or that there are a certain number of attributes, neither of which is true, or at least no experiments have been described to demonstrate such a proposed advantage.

6.2.3.6 Be quantitative and use a repeated trials design

The need for quantitative information is essential in descriptive analysis, as is a need for obtaining more than a single judgment from each subject for each attribute for each product. It is no different than any scientific analysis; whether it is physical or chemical or both, it must be replicated. Surprisingly, this lack of replication continues without much objection by requestors and by some sensory professionals, all of whom should know better. Because considerable emphasis is placed on results derived from a limited number of subjects, a large number of attributes, and, in some instances, a large number of products, one must be very confident about the reliability and validity of the responses that lead to actions such as product changes. Obtaining statistical significance is relatively easy given the large database; however, the real question is whether these differences have practical versus statistical significance. Earlier descriptive methods, such as the Flavor Profile, were criticized because of the difficulty in understanding the meaning of the words in the description and the lack of a true numerical system for assessing product differences based on those descriptions.

For QDA, the issue of quantification was addressed taking into account the literature on measurement, challenges to the Flavor Profile method, and recognition of differences in behavior and sensory skill. An unstructured graphic line scale was selected because it provided subjects with an infinite number of places in which to indicate the relative intensity for an attribute (within the constraints of the actual length of the line). Also, any scale that was selected would not have numbers directly associated with it. Subjects were instructed to place a mark on the line at that location that best represented, to that subject, the strength for that product attribute, keeping in mind that the far left of the line represented no detection, the first vertical line represented strength at or near threshold, progressing with increasing strength to the strongest among the products at the left end of the line. Initial subject reaction was and it still is typically one of concern regarding the correct placement; the request directed to the panel leader was "Tell us where we should place the line for some product attributes and then we will know what to do." Not surprisingly, this request is rejected and subjects are encouraged to proceed. After a few products, the process proceeds smoothly with little or no need to provide additional instructions other than a reminder to remember scale direction, which some subjects forget (leading to crossover interaction, which is discussed later in this chapter). With use of a repeated trials design and ANOVA, concern about specific scale location for a response is of less importance than are the twin issues of reliability and sensitivity to product differences. From a measurement perspective, use of a line scale was consistent with the concept of functional measurement and the graphic rating-scale approach of Anderson (1970) and in part with the trial-and-error process described by Stone *et al.* (1974). Although line scales, in various forms, have been used for product evaluation for many years (Baten, 1946; Hall, 1958), the philosophical approach developed by Anderson and co-workers proved to be especially useful.

The scale shown in Figure 6.1 is a 6-inch line with word anchors located 1/2 in. from each end. In metric terms, the line length is 15 cm with anchors located approximately 1.5 cm from each end. The scale direction always goes from left to right with increasing intensity; for example, weak to strong, light to dark, or some similar designated set of word anchors. The specific length of the line was selected based on a series of tests using lines of varying length. For example, it was observed that extending the line to the full length of an 8.5 × 11-in. paper did not increase sensitivity and there was a tendency for variability to increase. Making the line shorter than 6 in. (e.g. 5 or 4 in.) also appeared to reduce sensitivity; that is, the line restricted the number of places to put a mark. Although these observations did not constitute an exhaustive evaluation on the appropriate scale length, the evidence at the time and for several years after (examination of responses from several thousand judgments and many tests) clearly supported the continued use of a 6-in. or 15-cm line. On the other hand, shifting the anchors to the ends of the line had a more dramatic effect on responses; response patterns would take on the character of a true category scale and greater departure from linearity. For computational purposes, the distance along the

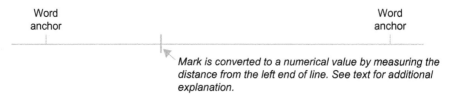

FIGURE 6.1

An example of a graphic rating scale (or line scale) used to obtain relative intensity measures from the subjects.

line to the mark is measured to yield a numerical value for computation. Successful use of this line scale depends on the instructions during training and the functional aspects of the measurement process. During language development, subjects are familiarized with use of the scale when evaluating products, starting with the third session. Effective functional measurement requires that the subjects experience what constitutes the extremes in intensities for many of the attributes in the context of the products that will be evaluated. This procedure helps to minimize end-order effects (avoidance of the extremes) and encourages full use of the scale to express product differences. Subjects are not penalized if they use less than the full scale. The emphasis is for the subjects to be as consistent as possible regarding location of the mark they place on the line and also to understand the scale direction. These twin ideas are reinforced throughout the language sessions as a result of sampling products, marking scorecards, and voting followed by a discussion of the results. If subjects can be reasonably consistent with regard to which product is stronger versus another, then a successful outcome can be expected. However, this is not a one-trial experience, and the process has to be repeated for the subjects to gain the confidence they need when actual product evaluation is initiated. Adding a constant does not change the relationship between products, so concern about where a subject marks a scale is of minor importance. Forcing a subject to use more of the scale or to move to a specific location is disruptive, it will alienate that individual, and, as mentioned previously, this action is suggesting to the subjects that there is a correct line placement and a correct response will be provided if they wait long enough.

Empirically, this procedure was found to be very easy to use, and subjects were able to complete a scorecard (containing as many as 40 attributes) within a few minutes with a high degree of reliability. In early applications, use of a third anchor at the scale midpoint was common. However, removal of the center anchor was observed to reduce variability by 10–15%, which represented a substantial improvement in sensitivity, and the current QDA scale has only two anchors.

As mentioned previously in this discussion, the elimination of numbers from the scale and from the subjects' responses eliminated two sources of bias—avoidance of or preference for particular numbers that have negative or positive connotations, respectively. The second source of bias was the subject who changed use of numbers over time. This latter situation was noted by Ekman and Sjöberg (1965) in their review on scaling, and it can be especially troublesome because one might not know if this change reflected a product difference or true bias. During training in the use of the QDA scale, numbers are not discussed and the subjects are not told that the scale has numerical content. They may infer that the marks are converted to numerical values, but that is not the same as being told so by the panel leader.

The use of the subject's own word anchors also contributes to the utility of the scale and the linearity of the responses. The responses are used directly in statistical analysis; that is, no transformations are necessary and the monitoring of response patterns is directly achieved.

A related issue is subject reliability, which is directly addressed by having each subject provide responses to the products on more than one occasion. The importance of this particular requirement cannot be underestimated. It represents an essential internal check on the response system, enabling the panel leader to determine individual subject and panel consistency within and across products, from trial to trial, and to establish numerical measures of performance. It has often been argued that incorporating replication into a test delays the time required for making recommendations. Although repeated judgments extend the time required to complete a test, this additional time is well worth the effort. The question of how much replication is sufficient depends on numerous factors, including the extent of product variability and magnitude of differences among products, the stage of product testing, and so forth. Empirically, we observed that four was a sufficient number of trials, but this could be reduced by one if there were eight or more products. Alternatively, the panel leader may decide that fewer trials would be satisfactory. For example, if the test is during the early stages of product development, then three replications could be adequate to provide direction. Alternatively, if the development effort were at a final stage, then more replication would be appropriate. We observed that beyond four replicates, little or no obvious improvement in sensitivity is achieved unless the product array is highly variable (e.g. agricultural products). For example, in a test of six products, each subject (assume 12) would evaluate each product four times, yielding a total of 288 evaluations for the panel for each attribute, which would be a sufficiently large database on which to reach conclusions with confidence about product differences for each subject and the panel.

On several occasions, it was possible to evaluate test results after 4, 6, 8, 10, and 12 replicates. The evaluations included typical statistical measures, such as means and standard deviations, individual and mean ranks, and ANOVA, and in each situation the conclusions have not changed. In some instances, the level of statistical significance changed; however, the conclusions about product differences were not altered. Observation of results after only one replication yielded a less clear result for some attributes (approximately 15–20 of them) but with most decisions unchanged for all other attributes.

The experienced panel leader will develop guidelines regarding the appropriate number of replications. These guidelines must take into account the nature of the problem, product variability, the anticipated degree of difficulty, subject skills, and the expected use of the results. However, one point should be very clear: It is extremely risky to rely on descriptive information for which there are no repeated judgments.

At various times, it has been suggested that subject performance can be improved through the application of feedback during the training sessions. This is not a new concept; psychologists have used this to improve performance of a variety of motor skills. It has also been applied successfully in sensory evaluation. More than half a century ago, Engen (1960) demonstrated that olfactory thresholds could be reduced with instructions and with practice. Similar results were obtained with taste thresholds. In both situations, the more familiar the subject was with the task, the better he or she performed; and sharing information with a subject directly after the test resulted in a lower threshold value. Feedback was effective in improving the sensitivity of the subject. However, no two subjects were alike; each learned at a different rate, and each person's threshold was different. In the case of descriptive analysis in which one is capturing intensities, assuming that the responses will be the same is naive and incorrect, regardless of the feedback. The taste of a product from the same container will not be perceived the same by the subjects. People do not have the same perceptions, which is why one has a panel of subjects and not the N of 1. There seems to be a point of view within sensory evaluation that considers differences between individual subjects as undesirable, as some kind of measurement error that needs to be removed or prevented from occurring. As we have noted at various places in this book and elsewhere, diversity of perception is real and fundamental to the science of sensory evaluation. In the descriptive analysis process, there is a fine line between encouraging subjects to verbalize their perceptions and telling subjects what words they should use to represent a specific perception. As noted previously, when QDA subjects develop a set of words to describe their perceptions and then assign intensities to those words, it becomes their scorecard and there is no suggestion that there are correct versus incorrect words to use and/or correct intensity values. The intensities reflect the individual's sensitivity and will be different from subject to subject, so there should be no question as to which is correct: All are correct. Attempts to modify behavior may give the responses the appearance of repeatability but at the cost of external validity and predictive validity.

6.2.3.7 Have a useful data analysis system

The ANOVA model is the most appropriate statistical procedure for analyzing scaled responses from a descriptive test. Use of the ANOVA model in sensory evaluation was discussed in Chapter 4, and the reader will find it helpful to review that information before proceeding. In a QDA test, the subjects evaluate all of the products on an attribute-by-attribute basis on more than a single occasion, one product at a time. In design terms, this represents a treatment-by-subject, repeated measures design. The use of replication adds additional time to a test; however, from a sensory point of view, it is essential in directly characterizing the reliability of response patterns, as noted in the previous section. This does not mean that sufficient replication from a limited number of subjects enables one to automatically generalize to the population but, rather, that there is confidence that the differences are due to true product differences and are not random occurrences. With this in mind, it is useful to list some of the more fundamental questions that need to be resolved before product similarities and differences can be considered and conclusions and recommendations made. These questions include the following:

1. Are individual subjects consistent from trial to trial? How much is each subject contributing to product differences on an attribute basis?
2. Is panel performance consistent or is some additional training required?
3. Are all attributes contributing to product differentiation or is additional training required?
4. Are responses to product differences consistent across all subjects?

These questions are answered through use of several different computations, such as means, standard deviations, rank order, and ANOVA. For example, a subject's ability to distinguish between products can be directly measured by use of a one-way ANOVA on an attribute-by-attribute basis. The magnitude of the difference between products and the consistency of the scores for the products from trial to trial is reflected in an F ratio and its corresponding probability value. The smaller probability values (e.g. 0.01 and 0.001) reflect greater differentiation among products and a higher level of consistency, and this can be considered as a measure of that individual's performance (for each attribute for that particular set of products). A subject whose probability values are small (e.g. 0.10 and lower) would be considered as contributing to the differences among the products and would be rated as a good discriminator of product differences compared with a subject whose probability values are higher (e.g. >0.75) and who would be considered as making a smaller contribution to the differences among the products. Probability values (associated with the F ratio) can range from 0 to 1.0; the smaller the value, the larger the contribution that subject is making to product differentiation. This process is not different from the computations involving all subjects; to wit, are the subjects in sufficient agreement to conclude that there are significant differences among the products? Regardless of the individual probability values, each subject is contributing to product differentiation, but that contribution is different for each subject for each attribute across all products. By comparing subject probability values, one has a useful measure of that panel's level of performance. This output also provides information about specific attributes; that is, some attributes contribute more to product differences and others provide very little and may be potential candidates for dropping from the scorecard at future tests.

As shown in Table 6.3, probability values derived from such an analysis are very informative and more useful when combined with that subject's individual product means, rank orders, and extent of interaction. Here, we focus on the use of the one-way ANOVA probability values. Responses from subject 831 yielded values of 0.00 for attributes A and B and 0.06 and 0.31 for attributes C and D. The corresponding values for subject 833 were 0.68, 0.59, 0.01, and 0.09, respectively. Subject 831 differentiated among all of the products on all four attributes; however, the contribution to differences among products was smaller for attribute D (where

Table 6.3 Probability Values for Individual Subjects Derived from a One-Way Analysis of Variance on a Scale-by-Scale Basis

Sensory Characteristic	Subject				Panel
	831	832	833	834	
A	0.00[a]	0.19	0.68	0.00	0.04
B	0.00	0.02	0.59	0.06	0.01
C	0.06	0.09	0.01	0.02	0.00
D	0.31	0.38	0.09	0.43	0.38

[a]*Entries are the truncated probability values; see text for additional explanation. These values were excerpted from a larger database and are used here for illustrative purposes.*

the value for D was 0.31). Alternatively, subject 833 contributed much less to the differences among products on attributes A and B and contributed much more for attributes C and D.

Further insight into performance is achieved by examining mean values, standard deviations, and the scores converted to ranks. The latter are obtained by transforming the intensity measures to ranks and examining them versus the panel ranks, keeping in mind that in situations in which the means are close (i.e. not statistically significant), it is reasonable to expect that there will be less agreement in the ranks than if the means were very different from each other. All this information, individually and collectively, is used to evaluate subject and panel performance and to identify those attributes that did not provide satisfactory product differentiation (e.g. see D versus C in Table 6.3). The latter problem may be due to the subjects not being in sufficient agreement as to the meaning of that attribute or using the scale in the direction prescribed; however, subject 833 was able to use the attribute to differentiate among the products. Prior to the next test and during a language warm-up session, the panel leader might want to request that subject 833 describe (to the rest of the panel) the basis for product differentiation using that attribute. This discussion could lead to additional information being added to the explanation, thus improving the criteria used by each subject and the panel and, finally, better product differentiation by the entire panel. For this attribute, product differentiation was not very good—the probability value was 0.38—and depending on the remainder of the test results, this attribute should be a candidate for discussion before the next test.

The panel leader examines the pattern of responses to develop a clear understanding of what should be done before the next test. To what extent is each subject contributing to product differences (e.g. more than half, less than half)? Is the lack of differences a result of increased variability from incorrect scale use (a cross-over interaction)? Is it a true lack of sensitivity or are the product differences so small as to be unimportant? Answers to these questions must be obtained before reaching product conclusions and reporting a course of action to the requester, and before initiating the next test. Table 6.4 provides additional examples of the

Table 6.4 Hypothetical Means and Standard Deviations for Assessment of Subject/Panel Performance[a]

| Subject | Attribute XYZ | | | |
	Product	Mean	SD	Probability[b]
A	1	14.8	9.4	
	2	16.9	9.1	0.41
	3	13.4	8.3	
B	1	14.8	4.8	
	2	16.9	4.7	0.05
	3	13.4	4.0	
C	1	24.8	18.5	
	2	20.5	14.0	0.25
	3	38.0	19.0	

[a]See text for detailed explanation of the tabular values.
[b]Probability values from the one-way ANOVA.

use of summary statistics (in this instance, the means and variance measures) to gain further insight into subject response behavior and future training emphasis. In this example, subject A exhibited minimal contribution to product differences primarily because of variability in the responses; the standard deviations were relatively high (8.3–9.4) compared with the means (13.4–16.9). Alternatively, subject B, with identical means, had smaller standard deviation values (4.0–4.8) and thus made a greater contribution to product differentiation. Subject C showed a different response pattern: The mean values were larger and not in the same order as reported for subjects A and B. The standard deviations were also much larger; however, product differentiation was better, as measured by the probability value of 0.25 versus 0.41 for subject A. Subject B exhibited the largest contribution to product differentiation, because he or she was less variable, and would be a useful tutor in the next training session. Subject C, on the other hand, was considerably more variable but contributed to product differentiation to a greater extent than did subject A. However, note that subject C also ordered the products differently, suggesting a possible crossover interaction. Here, too, attention must be given to attribute definition and scale direction.

Some additional comments about subject performance are appropriate here. The first is a reminder to the reader about not using performance data to justify removing that subject's data and reanalyzing results. Eliminating subjects (and their data) is not an acceptable course of action. The second is to keep in mind that what appears to be unusually high variability from a subject could reflect product variability and/or poorly controlled preparation and handling practices. The tendency to assign variability solely to the subject is usually unwarranted without other information. Each product type and manufacturer will have its own unique profiles and variability, and panel leaders must use their data files to formulate subject performance guidelines and not apply a rule arbitrarily and without fully appreciating the consequences.

In this way, the panel leader is able to isolate individual response patterns and from this information develop a sense of what needs to be done to improve panel performance as well as determine the relative quality of the database. This approach is considerably more demanding of the panel leader; however, the weight of the results justifies this degree of effort. Once there is satisfaction with the performance of the individual subjects and scales, then it is appropriate to evaluate product differences.

The data are analyzed using a two-way ANOVA, general linear model showing both subject and product effects. Subject-by-product interactions are also estimated when the experiment includes repeated judgments. These results can be summarized in typical ANOVA tables (Tables 6.5 and 6.6), with main effects tested against both error and interaction mean square. Table 6.5 contains the F ratios using the error or the interaction terms; the panel leader selects the appropriate F value.

For these results, interaction was not significant and no further consideration of interaction was necessary. Note that there was a significant subject effect, reflecting differences in scale use by the subjects; however, this difference in scale use was not unexpected and did not interfere with product differentiation. When interaction is significant (Table 6.6), indicating differences in the way subjects interpret or measure a given attribute, then one uses the interaction error term to test the product effect. It can be seen that this resulted in a loss of significance for products. The other analysis and the response patterns must be studied to determine the cause of the problem. It has been suggested that this switch from one value to another (error or interaction mean square) represents a shift in the particular mathematical model, and that there is some risk; however, we never rely on a single statistic, and the approach described herein is extremely helpful and inappropriate business decisions have yet to be encountered.

An alternative approach would be to consistently use interaction mean square for testing significance of product differences. One problem with this approach is that there will be instances in which product differences are not significant when tested against a nonsignificant interaction but are significant when tested against residual error. To have both available enables the panel leader to make the appropriate choice, always keeping in mind that the interaction significance is the gatekeeper.

Individual interaction values can also be computed for each subject (for each attribute). This provides still further insight about subject and panel behavior. As a measure of this, the interaction sum of squares is partitioned in such a way as to determine how much each subject has contributed to the interaction F ratio for the panel. In this case, a large F ratio (and small probability value) for a subject is indicative of subject–panel disagreement.

Knowledge about interaction is important in understanding the nature of disagreements between individual subjects and a panel. As depicted in Figure 6.2, there

Table 6.5 Analysis of Variance for a Specific Sensory Characteristic

Source of Variance	df	Sum of Squares	Mean Sum of Squares	Versus Error[a]		Versus Interaction[a]	
				F-ratio	Probability	F-ratio	Probability
Samples[a]	4	10718.89	2679.72	28.49	0.00	47.81	0.00
Subjects	8	8441.71	1055.21	11.22	0.00	18.43	0.00
Interaction	32	1793.55	56.05	0.59	0.95		
Error	126	11848.07	94.03				

[a]Main effect of samples is significant; main effect of subjects is significant; interaction (sample × subject) is not significant.

Table 6.6 Analysis of Variance for a Specific Sensory Characteristic

Source of Variance	df	Sum of Squares	Mean Sum of Squares	Versus Error[a]		Versus Interaction[a]	
				F-ratio	Probability	F-ratio	Probability
Samples[a]	4	1567.43	391.85	2.95	0.02	1.16	0.34
Subjects	8	14877.14	1859.64	14.01	0.00	5.54	0.00
Interaction	32	10730.41	335.32	2.52	0.00		
Error	126	16712.64	132.72				

[a]Main effect of samples is significant; main effect of subjects is significant; interaction (sample × subject) is significant; main effect of samples (with new error term) is not significant.

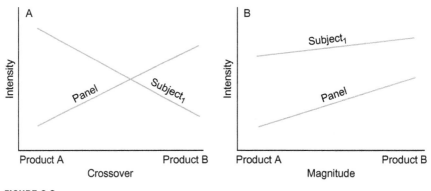

FIGURE 6.2

Examples of two different kinds of interaction that can occur in a sensory test: Panel A represents the classic crossover type of interaction; Panel B demonstrates a more subtle problem, called magnitude interaction.

are two types of interaction, but they are not equivalent. In the first type, crossover (Figure 6.2A), the subject or some subset of a panel reverses the products compared with the panel; that is, product A was perceived as more intense than product B, whereas the panel as a whole perceived the opposite. This may reflect a true difference in perception (i.e. bimodality), or it could reflect confusion in scale use by the subject, either of which would not be readily apparent without direct examination of response patterns. The other type of interaction, referred to as magnitude, is shown in Figure 6.2B. In this instance, there was no confusion about which product was most intense; however, the magnitude of the difference between products was relatively small for subject 1 versus the large difference for the panel as a whole. This latter type of interaction is of little or no consequence compared with crossover. Conversion to ranks minimizes the effect of that difference, allowing for still another view of subject response behavior. Magnitude interaction reflects differences in subject sensitivity and/or scale use and not in the ordering of the products. By differentiating these types of interaction, the panel leader is better able to consider the conclusions about the products for that specific attribute. If the statistics indicated that the significant difference among products was lost because of the magnitude interaction, then it could be ignored. Alternatively, crossover interaction is a more serious issue because it may reflect a failure to use the scale correctly or a true lack of sensitivity to the differences among the products. We use a proprietary statistical analysis to better identify subjects contributing to crossover interaction and to quantify the severity of the crossover on a subject-by-attribute basis. Significant product differences are first determined for the attribute. Subject scores are then converted to ranks, as are the panel means, and a sum of squares is computed for each subject. The computation reflects the accumulated rank distance for product means that are significantly different (based on total panel) and ranked by the subject in a different order than the panel. The example shown in

Table 6.7 Means, Ranks, and Crossover Scores for Sweet Aroma

Product	Panel Mean	Subject Means							
		1	2	3	4	5	6	7	8
Prod E2	35.79[a] a	48.3	26.3	12.8	41.0	46.7	34.0	45.5	31.7
Prod E1	30.15 b	42.5	26.5	17.3	33.8	32.3	28.2	40.5	20.0
Prod S	27.90 bc	36.2	22.2	22.7	34.0	26.5	27.5	38.0	16.2
Pilot 2	25.85 c	36.0	16.8	15.2	33.7	22.7	27.3	37.5	17.7
Mean	29.92	40.8	23.0	17.0	35.6	32.0	29.2	40.4	21.4
N	192	24	24	24	24	24	24	24	24
		Subject Ranks							
		1	2	3	4	5	6	7	8
Prod E2		1	2	4	1	1	1	1	1
Prod E1		2	1	2	3	2	2	2	2
Prod S		3	3	1	2	3	3	3	4
Pilot		4	4	3	4	4	4	4	3
Crossover score		0	32	71	0	0	0	0	0

[a]Panel means designated with the same letter not significantly different based on Duncan multiple range test.

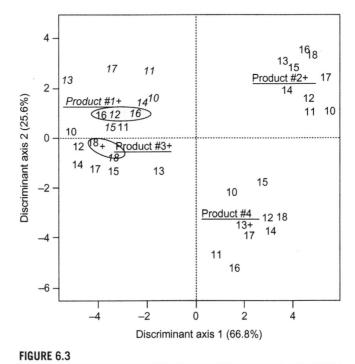

FIGURE 6.3

Subject discrimination of products across attributes.

Table 6.7 identifies subject 3 as having the largest crossover score, and this is consistent with scoring product E2 lowest for this attribute, whereas the panel scores it significantly higher than the other products. Subject 2 also has a crossover score, albeit lower (32 vs. 71), because product E2 is ranked second highest for this attribute. The remaining subjects score product E2 highest and reverse product E1 with S, or S with Pilot, comparisons that the panel result does not identify as significant. Consequently, the calculated crossover scores for these subjects are 0.0. Summarizing crossover scores during training provides the panel leader with useful information about subject and attribute performance, and during final data analysis it is helpful for identifying potential problem areas when assessing consumer behavior. Interaction in relation to multiple range tests and significant differences between products is discussed in more detail later. For a more complete exposition on the statistics of interaction, see McCall (1975), McNemar (1969), and Winer (1971).

QDA data lend themselves well to multivariate analyses to better understand subject performance and attribute relationships. The data in Figure 6.3 are examples of the former, and the data in Figure 6.4 are examples of the latter. Examining Figure 6.3, we see that the subjects discriminated the test products quite well. The panel scored products 1 and 3 most similar to one another, where subject 16 scored product 3 more similar to product 1 and subject 18 scored product 1 more similar

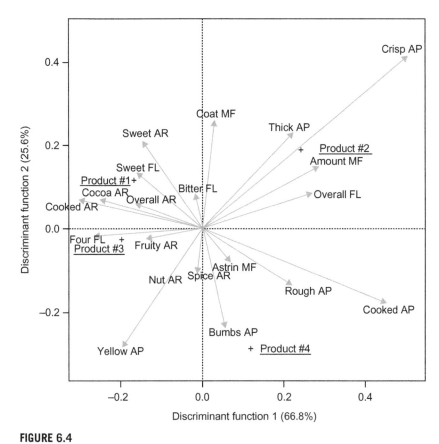

FIGURE 6.4

Attribute factor loadings.

to product 3. The information provides insight about what may occur in consumer testing, and an area for follow-up with the panel. Figure 6.4 allows identification of those attributes that most differentiate the array of products to the panel. For example, product 1 can be characterized as aromatic with sweet, cocoa, and overall aroma scores; product 2 is characterized for its thick and crisp appearance and high scores for mouthfeel; product 3 has a fruity aroma and flour flavor; and product 4 can be differentiated on several appearance measures.

The ANOVA model also enables the panel leader to determine whether the mean scores for several products differ from one another in sufficient magnitude to justify considering them different at some stated (and low) level of risk. The analysis does not specify which products are different, and one must do additional computations after the F test, sometimes referred to as multiple-range tests. There are a variety of these tests, and some needless controversy has developed concerning which one is most appropriate. The tests are not interchangeable, and different tests can lead to different conclusions regarding significant differences among products. This does not mean that one test is more or less correct but, rather, that they reflect different philosophies with regard to risk and minimizing decision errors. These tests include Duncan, Newman–Keuls, Tukey (a), Tukey (b), Scheffé, and Dunnett. The first three tests and the Scheffé test are most commonly used in sensory tests. In effect, one computes a critical value, and those mean differences that are greater are considered significantly different. The formulae are quite similar to those shown here. For additional discussion on these computations, see Bruning and Kintz (1977) and Winer (1971).

6.2.3.7.1 Duncan
The computation is as follows:

$$\text{Critical differences} = k_r \sqrt{\frac{ms_e}{n\,(\text{per group})}}$$

where k represents values derived from table, and r is the number of means being compared (number of steps apart). The Duncan procedure uses a protection level for α for the set of samples being compared rather than an α level for the individual pairs of samples within the test. This approach minimizes the risk of a Type 2 error, but it may concomitantly increase the risk of a Type 1 error.

6.2.3.7.2 Newman–Keuls'
The computation is as follows:

$$\text{Critical difference} = q_r \sqrt{\frac{ms_e}{n\,(\text{per group})}}$$

where q represents values derived from table, and r is the number of means being compared. The Newman–Keuls' procedure emphasizes ranges rather than differences; the computations are the same as in the Duncan test, but the values q derived from the table are different and the result is that larger differences (between means) are required before statistical significance is achieved. In effect, the procedure provides greater protection from risk of a Type 1 error.

6.2.3.7.3 Tukey (a)
The computation is as follows:

$$\text{Critical difference} = q_r \sqrt{\frac{ms_e}{n\,(\text{per group})}}$$

where q represents values derived from table, and r is the number of means. The Tukey (a) or HSD procedure uses a single value for all comparisons based on the total number of means involved, and larger differences are therefore required before statistical significance will be achieved. As Winer (1971) noted, this test has lower statistical power than the Newman–Keuls' and Duncan tests.

6.2.3.7.4 Tukey (b)
The computation is as follows:

$$\text{Critical value} = \frac{\text{critical value } N - K + \text{critical value Tukey (a)}}{2}$$

This procedure represents a compromise between the two previous tests; the critical value changes with the number of means being compared and provides protection from Type 1 error.

6.2.3.7.5 Scheffé
The computation is as follows:

$$\text{Critical difference} = (a - 1) F \sqrt{\frac{2ms_e}{n \text{ (per group)}}}$$

where a is the number of groups to be compared, and F is the tabular value for F with appropriate degree of freedom (df). The Scheffé test yields a single value for use in determining the significant difference between any two pairs of sample means and, as can be seen, uses the F value and further minimizes risk associated with Type 1 error, to the greatest extent.

6.2.3.7.6 Other tests
Other possible tests include the Fisher modified least significant difference (LSD) and Dunnett tests. The former test is, according to Winer (1971), suitable for comparisons planned prior to data evaluation, whereas the latter test is designed for use when one is comparing products to a control.

Winer (1971) prepared a tabulation of these alternative tests using a common data source for comparative purposes. The ordering was Scheffé, modified LSD, Tukey (a), Tukey (b), Newman–Keuls', and Duncan. The author believed that the Tukey (a) test had much to commend it, including broad applications and relative ease in computation. A similar type of comparison was made by Bruning and Kintz (1977), but no recommendation was made. Rather, these authors simply identified significant and nonsignificant occurrences using a common database.

Another test that can be applied is the Bonferroni t (Myers, 1979). This test, like Tukey (a) and Scheffé, is especially protective of Type 1 error. It takes into account the problem of the change in Type 1 errors with multiple contrasts, and it is especially applicable where group sizes and variances are unequal. Thus, if a test objective were to identify the product that is perceived as different for an advertising claim, then the Bonferroni t may be the method of choice. The emphasis of these tests is on minimizing Type 1 errors; however, this objective needs to be considered relative to the test objectives encountered in many business situations. As discussed in Chapter 5, most product work (e.g. ingredient substitution, cost reduction, and matching a target) is concerned with minimizing the risk of perceiving a difference between products and reducing the risk of a Type 2 error. Therefore, it will be more appropriate to use either the Duncan or the Newman–Keuls' tests for determining which means are different. However, even these latter procedures entail some risk, which is only appreciated after a thorough examination of all data, including subject response patterns.

If we consider the results shown in Table 6.8, this concern will be more fully appreciated. These data were derived from a QDA test involving 12 experienced subjects, four products, and 30 attributes; each product was evaluated once in each of four sessions (a.m. and p.m. for 2 days), with a total N of 48 judgments per product.

Based on the Tukey (a) test, products A, B, and C are not considered different from each other, and products B, C, and D are not considered different from each other (the Duncan brackets are used here for ease of communication). Based on the Duncan test, products A, B, and C are not considered different from each other, which is consistent with Tukey (a); however, products B and C are considered different from product D. Further insight into this "discrepancy" can be gained by examination of the way the individual subjects rated products B versus D and C versus D. For example, considering products B versus D, 11 subjects rated B as more intense than D, and the magnitude of the difference, on average, was 5.9 scale units, whereas only one subject rated D as more intense than B and the magnitude was 0.7 scale units. It is difficult to accept a Tukey (a) no significant difference decision in this situation, considering the responses from the individual subjects and the average magnitude of that difference.

In this particular instance, which involved ingredient substitution, a primary objective was to minimize detection of a difference. Stating that there was no significant difference between the control and 50% replacement would be very risky (because 11 of the 12 subjects stated there was a difference). Examination of the C versus D comparison revealed a somewhat less clear picture. Seven subjects rated

Table 6.8 A Comparison of Duncan and Tukey (a) Results for Significant Differences

Characteristic[a]		Mean[b]	Duncan	Tukey (a)
A	100[c]	29.7	⎤	⎤
B	50	27.1		⎤
C	25	26.9	⎦	⎥
D	0	21.7		⎦

[a]The particular characteristic is not germane to the problem (see the section on language in this chapter).
[b]Entries are mean values in a scale of 0–60. The two-way ANOVA revealed a significant F ratio, $p < 0.005$, and there was no significant interaction.
[c]Added ingredient at 100, 50, 25, and 0% of normal; that is, the current production was the 0 product.

C as more intense than D, and the magnitude of the difference, on average, was 11.6 scale units; five subjects rated D as more intense than C and the magnitude was 3.8 scale units (on average). Recommending a significant or nonsignificant difference was more difficult. The number of subjects favoring either product was almost equal, and the only source of hesitation in reaching a recommendation was the magnitude of the difference. The sensory professional would need to discuss the situation with the product specialists as well as evaluate any other sources of product information before reaching a meaningful business decision.

This particular situation occurs often enough that one should never be complacent about results, and it is a reminder that no one statistical measure will be totally satisfactory in explaining product similarities and differences. It can be argued that this approach to the evaluation of data is time-consuming and not practical in a typical busy, company testing environment. This argument is, at best, a weak one because it fails to acknowledge the purpose for a test and the importance of making correct and supportable recommendations. Sensory professionals cannot afford to make recommendations about products that are not consistent with all the results. Also, the time required to run and analyze QDA data with the appropriate software is not likely to be more than 1 hour, a small enough time period to be confident when making recommendations. At this stage, a conclusion is reached about product differences, and a written document is prepared summarizing the results in both a technically and a visually acceptable format. Technical documentation typically follows a specific report format (appropriate for that company) that most often includes a listing of means and variability measures. However, this method of communication is rarely satisfactory because it is relatively easy for numbers to be misinterpreted.

This problem represented a challenge (to the authors), leading to development of the now familiar "spider webs" or sensory maps/pictures. The intention was to display results that did not rely directly on numerical values (Stone et al., 1974). As shown in Figures 6.5 and 6.6, the visual display is achieved by plotting the relative intensity values for the various attributes on a series of lines (or spokes) that radiate from a center point. These radiating lines can be envisioned as the attribute scales without the anchors. Currently, the most useful approach is to display attributes in groupings—for example, all the appearance attributes regardless of whether product differences were significant. The problem of deciding which attributes should be grouped could be resolved through factor analysis or some other procedure (see, for example, Petty and Scriven, 1991; Powers, 1988); however, subsequent testing of a related group of products could lead to quite different results, including significances for attributes that were previously nonsignificant. Some displays have included confidence levels. It should be clear that there is no one way that is best for displaying results. The primary objective is to convey, in as simple a way possible, the specific differences in a set of products. Visual displays such as the one shown in Figure 6.6 have proven to be extremely useful without compromising numerical significances. As a further step in the process of displaying and communicating product information, products can be added, one at a time, in a layering

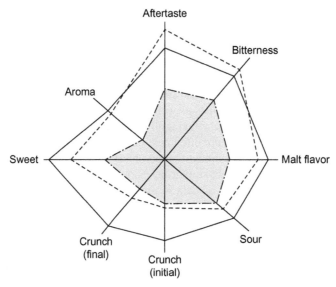

FIGURE 6.5

Visual display of the sensory attributes based on the results of the QDA test. For each attribute, the relative intensity increases as it moves farther away from the center point. See Figure 6.4 and text for additional explanation.

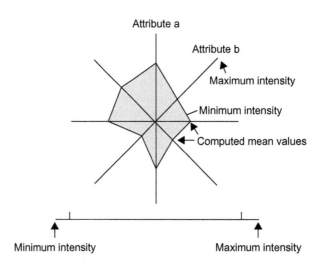

FIGURE 6.6

A schematic diagram with explanation of the procedure used to prepare the visual display (sometimes referred to as a spider web) of QDA results.

approach. By layering the plot, all possible combinations can be presented and discussed. Figures 6.7 and 6.8 are results from an evaluation of appearance and aroma and also mouthfeel and aftertaste attributes for cognac and brandy (the French and American designations, respectively). This display makes it easy to convey information to requestors, and the plotting program now includes options for displaying a significance designation for each attribute (Stone and Sidel, 1998). Also, by simply "clicking on the plot," one can add or remove any product quickly. To enhance the communication process, the research team is provided with the scorecard, definitions, and products, enabling the team to experience the results in a meaningful way. It is encouraging to note how frequently this approach is used to display sensory information other than that obtained using the QDA method.[1]

6.2.4 **Spectrum descriptive analysis**

This method was developed primarily from the Flavor Profile and Texture Profile methods, and a description can be found in Meilgaard *et al.* (2006; see also Rutledge and Hudson, 1990). The selection and screening for subjects follow those described previously for the Flavor Profile; that is, there are standard threshold, odor, and taste recognition tests along with personal interviews to determine the individual's interest in being a part of the panel. Discrimination testing was also proposed as an alternative or as an additional part of the subject screening process.

The training activities, as described, are quite extensive, reflecting the basic Flavor Profile and Texture Profile procedures, with particular reliance on the Texture Profile method of training subjects with specified standards of specified

---

[1]In 1974 Tragon co-founders Joel Sidel and Dr. Herbert Stone developed what has become the leading descriptive methodology for measuring a product's sensory attributes. In those earlier years, it never occurred to them to trademark or otherwise legally protect the Quantitative Descriptive Analysis (QDA) methodology. Nearly 30 years later, the term the methodology is most widely known by (QDA) has become synonymous with sensory evaluation. Tragon Corporation, therefore, found it necessary to seek a legal way of distinguishing its approach to the methodology (as originally developed) from the many off-shoots that exist today.

In 2007, Tragon filed for and received protection from the USPTO as well as the European Union for the following terms: Tragon QDA; Tragon Quantitative Descriptive Analysis; Quantitative Descriptive Analysis (QDA); and Tragon Quantitative Descriptive Analysis (QDA). Tragon licenses the above terms to companies that adhere to a strict licensing agreement and vigorously defends against any unauthorized use.

In 2011 Tragon released its first major version update to its industry leading descriptive analysis software in over 10 years. Developed to operate in the cloud (and as a Windows 8 Metro Page) upgrades include a GUI interface allowing for ease of use, panel performance metrics, ability to exclude data from the analysis, as well as the ability to easily restore data. It also creates principle component analysis (PCA) plots with ease. The software is ideal for key decisions of significance, interactions, correlations, and principal component analysis as well as resulting maps, panel performance, attribute performance, and product similarities and differences. Results are easily used in conjunction with other data for advanced multivariate analysis techniques. And because the entire application runs in the cloud, it is easy to access, share, and analyze data from any computer at any time.

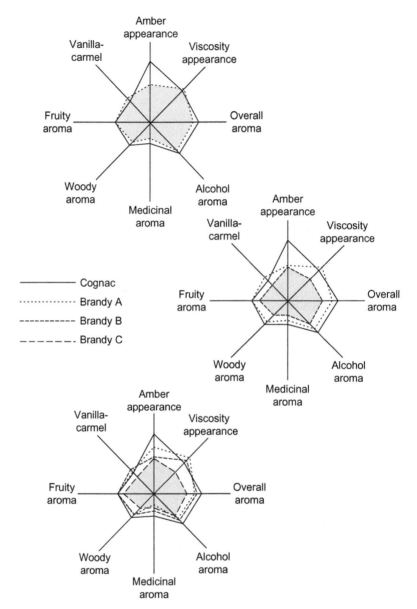

FIGURE 6.7

Appearance and aroma attributes used by a QDA panel to score the four cognacs and brandies. The top figure shows two of the products, the middle figure adds a third product, and the bottom figure allows for examination of all four products that were evaluated. See text for further explanation.

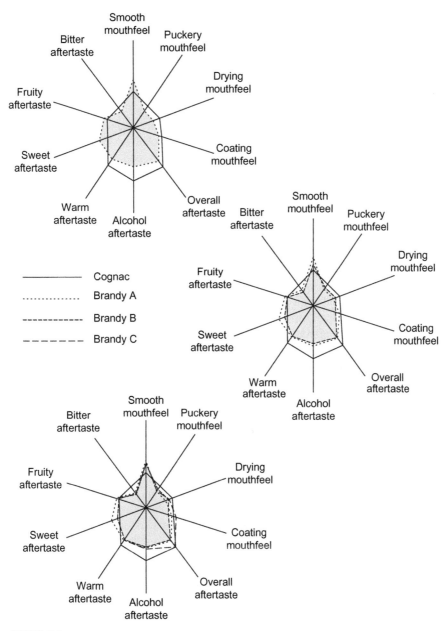

FIGURE 6.8

Mouthfeel and aftertaste attributes used by a QDA panel to score the four cognacs and brandies. See Figure 6.5 and text for further explanation of the display.

intensities. The training process is lengthy, requiring 6–8 hours per week for a period of 14 weeks, with 100 hours or more of training time per modality. This training time is described as necessary so as to enable the panel to be universal—that is, to be able to evaluate all types of products. Whether one can, in fact, train subjects to be qualified as an "expert" in all products is an interesting idea, but it remains to be demonstrated whether it can or should be done. It is also possible that this extensive training time is related to the use of reference standards and specific intensities for those standards in an effort to develop "universal scales."

Although the concept of universal scales may appeal to some, the scientific reality of individual differences in sensory physiology and behavior runs counter to the notion of universal perception and related scales. Boelens and Boelens (2002), discussing olfaction from the perfumers perspective, stated that "it is highly unlikely—in fact, practically impossible—that two human beings would describe the odor of a pure chemical compound in exactly the same way." We concur with Lawless *et al.* (2000, 2010), who observed that "it is widely believed, although rarely [if ever] demonstrated, that trained panelists in applied descriptive analysis techniques, such as the Texture Profile or the Spectrum Descriptive Analysis technique, can be calibrated and anchored to physical examples to stabilize and equate their scale usage." That subjects can be taught to associate a specific stimulus with a specific response demonstrates nothing more than a learned activity, a learned "trick." It may be an effective way for demonstrating repeatability of a behavior; however, it tells us nothing about what is perceived or the validity of the activity for evaluating products.

As noted in the introductory segment of this chapter, the variable nature of standards and the efforts of the subjects to adjust their perceptions to the assigned numerical values make it clear as to why the required training time is so long. The panel leader has an active and direct role in this training effort, which it is claimed is reduced as the panel develops its confidence in its task and requires less direction.

To evaluate an array of products, this previously trained panel goes through a language development process similar to that described for QDA but with considerable attention given to the identification and use of specific intensity references to represent the various attributes and intensities of each attribute. Here, too, the panel leader has an active role in directing what the panel does and what are acceptable responses. Depending on one's perspective, the strengths or weaknesses of the method are in the number of descriptive words and associated references taught to subjects.

Recommendations for test design, data collection, and analysis are similar to those already described for the QDA method, although consensus results are reported.

6.2.5 Free-choice profiling

Williams and Langron (1984) described a radically different approach to descriptive analysis in which no subject screening or training times were required and subjects could use any words they wanted to describe the products being evaluated. In addition, subjects' words were unique to each subject, as were their scorecards. This approach was presented as an alternative to the protracted time required to develop a language and to achieve agreement among the subjects using conventional descriptive methods. The authors claimed, in discussing other methods, that "complete agreement between assessors is often very difficult, if not impossible, to obtain." The issue of protracted time to achieve complete agreement is reasonable if all methodologies required 3 or more months as prescribed for the Flavor Profile, Texture Profile, and Spectrum Descriptive Analysis methods. The QDA method requires 2 weeks from screening (4 or 5 hours) through language development (8–10 hours), so the difference in timing is small, relative to free-choice profiling. Differences in data collection time depend on the amount of replication, which is essential if one is to provide reliable and actionable information. For an existing panel, there are no screening or language sessions. The second issue, concerning the need to have subjects achieve "complete agreement," is speculative, at best. As noted previously, it is highly unlikely that all subjects will be equally sensitive to all attributes. No data have been presented that support this statement. Such an achievement would enable each test to require only one subject (the N of 1). This particular issue has also been raised by O'Mahony and co-workers (1990) and Civille and Lawless (1986) in their discussions about concept development and concept alignment. This may be another example of a question that may be of interest to the academic community but has little relevance in practical or business terms. If subjects are identifying product differences on a repeated trials basis and these results help clarify consumer behavior and yield marketplace success, then perhaps the alignment issue is not an issue for descriptive analysis panels.

The free-choice profiling method, as described by the authors, requires the experimenter to spend time explaining the testing procedure to the subjects, including instructions on the scoring of the attributes. Unfortunately, the example used by these authors identifies 7 of the 10 subjects as individuals with years of experience tasting the product, in this instance, port wine. Therefore, the time advantage of the method may be nonexistent. It is interesting to note that soon after its publication, Marshall and Kirby (1988) described the method in terms of training time, which included a pretest of the subjects' texture vocabulary followed by 10 half-hour training sessions—5 hours. The individual results and the panel results are then analyzed using a generalized Procrustes analysis. This type of factor analysis analyzes data until an acceptable result is achieved, usually by excluding selected data. This procedure is used often, but it is not clear that researchers understand the implications of selectively excluding data until a satisfactory result is obtained. As Huitson (1989) noted, the analysis always produces a result that might not be justified based on examination of the database. Using a random number generator, this investigator obtained a result as good as that obtained by Marshall and Kirby (1988). Taking into account that the procedure does not yield attribute means (or variance measures) for products without training time, and the aforementioned question about the use of Procrustes analysis, it is difficult to see any advantage derived from this method.

6.2.6 Other methods

Other descriptive methods have continued to be described in the literature since the previous edition of this book was published—evidence of its continuing popularity and its value in making better informed product decisions. Most of these appear to be adaptations of existing methods, for example, to expedite the language development process as described in free-choice profiling, or providing subjects with feedback as to whether they were in agreement with the panel to minimize variability and/or eliminate subjects whose responses are not consistent with those of the panel. This latter approach is not new, but the ability to capture information electronically and provide almost immediate feedback based on a principal component analysis (PCA), for example, has considerable appeal. To date, the focus appears to be directed at changing subject response behavior to better conform to panel response behavior. It is claimed that by sharing this information with subjects, their behavior will change to conform with the panel. It is interesting that in a scientific field so much effort is invested in changing behavior rather than understanding behavior.

Cross *et al.* (1978) described a procedure whose objective was to provide more standardization for investigators measuring meat quality and at the same time "could identify certain textural properties of meat," but also would have more general applications. A comparison with the Flavor Profile and the QDA methods shows many common features in terms of subject screening and qualifying, the use of references, subject feedback during training, and data analysis. However, the method had the panel leader participate in product evaluation, and the attributes that were used were selected in advance by the experimenter without the opportunity to change based on subject input.

Larson-Powers and Pangborn (1978) described two approaches to descriptive analysis as part of a larger investigation on the scoring of sweeteners. The first test included the use of a reference product against which the other products were evaluated. An unidentified reference was also included as an internal measure of subject reliability, and a degree-of-difference scale was used with samples scored relative to the reference. The second test used a unidirectional line scale anchored from none to extreme. Unfortunately, the latter approach was used after four sessions with the anchored procedure so that a direct comparison of the two methods must be viewed with caution. The authors reported more significant differences with the anchored procedure. Because there were fewer responses with the latter method, it could account for the difference. The products were presented simultaneously, and some additional variability might also be expected (hence fewer significant differences). Although one may consider the presence of a reference as contributing to stability of the judgments, this assumes that the references do not change (from the start of testing to the end) or affect sensitivity, and that subjects will not be confused by the presence of this additional sensory input.

The introduction of so many additional sources of variability must be carefully considered relative to the reported advantage in subject skill and product differentiation.

For some product categories, such as salad oils, spicy foods, salad dressings, and cigarettes, the problems of sensory fatigue and the removal of residual flavors make this particular approach impractical. Whether this is superior to direct scaling with no references was not demonstrated in the study; however, it does identify further opportunities for research. In our experience, testing without a reference has always shown a higher degree of reliability and optimal differentiation between products, but only when qualified subjects are used. The results by Lyon *et al.* (1978) and Zook and Wessman (1977) using the QDA methodology were consistent with this conclusion. Nonetheless, research of this type must be continued if the descriptive methodology is to be optimized.

The need for modifications allowing statistical treatment of Flavor Profile data has been recognized since its introduction (Hall, 1958; Jones, 1958). Supporters of the profile method (Miller, 1978) addressed the issue of statistical treatment by suggesting a mathematical procedure for transforming profile data into numerical values. The reason for the transformation was to permit calculation of an index (Z) derived from amplitude and total intensity. The index was reasoned to provide a single value representing overall product quality. We strongly agree with the author's caution against using this index to oversummarize a complex sensory phenomenon, and we would add that the assumptions underlying this index, the method for transforming profile symbols, and the resulting computations should be treated with equal caution.

The developers of the method also discussed a new version of the methodology (Hanson *et al.*, 1983) called Profile Attribute Analysis. The method addressed the lack of scaling by incorporating 7-point numerical scales that permits statistical treatment of responses. This has long been needed because those still using the method have made numerous modifications to obtain intensity judgments. However, if the change is nothing more than the addition of a scale and a procedure for calculating a quality index, the method will continue to be of limited use as a descriptive tool. To reflect contemporary knowledge of behavior and measurement, consideration should be given to modifications in subject screening, language development, the evaluation procedure, and the role of the panel leader.

We previously noted that the Spectrum method is often positioned as a hybrid based on Flavor Profile, Texture Profile, and QDA; in a similar vein, Stampanoni (1993) introduced another hybrid of the two, which she refers to as "Quantitative Flavor Profiling." Several other authors, including Lawless and Heymann (1999), refer to generic descriptive analysis for those methods having elements of Spectrum and QDA, without formal training by the developers of the methods.

As mentioned previously, other methods focused on greater use of statistics for identification of subjects whose scoring patterns were inconsistent with the panel and for identification of attributes that were redundant (measuring the same sensation). The technical and statistical details of these procedures are described by Powers (1988). As already noted, it is logical to use various statistical procedures to analyze response patterns as a means of assessing subject agreement; however, such procedures have their own risks when they are used to eliminate subjects, particularly when done during training.

In the first instance, subjects will exhibit inconsistencies in their responses during training, not only within themselves but also between each other, and one does not need any analysis to recognize this. A purpose for the training sessions is to enable subjects to practice scoring products, to identify and to resolve, as much as possible, the words used to represent the sensations as well as conflicts in scoring. This resolution comes about through discussions led by the panel leader. For QDA training, the subjects typically score two products (one at a time) and then, with a show of hands, vote on which product is further to the right for each attribute. After one or two sets of products have been scored and the votes tabulated, disagreements are more easily observed and the panel leader can direct the subjects' attention to those attributes. For some subjects, the resolution of disagreements is easy, for others it takes longer, and some subjects may never resolve a scoring difference for some attributes. This lack of resolution for some attributes should not be used as a basis for removing a subject. That same subject may be the most sensitive on many other attributes. After all, during training the subjects are still deciding what attributes are needed, whether some will no longer be needed, and so forth. These questions cannot be answered with confidence until the range of sensory differences for that array of products has been evaluated (during training). Using responses obtained during training to identify subjects who are outliers and dropping them from a panel is not recommended. This could result in a group of subjects who agree with one another but are less sensitive to product differences. A panel leader must be careful to not use procedures that suggest there are correct responses, which will lead to development of errors of central tendency—to be safe, subjects will tend to score in the mid-range of the scale. Finally, some disagreements will arise whenever product differences are small.

After data collection, these multivariate analyses will be helpful. However, we observed that the simple one-way ANOVA, the individual product ranks, and the subject rankings provide considerably more information on an attribute-by-attribute basis about each subject than does a factor analysis. Not only is a direct measure of performance obtained for each subject but also one can determine whether a subject is causing interaction and whether that interaction is crossover or magnitude. As previously noted, this information has a direct impact on conclusions about product differences but should not be used arbitrarily to eliminate a subject so as to obtain a better result.

These same arguments develop when using multivariate analyses to eliminate attributes during training. It is conveying a message that there is a "correct" list of attributes, but the groupings of attributes (correlation of each attribute to each factor) may not be based on a causal relationship. Thus, the two or three attributes highest correlated to a factor could be selected, yet these attributes, in all likelihood, would not have been used by all subjects with the same degree of sensitivity to differentiate the products. There is no question that a scorecard with 50 attributes contains redundancy; however, all subjects are not equally sensitive to all attributes and the scorecard must reflect this diversity. In the QDA method, eliminating attributes is considered only after three or more tests to be sure that no product differences were obtained and most responses are at the far left of the scale; that is, little or none of that attribute was detected. Here, too, the subjects, not the panel leader, decide whether the attribute should be eliminated. In addition, the number and diversity of the attributes reflect the multivariate nature of the evaluation process and of the products. It is risky for a panel leader to focus on results from multivariate analyses without first taking into account more basic issues such as subject ability and performance criteria.

6.2.7 Experts and expert panels

As noted in the beginning of this chapter, the development of sensory evaluation and descriptive analysis occurred as a result of the difficulties and/or challenges of making use of expert information in today's increasingly complex and competitive marketplace. However, these experts and expert quality panels continue to function, in most instances, independent of formal sensory testing activities. In some instances, efforts are made to compare results from these panels with those from a sensory panel.

We begin with a brief explanation of what is meant by an expert or expert panel. These are individuals in a company with specialized sensory capabilities or members of an association recognized by the government as responsible for designating a product's quality. In food companies, these would be individuals who have sampled raw materials and finished products daily for many years, thus developing a keen sense of what a product from their company should taste like before it goes to the market. In the distilled spirits business, they may be referred to as "the nose" or some such designation. Their abilities are highly regarded, as is their impact on various technical matters (and, not surprising, they are compensated accordingly). Throughout the years, these individuals developed specialized knowledge about their company's products; they served as a primary basis for specific recommendations—for example, to purchase ingredients, to assess the impact of new technology on finished product quality, and to assess the marketability of a particular product (an area of some dispute). In the wine industry, there are groups who were responsible for sampling the wines of their region and awarding the owners the right to assign a particular designation to those wines. This ensured that the wines were of a certain quality, thus providing a seal of approval. For some markets and for some consumers, such designations were and still are very important for their marketing and sales efforts. For the most part, these individuals have been successful, and their actions served as the forerunner of sensory evaluation. The designation of expert was acquired through practice on a regular basis, working with an acknowledged expert who was usually older and more experienced. This practice (i.e. apprenticeship) was, and in most instances continues to be, done on an *ad hoc* basis—that is, through participation in the evaluation process with the awareness that their judgments will not necessarily be included in any decision until such time as determined by the acknowledged expert. The training varied from one individual to another, and the criteria by which an individual was certified also could change. The difficulty with this qualifying process was and still is the lack of a formal description of the qualifying criteria—that is, the scores expected for specific products, the numbers of tests and basis for qualifying or for continuing to train, and so forth. Despite these shortcomings, this system continues to be used in many companies; however, the primary role of the expert or apprentice is most likely associated with assessments of product quality and, secondarily, the initial stages of product creation. In effect, their impact has been lessened and more directed toward specific problems. The other challenge that these groups face has been the changing nature of the marketplace, leading to a variety of problems including declining sales; that is, designating a specific quality no longer equates with consumer preferences and sales. This does not mean that such practices should be discontinued; rather, their function needs to be changed to reflect how best the information should be used in today's global marketplace.

Experts functioned primarily in making judgments about products in terms of quality—a quality consistent with what they had been judging in the past. Each product category had a language and a focus, and not surprisingly, many of the words connoted quality or a quality defect. This assumed that a quality defect to an expert was a defect to the consumer—an assumption that has not been demonstrated. Sensory panels provide measures of the strengths of product characteristics or differences between products but do not directly assess product quality (discussed in Chapter 8). Comparisons of experts and sensory tests are usually not helpful for sensory programs and should be avoided because generally they are focused on different business objectives. Unfortunately, situations can occur in which the comparison is inevitable, and careful planning is warranted. The request to compare the two (often stated as using both tests to increase the product information base) arises because past sensory results suggest product changes were necessary and the technical group, for whatever reason, does not accept or believe the recommendation. Initial actions often include challenging how the results were obtained, whether the subjects were appropriate, etc. When these actions fail, the next course of action is to bring the expert or expert panel into the "picture." Once the actual decision is made, the sensory professional needs to vigorously engage the requestors in extensive dialogue to delineate and document the objective, including the criteria by which results will be judged, what products will be evaluated, and the actual evaluation protocol. Because both panels evaluate products from different perspectives, the results are not likely to be the same, and agreement must be reached with regard to how the information will be interpreted and applied. In most instances, the comparisons involve some type of descriptive analysis. Experts typically work from a list of terms that have been developed over many years, in association with their respective technical peers in research and product quality control. Most of these terms are technical in origin and represent existing knowledge about the chemistry, ingredients, and formulation of the particular product. In some instances, the terminology will include language obtained from an industrywide effort. It is interesting to note that experts usually have a primary modality that they use in their evaluations; for example, for wine it is often a list of 30 or more aroma words, some of which relate to defects and have not been discussed to eliminate the obvious redundancies in such a long list.

The QDA descriptive panel uses a language that is nontechnical; it is consumer based and encompasses all modalities. Thus, efforts to directly compare the terminologies will be difficult. In some of our unpublished research, we have had the opportunity to compare results from expert, consumer, and especially expert versus QDA panels. That research provided some useful findings.

Consumers, experts, and a QDA panel evaluated sensory attributes for an array of 18 beverages; all except three were commercially available at the time of the test. As a result, there was a wide range of product differences, a not unexpected finding. Although the primary objective was not comparing the two "panels," the database and technical interests necessitated such a comparison using some typical multivariate analyses. Separate PCAs of products across attributes demonstrated that the underlying grouping of the products and the attributes were different for the two panels. Experts organized products into two groupings, one defined by their company's products and the other by competitor products (products were unbranded and coded for these evaluations). The underlying structure appeared to be based on their perception of quality or, alternatively, recognition of their own products. The descriptive panel results provided more than two product groupings, which were defined by style (or flavor strength), independent of manufacturer. Product groupings for the descriptive panel were consistent with the consumer preference patterns.

It is not surprising that the results were different for the panels, especially because most expert panels (e.g. coffee, beer, and wine) are used for quality control purposes and, as a result, they have limited exposure to competitive products. Frequency and type of exposure to specific products will influence the way subjects perceive products, which will also influence their evaluations.

In addition to using the aforementioned data reduction techniques, regression equations were developed using results from the expert and descriptive panels and results from the physical and chemical analyses. We concluded that (1) multivariate analytical models predicted sensory attributes substantially better than did univariate models, (2) different models were required for the different panels, and (3) the model for the expert panel contained more variables than did the model for the descriptive panel. These results (and others obtained in similar tests) demonstrated that the same descriptor had different meanings to the different types of panels, even when the descriptor was for a basic taste such as sour or bitter. The different meanings appear to be based on different underlying constructs (or different physical components represented by the analytical measures in the prediction equations). For many of the attributes, the underlying perceptual structure for the experts was more complex than for the nonexperts, a not surprising conclusion. After all, the experts spend considerable time each day evaluating their company's products with knowledge as to the ingredients, the process, and what the products should "taste" like. Expert evaluations are accomplished from a different context than that used by the descriptive subjects. The latter have no technical knowledge about the ingredients or the process, and their evaluations are based on the context of the array of products, not what a particular product is expected to taste like. Perrin *et al.* (2008)

described results from a comparison of conventional versus free profiling using a panel of wine tasters with 3–6 years' training and testing experience compared with a panel consisting of wine makers, enologists, and merchants. Not surprising, the authors concluded that very similar words were used to describe the wines; however, the authors also concluded that the free profiling did not "indicate the subtle differences among products as in conventional profiling, since data is not sufficient to perform analysis of variance." So what would be an advantage to such a method? The power of descriptive analysis is to provide information about a product's differences to better inform technologists and marketers. To determine that wine tasters, enologists, and wine merchants can agree with each other about the taste of a wine is interesting, but knowing what style of wine the consumer likes is far more important.

As previously mentioned, much of the focus in recent years has been on reducing the time to develop a language to evaluate an array of products. In fact, one could significantly reduce the time by providing the list of words to the subjects based on a literature search. Empirically, it can be shown that subjects will complete such a scorecard and provide information that can be presented to the requester. Because other research has shown that subjects need practice and dialogue to be sure that their judgments are based on the same criteria, the issues of reliability and external validity cannot be determined. The only way to know is for the technologist to make recommended changes and see what happens in the marketplace. The current product failure rate is greater than 90%, and it had been much higher before formal sensory evaluation programs were developed, so increasingly faster methods with no measures of reliability and validity does not seem to be the best direction to be heading.

6.3 Applications for descriptive analysis

There are numerous applications for descriptive information, including monitoring competition, storage stability/shelf-life, product development, quality control, physical/chemical and sensory correlations, and advertising claim substantiation, which will be described in more detail in Chapter 8. Before briefly describing them here, it is useful if we reemphasize the role of the panel leader in deciding how best to use this resource.

Once the decision is made that a descriptive test is appropriate, it is necessary to plan the test, bench-screen the products in the context of the test and the project objectives, and then decide on the size of the database that will be developed. In product development, a large array of products (e.g. 10 or more) could be evaluated in a single test with less replication (three) than if there were only three or four products to be evaluated. The number of replicates would be sufficient to assign a level of confidence to the results and the conclusions and to be able to differentiate the products based on the panel leader's judgment. As a result of formulation efforts, new versions of the products would be submitted for evaluation; however, there would likely be fewer of these products. For this latter test, more replication would be incorporated in the expectation that the differences would be less obvious and that the relative importance of the decisions would be that much greater. Similarly, a storage study might involve only a few products, but in each test a high degree of precision, as well as reliability with each result, would be very important. Therefore, additional care would be taken in the test plan to ensure there was sufficient replication so that any comparisons would be made from a larger database. In addition, individual subject performance would be critical to the decisions about specific product changes, and therefore the replication per subject would also be important.

The list of applications should be considered as a general guide; a similar list could be prepared for discrimination and affective tests. For the sensory professional, the decision regarding what methodology (discrimination, descriptive, or affective) will be used is addressed once the problem is understood and the objective has been stated. Because the output from a descriptive test is so extensive, there is a tendency for it to be considered for all applications; however, this should not be done to the exclusion of the other types of test methods. The following list should be considered in that context:

1. *Monitor competition.* It is especially important to know in what ways competitive products differ; such information can be used to anticipate changes and

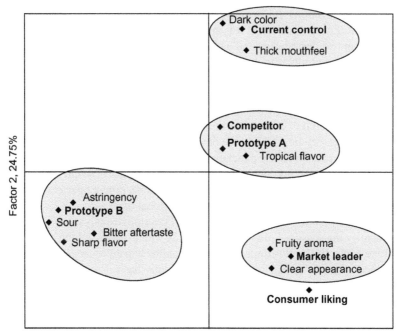

FIGURE 6.9

A map of sensory and preference differences. The two factors accounted for 69.91% of the differences that were measured.

to identify product weaknesses. Without such information, it is relatively easy to initiate product reformulation efforts based on circumstantial evidence that a change in market share reflects a changed competitive product. The descriptive information provides a primary basis and more precise direction for any proposed changes. In addition, one can map the relationships of preferences with specific attributes (Figure 6.9) to gain more insight to how products are "perceived" by the consumers relative to their sensory differences.

2. *Storage testing.* In a storage test in which product changes occur over time, descriptive analysis at the start of the test provides a basis on which changes can be compared. A major problem in all storage tests is the availability of control product throughout the study. Product placed in some type of controlled environment, such as a freezer, does not prevent change, and providing fresh product for comparison with the stored product introduces other sources of variability. The QDA method eliminates the need for the control product. This particular application and the following three are all discussed in more detail in Chapter 8.

3. *Product development.* Descriptive analysis is used to delineate a target product, determine whether experimental formulations match that target, and provide precise sensory information about the finished product. This latter function could be used to support advertising as well as be of value in setting quality control specifications. It can also be used in evaluating the usefulness of a new ingredient. For an example of this application, see Clark (1991). Chapter 8 of the current text describes use of a trained sensory panel to provide a descriptive analysis of the products included in optimization research. The statistical analysis relates the consumer, trained panel, and analytical (if included) data sets to determine combinations of sensory and analytical attributes that best predict optimal consumer liking. The descriptive panel is also used for the follow-up research to evaluate test products until one or more satisfy the optimized sensory target. With the development of portable data collection devices, it is easy to have a panel evaluate products in typical use situations. Having a panel take product home or to a typical product use environment allows for considerably more detail regarding how a product performs. In fact, we use this approach to data collection on a global basis.

4. *Quality control and sensory specifications.* Descriptive analysis could be used to identify the sensory limits for a product as well as to track long-term trends (Stone *et al.*, 1991). It is impractical as an online capability because of the large numbers of products that are evaluated on a daily basis in a manufacturing facility. However, it can be adapted for use in quality control applications (discussed in Chapter 8).

5. *Physical/chemical-sensory relationships.* Descriptive analysis is especially helpful in identifying specific product differences that can be related to differences in various instrument and chemical measures. Once identified, these differences can be explored in more detail, again using the descriptive model (discussed in Chapter 8).

It is possible to identify other applications for descriptive analysis; our primary goal here is to increase the reader's awareness of the options.

6.4 **Conclusions**

In this discussion about descriptive analysis, the development of the methodology has been traced from the use of a product expert through the more formal and rigorous approach applied with methods such as QDA. It is reasonable to expect that there will continue to be changes and growth in the methodology and in its applications. In particular, emphasis must be placed on the importance of the quantitative aspect of descriptive analysis.

Quantitative descriptive methods are not intended to solve problems in an hour. Such problems, in all likelihood, require neither descriptive analysis nor any other sensory test. Descriptive methods were designed to analyze products with a high degree of reliability and precision. Although the opinions of experienced staff obtained in bench screening may provide clues to a particular problem and in some instances will solve a problem, this would not constitute a basis for rejecting formal and rigorous sensory analysis of a problem. Similarly, efforts to replace the trained panelist with the consumer, as suggested by Cardello *et al.* (1982), do a disservice to both sensory evaluation and consumer testing. Neither is intended as a substitute for the other, and it is quite risky to confuse their respective tasks and their skills. These latter investigators observed that there were clear differences between the two types of panels, a not unexpected finding. Similarly, and equally not unexpected, differences can be expected when different descriptive training methods are applied (Lotong *et al.*, 2001). This issue is not trivial for the sensory professional, and it requires a clear understanding of the test objective and an appreciation for those screening, training, testing, and data analysis procedures most likely to achieve the test and business objective.

Descriptive analysis is a sophisticated concept in the sensory analysis of products. It has evolved from the opinions of the expert to a more rigorous and scientific approach to measuring perceptions. The QDA methodology, in particular, emphasizes the behavioral basis of the judgment process, and combined with a rigorous approach to assessing response reliability, it provides a level of sophistication heretofore not possible in sensory evaluation. However, it would be naive to assume that descriptive analysis is the only category of test methods or that the methods are optimal. Certainly in the case of the QDA method, changes continue to be made as more experience is gained in the use of the method as well as from new developments in measurement and related fields of investigation.

Continued and successful application of descriptive analysis requires an appreciation for the basic principles and test guidelines described herein. In our experience, this will result in a higher level of awareness of the methodology and a greater role for sensory evaluation in the decision-making process.

Affective Testing

7.1 Introduction

Acceptance testing is a valuable and necessary component of every sensory program. In the product evaluation process, acceptance testing usually, but not always, follows analytical sensory testing using discrimination and descriptive analysis. Analytical results can effectively reduce the number of product alternatives to a more limited subset. The selected subset should have demonstrated sensory

H. Stone, R. Bleibaum, H. A. Thomas: Sensory Evaluation Practices, fourth edition.
DOI: http://dx.doi.org/10.1016/B978-0-12-382086-0.00007-8

differences that can be recommended to proceed to larger scale consumer testing, generally conducted by marketing research or consumer insights. This evaluation or screening task is referred to as acceptance, preference, guidance, or consumer testing—labels that can have different meanings depending on one's experience and professional responsibilities. From a sensory science perspective, acceptance testing should have a specific meaning with regard to the research objective, methodology, participant qualification criteria, and key deliverables.

By acceptance testing, we mean measuring liking or preference for a product. Preference is that expression of appeal of one product versus another. Preference can be measured directly by comparison of two or more products with each other; that is, which one of two or more products is preferred. Indirect measurement of preference is achieved by determining which product is scored significantly higher (more liked) than another product in a multiproduct test or which product is scored higher than another by significantly more people. There is an obvious and direct relationship between measuring product liking/acceptance and preference. To be most efficient, sensory science should emphasize measuring product liking/acceptance in multiproduct tests and, from these data, indirectly determine preference. Scaling methods allow us to directly measure degree of liking and to compute preferences from these data. Although the reverse is possible (Guilford, 1954), it is more complex, time-consuming, and costly. Later in this chapter, we further discuss the relative merits of these two measures of acceptability. The hedonic continuum is another frequently used expression for product liking and may be considered the more generic representation of the affective process. As Young (1961) noted, the hedonic continuum represents "the sign, intensity, and temporal changes of affective processes," and it is these changes for a set of products that we are interested in measuring. The hedonic continuum is discussed in more detail in the section on specific methods.

The measurement of unbranded liking is a logical and necessary step in the development process before substantial capital has been invested in equipment, packaging, production, distribution, advertising, and so forth. Branded versus unbranded testing is a common debate between sensory scientists and marketing. At what point should brand be included to help determine potential success in the marketplace? Unbranded sensory testing has been challenged by marketing and sales, stating that it is unrealistic because the product will be purchased by consumers with the full experience (Garber *et al.*, 2003). However, one of the values of sensory research is to better understand potential for repeat purchase. This unbranded research allows the sensory scientist to provide critical feedback to product development on the sensory experience of the product, without the biases inherent in the marketing materials.

We would be very reluctant to invest in a product if we knew that it was not liked because of a sensory deficiency. Therefore, we organize a test methodology that will give us an estimate of product acceptance based on what Cardello and Schutz (2003) refer to as its intrinsic (e.g. ingredient and process related) sensory properties. This measure of sensory acceptance does not guarantee success in the marketplace because extrinsic variables (Schutz and Cardello, 2003) such as packaging, price, advertising, and market segmentation will have an effect. However, it does provide us with a good indication of the product's potential as a product without any other of these accompanying features that we expect will enhance its acceptance in the marketplace. Also, the sensory acceptance test does not infer share of market because this topic is beyond the scope and responsibility of sensory evaluation. In that sense, the sensory response is a passive measure; that is, no action is inferred from the respondent. Because most sensory acceptance tests involve limited numbers of consumers, they cannot represent all segments of interest to marketing—it would not be strategically useful to the business. It is possible to obtain a more action-oriented measure by use of the food action rating scale (Schutz, 1965). This scale requires the subject to estimate frequency of consumption (or use) of a product. Once more, the information will be helpful as one proceeds from the smaller scale to the larger scale testing models.

Sensory acceptance can also and should be measured as a function of product usage in a normal consumption situation, as would occur in a home-use test. If development efforts have yielded a large number of formulation alternatives, it is a responsibility of sensory evaluation to identify which product(s) warrants further attention—that is, which one (or more) most closely matches the project objective. In addition to reducing the number of alternatives, the results of the test will also provide an indication of expected reaction with consumers relative to commercially successful products. This latter information can be used to prevent larger scale tests or at least alert project staff that these tests may not be warranted without reformulation. The value of sensory acceptance testing is obvious, and over time a reliable and valid database should be developed for key products and their competition. With such a database, acceptance norms can be established for key products, further enhancing the usefulness of the sensory acceptance model.

It is also important to emphasize that the sensory acceptance model is not a substitute for larger scale consumer tests nor is it a competitive alternative. The sensory test should not be used in place of the larger scale consumer test when the latter is needed. A frequent misconception is that use of one type of test will replace the other. Often, however, other factors influence the decision as to use of the sensory acceptance test. Of these, the most important are budgetary and timing. Sensory acceptance tests can be organized and fielded at a fraction of the cost and time of larger scale tests. When budgets are restricted or are depleted, the temptation to substitute the sensory acceptance test (for the larger scale test) is very strong. This is a more common occurrence as a sensory acceptance database is developed and its credibility is established. Its very success can and will lead to some misuse.

Of course, there will be situations in which the sensory test is the only test that can be done—for example, the product's sales cannot justify the expense of the larger scale test. It is a responsibility of the sensory professional to review each request, to ensure that the specific problem warrants use of the sensory acceptance test, and to provide a realistic framework for any conclusions derived from this latter, smaller scale test. Through communications with requestors and with brand managers, sensory should make every effort to limit the extent to which there is misuse. The sensory acceptance test is a small-panel guidance test—that is, usually 50–75 participate—whereas the marketing research test involves more than twice that number, often in more than one geographic location. The sensory test focuses on product and determining the best product(s) for the consumer target, whereas marketing research focuses on populations and identifying consumers to whom the product will have the most appeal and also developing the means to influence those consumers (Carter and Riskey, 1990; McDermott, 1990). Clearly, these are two separate but related activities that complement one another and rely on different testing procedures and different research objectives.

In subsequent sections, we describe the various types of acceptance tests—central location, home use, and so forth—within the context of accepted sensory evaluation principles and practices. This information should make it clear as to the differences between sensory acceptance testing and market research acceptance testing and the means by which they complement one another.

Considering the relative value of this information, it is surprising that there is controversy with regard to who should do this testing. In some companies, sensory evaluation is not allowed to do acceptance testing; such testing may be mistakenly positioned as consumer testing and designated as the responsibility of marketing research. If sensory evaluation intends to use employees as the subjects, management may argue that the employees are not representative of the target consumers or are biased, and therefore sensory professionals cannot do acceptance testing. The most frequently cited example is the result from a sensory acceptance test that was reversed when the products were evaluated by "real" consumers. Implicit in this statement is the notion that a test to measure acceptability with consumers is always valid but a test to measure acceptability with employees is not valid. Of course, results from two tests can differ for many logical reasons: products were manufactured at different times or locations (production vs pilot plant), test procedures, criteria for statistical significance were different, scales differed (a less critical issue), and so forth. In some instances, statistical significance is achieved in one test but not in the other; however, this result should not be surprising because the numbers of consumers tested differed. Obviously, these problems must be given a professional review, and their resolution must be brought to the attention of management. Although these concerns may seem trivial, for long-term development of one's sensory program, such differences must be resolved. Sensory professionals must demonstrate, to their own and to their management's satisfaction, that the procedures yield reliable and valid information. For selected competitive products, there should be consistent response patterns, and results should be in agreement with results previously obtained. For example, if one product is consistently scored significantly more acceptable than another in central location tests, a similar result should be expected with an employee panel. We discuss the issue of comparative testing as part of the sensory evaluation strategy for establishing and maintaining the sensory acceptance testing resource.

Alternatively, sensory may use local residents to overcome the criticism of using employees but still find that results lack credibility. Often, this can be traced to a lack of understanding of the uses and limitations for sensory acceptance information, which derives, in part, from confusion about marketing decision criteria. Almost all companies have developed management-approved consumer testing guidelines as a means of standardizing procedures and the decisions based on the results. These guidelines stipulate the number of consumers, the research methodology, the number of locations, and so forth, all of which are intended to minimize risk in the decision-making process and to enable managers to more easily make comparisons. Unfortunately, over time these guidelines undergo a metamorphosis into inflexible rules with no easy means of change to take into account advances in measurement, methodology, or in recruiting the right consumer. Often, the decision criteria (developed several decades before development of appropriate scales or recruiting criteria) are so onerous that product changes are very difficult to achieve, and this leads the clever manager to select products and test protocols that improve the odds for success. The situation becomes even more complicated by the unwillingness to remove questions that are no longer relevant or actionable or obtaining responses to sensory attributes that are best obtained from a descriptive panel. As a result, questionnaires become cumbersome and the results are less sensitive, thus making comparisons even more difficult. Although it is not our purpose to analyze these practices in detail, it is important to make sensory professionals aware of the complexities and the challenges that can help explain, in part, the issue of credibility for the sensory acceptance model.

In this chapter, we describe acceptance testing using the same format that was used in the two previous chapters. As a part of this discussion, we also address a number of issues inherent in the relationship that exists among sensory evaluation, marketing, market research, and consumer insights. These issues include the inappropriate application of sensory test methods, the use of large numbers of respondents, and the use of multiple-page scorecards. In each instance, the potential for conflict or problems in interpretation can be minimized through development of a sensory approach to acceptance testing based on the guidelines described herein. Experience has demonstrated that sensory acceptance testing is cost-effective and a very useful screening tool before large-scale project commitments are made. If the sensory professional staff has established and implemented the guidelines by which the sensory program operates, then sensory acceptance testing will very likely receive no more criticism than any other methodology.

There can be severe consequences for a sensory program in which the types of consumer tests appropriate for the science are not clearly delineated, the value of the sensory affective test is not clearly understood, or the guidelines for those tests are not followed. Consumer testing is not an allowed activity for some sensory programs, whereas other programs have been assigned to business groups that neither appreciate nor recognize the broad range of support that sensory evaluation provides to R&D, marketing, and manufacturing. In some cases, this has led to the eventual dismantling of some sensory programs.

7.2 **Methods**

The two methods most frequently used to directly measure preference and acceptance are the paired-comparison test and the 9-point hedonic scale, respectively. Other methods are described in the literature, but many of them either are modifications of these two methods or are types of quality scales—for example, excellent to poor and palatable to unpalatable. We discuss some of these methods later with particular emphasis on those that represent opportunities (Cardello and Schutz, 1996; Lim, 2011; Schutz and Cardello, 2001) and those that represent risks for sensory evaluation. The review by Lim provides a summary of the merits of the various methods that have been used as well as current developments in measuring liking.

Before proceeding, the reader may find it useful to review the material on measurement presented in Chapter 3. The guidelines and rules regarding the types of scales and the permissible mathematics are equally applicable in acceptance testing. In addition, the admonition to not combine different sensory tasks, such as discrimination with preference, is especially applicable. Not only will the odd-sample bias be especially strong but also subjects for sensory acceptance testing are not screened for their sensory acuity nor trained, are relatively naive about the sensory task, and may be used more frequently than consumers in marketing research tests. In the larger scale marketing research test, consumers are current or potential users and are qualified if they have not participated in any testing in the previous 3 months (the time will depend on the particular test specification), meet various demographic and product usage criteria, and, in some situations, are concept acceptors.

In addition to the combining of difference with preference, having subjects evaluate specific product characteristics as part of the acceptance-preference test is frequently done, but it also is not recommended. The evaluation of sensory characteristics, whether for intensity or for acceptance–preference, is quite risky for sensory evaluation because the test population is often too small to compensate for semantic differences. Also not recommended is requesting the reasons for a respondent's decision—that is, what he or she liked or disliked about a product—in an open-ended format. As Payne (1965) observed more than four decades ago, open-ended questions are generally not worth the effort. Subjects are very likely to justify their decisions, and this is often a recital of advertising information or their expectation of what the product should be like. Although such information may have meaning for marketing and advertising, it is generally inappropriate for sensory evaluation.

The problem with the combining of tasks in a single test is not that a subject will refuse to complete the scorecard or indicate he or she does not know how to respond but, rather, it is highly likely that responses will be confounded. Each response may consist of a component that represents intensity and one that represents affect, and each subsequent response is influenced by the response to the preceding question (often referred to as a halo effect; see Chapter 4 for more information on this effect). This confounding typically results in higher than expected variability (and decreased sensitivity). We further discuss these issues later; however, our point here is to emphasize the importance of planning and limiting sensory acceptance tests to measure only acceptance (or preference).

7.2.1 Paired comparison

The paired comparison is probably the first formal sensory test method developed to assess preference (Cover, 1936). Therefore, it should be no surprise that there is extensive literature about the method and especially on the topics of test design, statistical analysis, and mathematical models to help explain choice behavior in the paired-comparison situation (Bradley, 1953, 1975; Bradley and Terry, 1952; Day, 1974; Gridgeman, 1955a, 1959). The test may involve one or more pairs of products, and the subjects may evaluate one or more pairs of products within a single session.

The method requires the subject to indicate which one of two coded products is preferred. A frequently used option allows the inclusion of a "no preference" as a third choice, whereas another option allows inclusion of a fourth choice, "dislike both equally." Examples of types of scorecards with instructions are presented in Figure 7.1. The product setup for the paired test is demonstrated in Figure 5.2. The test is relatively easy to organize and to implement. The only two orders of presentation are A–B and B–A, and subjects usually evaluate only one pair of products in a test with no replication. The preference test is usually a two-tailed test because *a priori* we do not have knowledge regarding which product is preferred. However, there will be situations in which the interest is only in the preference for one of the products, in which case the analysis will be based on a one-tailed test. The hypothesis assumes one of the products will be preferred.

Analysis of the results is accomplished using the same computations described for discrimination testing, with $p = 0.05$, or use of Tables 5.11 and 5.14. In advance

Name _____Code _____Date _____

Option A
  Evaluate both products starting from the left. Check the box for the product you prefer. You must make a choice.

  347 ☐   602 ☐

Option B
  Evaluate both products starting from the left. Check the box for the product you prefer. You must make a choice.

  347 ☐   602 ☐   No preference ☐

Option C
  Evaluate both products starting from the left. Check the box for the product you prefer. You must make a choice.

347 ☐
602 ☐
Like both equally ☐
Dislike both equally ☐

FIGURE 7.1

Example of the scorecard for the paired-preference test, showing option A, which limits the subjects to two choices; option B, which includes a no-preference choice for the subject; and option C, which includes two additional choices.

of a test, it is possible (and recommended) to compute the necessary number of respondents and the number preferring a particular product for a given confidence level. However, it does require the sensory professional to make assumptions about subject sensitivity and the magnitude of the difference. The computations, described in Chapter 5, are derived from Amerine *et al.* (1965), Radkins (1957), and Roessler *et al.* (1978). These authors provided a statistical basis for considering the risks and especially the risk associated with a Type 2 error in the preference test model. Although we have emphasized this issue in the discrimination model, it is equally important in the paired-preference test.

If the test design permits a no-preference decision or the additional category of dislike for both (options B and C, respectively), then the analysis will not be as straightforward as the binomial paired-preference model. Gridgeman (1959) observed that the inclusion of ties (no-preference choice) in the preference test is useful in providing more information about the subject's reactions to the products. However, with small numbers of subjects, permitting ties reduces the statistical power (i.e. reduces the likelihood of finding a difference). Whether to permit the no-preference option is not always an easy decision. In the sensory test situation in which the number of subjects is usually fewer than 50, the no-preference option is less desirable. The more subjects who do not or cannot make the choice of prefer A or prefer B, the smaller will be the database and the larger the difference in prefer-ence needed in order for it to be statistically significant. In tests involving a larger number of subjects (100 or more), a substantial number of no-preference decisions will be important, for example, indicating that the products are equally preferred. Within the sensory community, especially within ASTM E18, there continues to be discussion and revisions regarding the no-preference option and the appropriate data analysis of the no-preference ratings. The reader is referred to ASTM E1958 on sensory claims substantiation for a discussion of the no-preference option and its impact on legal and advertising claims.

Use of the additional category (dislike both equally) is an option included to ensure that the products are liked; that is, A could be preferred over B, but both could be disliked if the subject's choice was limited to a single no-preference cat-egory. The statistical treatment of results from the test could be based on the rea-soning of the previous option (ties permitted); that is, the analysis would be based solely on the number of preference responses for the two products as recommended by Gridgeman (1959). The expectation would be for minimal numbers of respond-ents in the "dislike-both-equally" category—for example, approximately 5–10%. When this test plan is used, the proportion of subjects expected for the dislike-both-equally category, as well as for the no-preference category, should be computed for comparative purposes. Of course, having a large number of responses in the dislike-both-equally category should raise questions regarding the appropriateness of the products for that test or, more important, questions regarding the subjects' qualifi-cations for that particular product.

Marketing research has made the greatest use of the paired-comparison prefer-ence method, either as a two-product test or as a much larger effort involving many products, which is referred to as multiple paired-comparison tests. The appeal of the test method is quite strong because of the unambiguous design; it eliminates transitivity, and the ease with which various marketing hypotheses can be assessed. For example, given different product positioning statements, will preference shift? From a sensory evaluation perspective, however, the paired-preference test (either as a single pair or as multiple pairs) is less informative and less efficient than direct scaling using methods such as the 9-point hedonic scale. The test is less informative because the response (prefer A, prefer B, no preference) provides no direct measure of magnitude of the preference; both products may be disliked. The test is less efficient because there is only one response for each pair of products, compared with a scoring method that yields one response per product. Although the Scheffé method of multiple paired comparisons (Scheffé, 1952) yields magnitudes of preference in a multiproduct test, there is still a single judgment per pair of products.

In multiproduct food and beverage tests, there will be considerable sensory interaction because of flavor carryover from one product to the next. This can be a serious deficiency for the paired-preference model, in which simultaneous (side-by-side) product presentation is the most typical paradigm. Alternatively, memory may become a confounding variable if a sequential presentation is selected for the paired test.

In the multiple paired test, it is possible to minimize the sensory interactions through control of the time interval between pairs. Whereas the time interval between pairs can be controlled, the extent to which the subject samples and re-samples each product cannot. As a result, there can be greater indecision, leading to a greater percentage of no-decisions, which could work to one's advantage in certain circumstances (e.g. competitive advertising). Multiple paired comparisons are also especially cumbersome from a testing standpoint and particularly with regard to product handling errors. For example, there are six pairs for a four-product test, 10 pairs for five products, 15 pairs for six products, and so on. The time required for the test is also extended. Unless one is prepared to have these consumers participate over a longer time period (several hours or on several days), then the alternative is to use incomplete designs.

Statistical analyses of multiple paired comparisons, including the ability to convert the comparative responses to partially metric/order data and to undertake a variety of multidimensional analyses, are described in the literature (e.g. Greene and Carmone, 1970; Nunnally, 1978; Schiffman *et al.*, 1981). With these resources, it is not surprising that the paired test has remained the method of choice for individuals in marketing research, advertising, and so on. Many of the methods for analysis are based on the binomial theorem and (except for the Scheffé procedure) do not permit the inclusion of no-preference ties. Baumgardner and Tatham (1988) recommend a trinomial approach (and Z test for significance) in which no-preference is permitted. They concluded that the trinomial approach is fundamentally (more) correct, even though they demonstrated little difference in Z scores when applying a trinomial approach or using a binomial distribution and eliminating the no-preference judgments. The statistics literature also contains numerous references to multinomial discrete distributions for those situations requiring them. This is a reminder to the individual contemplating such tests to ensure that the design and analysis are consistent with the model.

In this discussion, considerable attention has been directed to characterizing the relative value of the paired comparison model for sensory acceptance measurement. Despite weaknesses, the paired test is useful in certain situations. If an advertisement claiming a preference based on a direct comparison is contemplated, then use of the paired comparison is necessary. The paired test is also useful in situations in which the chemical senses (taste and smell) are not involved—for example, evaluation of a large array of visual or tactile stimuli.

If there is an interest in obtaining a direct measure of acceptance, then the use of a scaling method is appropriate (Pilgrim and Wood, 1955). We now direct our attention to some of these methods, with particular emphasis on the use of a 9-point hedonic scale.

7.2.2 Hedonic scale

For measuring product liking and preference, the 9-point hedonic scale is probably the most useful sensory method. As noted in Chapter 3, the method occupies a unique niche for sensory evaluation. Since its development (Jones *et al.*, 1955; Peryam and Haynes, 1957), it has been used extensively with a wide variety of products and with considerable success. An example of the scale is shown in Figure 3.5. The scale is easily understood by naive consumers with minimal instruction, results have proven to be remarkably stable, and product differences (in liking) are reproducible with different groups of subjects. Not surprisingly, the 9-point hedonic scale is used extensively by many companies with considerable success in terms of the reliability and validity of the results. Because most of these studies are not published (but are discussed informally at conferences), questions about the scale continue to be asked or studies are presented purporting superiority of or to some alternative method. For example, in carefully designed studies comparing the method of magnitude estimation with the 9-point hedonic scale, Moskowitz and Sidel (1971) and Warren *et al.* (1982) independently concluded that magnitude estimation was not superior. In contrast, a study by McDaniel and Sawyer (1981) suggested that magnitude estimation was more sensitive because it yielded more significant differences. However, test design limitations raise doubts as to the significance of that conclusion. In recent years, interest in these comparisons has not demonstrated any evidence for the superiority of the magnitude estimation method, and most professionals make use of the 9-point hedonic scale.

Criticisms of the hedonic scale method are primarily the same issues that are directed to many methods: the use of parametric methods for analysis with a scale that is bipolar, the lack of definitive evidence of the equality of the intervals, and avoidance of the neutral category (Day, 1974). These criticisms were discussed previously (see Chapter 3); however, the practical value of the method continues to be demonstrated. The reader is reminded that the issue of bipolarity, for example, is a technical one and not a consumer issue; that is, the consumer is unaware of this and uses the scorecard in a manner that suggests a scale of continuous measurement of liking. The explanation for this is probably manifested in the simplicity of the subject's task (checking a box for the appropriate expression) and the absence of numbers on the scorecard that might otherwise confuse or bias the subject. However, this ease of use must be preceded by an explanation given to the participants before the actual evaluations begin of how to use the scale.

For sensory evaluation, the results from use of this scale are most informative. Computations will yield means, variance measures, and frequency distributions, all by order of presentation and magnitude of difference between products by subject and by panel, and the data can be converted to ranks as well, which yields product preferences. Vie *et al.* (1991) recommended calculating R-indices for 9-point hedonic scale data. Additional information about product differences is obtained from the analysis of variance or the t test, depending on the number of products and number of responses per product per subject. When one considers all of this information relative to the information from the paired test, the usefulness of the former method to sensory evaluation is understandable. There is growing evidence of the labeled affective magnitude (LAM) scale described by Schutz and Cardello (2001), or some similar labeled magnitude scale, that may prove to be a significant improvement to the original 9-point hedonic scale. This needs to be confirmed through time and with broader use by sensory professionals, but it has been demonstrated that the LAM scale may have a slight advantage over the 9-point hedonic scale, especially for well-liked products. This scale is anchored with the high end as "greatest imaginable like" and the low end with "greatest imaginable dislike." There is discussion that these extreme anchors will result in compression of the ratings and alternative anchors such as "greatest imaginable like/dislike for products like this" would be better to encourage greater use of the scale categories. These researchers continue to explore this topic as a means for extending our knowledge of consumer preference behavior (Cardello *et al.*, 2008; Lawless *et al.*, 2010).

7.2.3 Other methods

In addition to the 9-point hedonic scale, other scoring methods have been used and continue to be used to measure product liking and preference. Many of the most common of these methods have been described by Amerine *et al.* (1965), Lawless and Heymann (2010), and Lim (2011). Many scales we have encountered tend to be associated with particular product categories or with a specific company. For example, meat product evaluation usually relied on an 8-point scale measuring product desirability or an 8- or 10-point product quality scale (excellent to poor); other products also made use of such scales (and continue to be used). Some companies have developed scales that they believe are "special"—ones that they consider as proprietary but unfortunately do not have any scientific basis. Another variation uses the 10-point numerical scale anchored at one extreme with a frowning smiley face and anchored at the other extreme with a happy smiley face. The ends of the scale, 1 and 10, are also defined with word extremes such as "doesn't appeal/very appealing" for appearance and "particularly bad/particularly good" for taste. Other examples include categories such as the best "food" ever, etc. These are nothing more than poorly disguised quality scales that are inappropriate for measuring acceptance-preference. Quality scales, whether devised to establish an excellent-to-poor or a palatability-to-desirability measure, should never be used with untrained subjects (i.e. consumers) let alone used with trained subjects; such individuals are highly influenced by each participant's attitude about what constitutes "excellent" quality or the more cognitive concept of desirability. The weaknesses associated with the use of quality scales in general have been discussed previously in Chapters 3 and 6, and the interested reader should review that material.

Another alternative is to rank products from best liked or most preferred to least liked or least preferred. Although these are among the least costly and most efficient multiproduct tests to administer, they involve the same compromises described for paired-preference tests (i.e. sensory interaction or memory).

The face or "smiley" scale is another example of a scale (including several versions) that has been used to measure product acceptance and is listed in several recommended test guidelines (American Meat Science Association, 1978; Anonymous, 1981b). The scale consists of a series of faces with expressions in gradations of smiles to frowns that are intended to reflect gradations of preference. Two examples are shown in Figure 3.7. The smiley scale was developed for use with children and/or individuals who could not read or understand the meaning of written words as presented in scales such as the 9-point hedonic scale. The origins of the scale are obscure, and there is little or no evidence that such individuals can, in fact, associate a particular face with their degree of liking for that product. For a child, this can be a challenging intellectual task, especially for relatively young children (8 years or younger). The issue is not whether a child can or cannot read or can visually discriminate the different faces but, rather, whether the child has sufficient intellectual skill to transfer sensory reaction to a product, to a specific face. The potential for misinterpretation is so great that it hardly seems worth the effort. Although Roper (1989) recommended against using rating scales with children younger than the age of 7 years, Kroll's (1990) data supported their use but demonstrated that the face scale was not better than other rating scales. Young children are often selected as subjects for sensory acceptance tests. They are consumers, they spend money and influence adult purchases (Clurman, 1989), and consequently their preferences are of increasing importance to consumer product companies. Kimmel *et al.* (1994) recommended that children can be used for discrimination and acceptance testing but there were numerous difficulties and/or challenges that needed to be managed if the information was to have any value. For example, the cognitive abilities and the actual test protocol had a major influence, requiring experimenters to be able to adapt a protocol to obtain the information without unduly biasing response behavior. Chen *et al.* (1996) used three-face, five-face, and seven-face scales to measure food preferences with 3-, 4-, and 5-year-old children. These researchers concluded that the face scales could be used but that cognitive abilities, as noted by Kimmel and co-workers, would limit experimenters to scales of 3 or 5 points. Popper and Kroll (2003) reported on their research on children's testing and reached similar conclusions as those reported by Kimmel and others. Essential to a successful outcome is the information provided just prior to the test, the orientation, and the type of scale used—paired preference or scaled responses.

Our experience with testing children using the 9-point hedonic scale has been positive; that is, expected preference differences have been obtained. First, no assumptions are made with regard to the ability of the children (8–12 years old) to understand the absolute meaning of each phrase; just the relation to all the other phrases. The fact that children can read does not mean that they understand what they have read. The experimenter must explain the test procedure with examples of use, and then the children can proceed with the actual test without having to first interpret the relationship of what they tasted with one of several faces. Experimenters are available to assist; however, children older than 8 years usually want to proceed on their own. When children are too young to comprehend verbal instruction for the 9-point hedonic scale, a paired-preference procedure is recommended with parental participation.

We agree with Roper (1989) and Popper and Kroll (2003) that testing children provides many unique challenges in addition to the usual concern for product serving and questionnaire design. For example, to what degree should the parent be included in the test? Should the child be tested alone or in a group? Should different questions and tasks be given to different age children in the same test? We believe that use of known scales with children and pre-teens makes sense provided the scales are demonstrated before a test, and depending on the cognitive abilities, either the parent or an experienced server is available to help. Facial scales, which use cartoon-like faces to express like and dislike and were developed for those with limited reading and/or comprehension skills, have not been shown to provide advantages over verbal scales when used in consumer research. On the contrary, facial scales introduce their own complications. For children, the cognitive task of matching an emotion expressed by a face to their reaction to a product may require more abstract thinking than a response on a verbal scale. Another potential distraction for children is that the faces may not reflect the child's racial or ethnic identity and that the meaning of facial expressions is far from universal across cultures. For these reasons and the ones stated previously, face scales are not recommended.

Other scales, such as semantic differential and Likert scales, have also been used to obtain measures of product acceptance–preference (also see Chapter 3). Semantic differential and Likert scales are used extensively in studies of purchase decisions, attitudes, and other sentiments; they are especially applicable in the initial investigations of product concepts. Throughout the years, these scales have found increased use in product optimization research for linking sensory perception to various measures such as situation appropriateness, and so forth (see Chapter 8). That these scales are used to measure product acceptance is very surprising inasmuch as they were not developed for that purpose (Stagner and Osgood, 1946). This should not be construed to mean that scales devised for one purpose cannot be applied elsewhere, as we have seen with optimization studies. However, in the context of measuring the sensory acceptance of a product outside the realm of optimization, these scales are inappropriate. They are intended for cognitive issues and are considered less sensitive than scales such as the 9-point and LAM scales, particularly when used with only 50 or 60 subjects as is common with sensory acceptance testing. Finally, there is the challenge of interpretation of results. As noted in Chapter 3, responses from these scales can be summed and/or percentages computed, but other statistical treatments of this data are inappropriate. However, their value can be enhanced in marketing and optimization studies by incorporating cognitive and sensory information in a meaningful preference map.

The just-about-right (JAR) scale warrants comment here (see also Chapter 3) because it has been incorporated into the acceptance methodology and product optimization research with increasing frequency. The JAR scale incorporates both intensity and hedonic judgments, and it assumes that the consumer has a common understanding of the attribute that is being measured. Its use in product testing has been focused on identifying and directing technology to make attribute changes. Considering the relative insensitivity of the scale (see Chapter 3) and the limitations to common terms, it is a poor substitute for descriptive analysis.

A common feature of these scales (and others) is that they are usually adapted from other applications in the mistaken belief that what works in one situation will work in another. Sensory evaluation must take care to avoid blurring the differences between perceptual and conceptual measurement techniques. These scales should not be used in sensory acceptance–preference tests outside the realm of optimization. Although they may be useful in those situations, they are not appropriate for small-scale sensory affective tests described in this chapter.

7.3 Subjects

The subjects participating in a sensory acceptance test should be qualified based on typical demographic and usage criteria or preference scores from survey data, if the former cannot always be satisfied. It may not be possible, or even necessary, to select subjects based on demographic criteria when employees are the subjects, and a large majority of sensory acceptance tests do involve employees. Many companies now use local residents in place of employees, a topic that is discussed later in this section. All volunteers should be screened for their product usage and likes/ dislikes. Screening is best done using a survey format similar to the one described in Chapter 2.

For employees, the attitudinal responses should be compared with data from other populations to determine that the subjects are comparable, at least in their preference toward specific products. Meiselman *et al.*'s (1974) report on degree of liking for more than 300 products based on responses from several thousand military personnel can be used for comparative purposes. A summary of all food acceptance testing in the U.S. Army was updated in 2003 (Meiselman and Schutz, 2003) and provides a foundation for testing. However, external survey data should be supplemented and eventually superseded with data obtained from one's own product testing. This information is used as a basis for selecting subjects for acceptance tests. One selection criterion, in addition to requiring that they are likers of the product, could be to use only those subjects whose attitudinal scores fall within ±1 SD of the grand mean for all survey participants. This measure should be consistent with published data, such as the aforementioned survey. The decision regarding the permissible range of scores that includes or excludes subjects is the responsibility of the sensory staff. The overall objective is to select a relatively homogeneous group, all of whom are within the norm of likers of the particular product (or product category), and to be able to exclude individuals who exhibit extreme or unusual response patterns, such as always scoring products as 9's or 1's. The latter do not differentiate among products and therefore contribute nothing of value.

This entire database should be available electronically. As tests are completed, the database can be updated for each subject's file. This electronic system can easily be used to select subjects for a test, ensuring that an individual will not be used too frequently. For employees, this test frequency should probably not exceed two or three tests per month for the same type of products. There is no question that these subjects will become experienced test participants within a few months; however, monitoring of response patterns will identify atypical results—for example, if the variance measures show a systematic decrease or current products are assigned scores higher than what has been observed in previous research. In most instances, a 1- or 2-month interval of no testing should be sufficient to correct this problem. Specific guidelines must be developed on a product-by-product basis. Of course, if one has available many hundreds of subjects, then test frequency can be further reduced.

Individuals who are qualified for discrimination and descriptive tests should not participate in acceptance testing regardless of their willingness. The training process, especially for descriptive analysis, results in subjects who have an analytical approach to product evaluation that will bias the overall response required for the acceptance–preference task. Similarly, individuals who may possess technical or related information about specific products should not be used because of their potential bias. Objectivity must be maintained if the results are to be considered as reliable and valid.

The importance of subject selection and frequency of test participation are critical issues in relation to the question of the objectivity and face validity of the results from employee acceptance tests. There is also a risk that employees will recognize a product, especially if there are brand names or unique markings on those products. For companies that have a very limited product line, even these precautions may not be adequate. Employees may easily recognize those products that are different from their own, whether they are those of competitors or experimental. However, before a decision is reached regarding the potential bias (or lack of) of employee acceptance test results, sensory evaluation should initiate a series of comparative studies to test this hypothesis.

The validity of employee acceptance tests is an important step in the overall scheme of the sensory evaluation program. At issue is the question of whether employee panel results accurately reflect those from a nonemployee panel. Care must be taken to ensure that products, scorecards, test procedure, and so forth are matched. Products should be obtained from the same production lot; scorecards, serving orders, and test instructions must be identical. For the analysis, the use of a split-plot, one-within and one-between analysis of variance model is most appropriate to compare results from the two panels (see Chapter 4). In this model, products are treated as the "within" variable, whereas the two panels are treated as the "between" variable. Of special interest would be the significance measure for the product × panel interaction. If this interaction is not significant, we can conclude that the two panels scored the products similarly, and we accept the employee panel's results as a good estimate of the nonemployee panel's results with respect to the relative difference between products. On the other hand, a significant interaction would require sensory evaluation to determine whether that interaction was the result of panel differences for product ranks or due to the magnitudes of difference between product means. Conversion of scores to ranks followed by rank-order analyses may yet demonstrate a consistent ordering pattern between the two panels.

Data obtained from several comparable studies would provide the basis for a correlation analysis. Although correlations are not sensitive to absolute values, they are a good indicator of whether the relative differences between products are comparable for the two panels. Statistical validation of employee panel versus external panelist results should be conducted periodically.

Satisfactory results from these studies allow substitution of preference behavior for demographic criteria in selecting subjects for acceptance tests. This approach is especially useful because most employees (and most local resident panels) cannot totally satisfy demographic criteria for all products. For example, in the military, much of the progress in development of rations and menu items was achieved through use of nonmilitary personnel as subjects and then verified with military personnel (Meiselman *et al.*, 1974; Peryam *et al.*, 1960). This rationale is applicable to products such as breakfast cereals that are intended primarily for children. An employee panel whose acceptance scores accurately reflect relative differences similar to a children's panel could be used in the development process.

As previously mentioned, there has been a gradual trend toward use of local residents or contracting with a local fielding agency that provides a sensory acceptance test service. This increased use of nonemployees reflects the general lack of availability of employees as a result of industrial consolidations and, to some extent, concern about the credibility of results from employees. There are, however, several advantages of using nonemployees, including their availability with minimal delay, their availability for a longer period of time, and the fact that a more accurate estimate of the cost for each test can be obtained. This latter point is especially important for financial accountability for each test. When nonemployees are used for a single test, detailed information on attitudes toward a wide array of products is not expected, and the demographic information becomes the dominant issue. Traditional demographic information such as age groupings, gender, frequency of category usage, and brand usage may be used to determine whether an individual is qualified.

Because the sensory acceptance test uses small numbers of subjects, it will be important to select subjects who are within the norm of likers for the product and who can effectively use preference rating scales to differentiate among products (not all people can), if rating scales will be used in testing. When testing with consumers, additional selection criteria based on demographics, psychographics, user graphics, market segments, and lifestyle are recommended (Carter and Riskey, 1990; McDermott, 1990). This information should be available from marketing (or from sensory optimization research, a topic discussed in Chapter 8).

It is becoming more common to have a pool of nonemployees participate regularly in sensory acceptance tests in much the same way as employees, and this is often referred to as a convenience sample. In this case, the qualifying procedures described for employees will be applicable and their responses must be monitored on a continuing basis. In this way, the reliability and validity of the results will be maintained, and the sensory acceptance test will continue to be an integral part of every company's product sensory test program. Of course, when using nonemployees, some monetary compensation will be necessary. For employee participation, an indirect reward is necessary, such as credit toward lunch at the company cafeteria or some similar reward. As discussed in Chapter 2, employees must be rewarded for participation but not with direct cash payments.

7.4 Types of acceptance testing

Acceptance panels, as noted previously, are used primarily as part of the product screening effort. They allow for reducing the number of alternatives by eliminating poorly performing products and for maintaining greater security (when employees are used) than may be possible in other types large scale of consumer tests.

There are three primary categories of sensory acceptance tests: laboratory, central location, and home use. In Table 7.1, these three test categories are listed in terms of types of subjects and their respective advantages and disadvantages. To some extent, the distinction between testing in a laboratory and testing in a central location is not as clear as that versus home-use testing. In the former situations, the subjects are brought to a central location for testing and the primary differences relate to the extent of the controls applied to the testing process and the environment of the test. Additional discussion about each of them follows.

7.4.1 Laboratory testing

Acceptance testing in the laboratory environment is the most frequently used of the various types of sensory acceptance tests. It is particularly appealing because the laboratory is usually very accessible to employees, the greatest possible control over

Table 7.1 Characteristics of Different Types of Sensory Acceptance Tests

	Laboratory	**Central Location**	**Home Use**
Consumer type	Employee or local resident	Public (general or selected)[a]	Employee or public
Responses per product	25–50	100+	50–100
Product number	Maximum of 5–6 per session	Maximum of 5–6	1–2
Test type	Preference, acceptance but not quality	Same as laboratory	Preference, acceptance, performance (intensity and marketing information)
Advantages	Controlled conditions Rapid data feedback "Test-wise" subjects Low cost	Large number of subjects No company employees	Product tested under actual use conditions All family's opinion obtained Marketing information (pricing, frequency of use, etc.)
Disadvantages	Familiarity with product Limited information Limited product exposure	Less control Limited information No lengthy or distasteful tasks Limited instructions Large number of subjects required	Little or no control Time-consuming Expensive

[a]General groups (subjects solicited at fairs, markets, retail outlets, etc.); selected groups (churches, clubs, schools, etc.).

all aspects of the test is ensured (lighting, environmental control, product preparation, etc.), and rapid feedback of results is possible. The number of responses per product is listed as between 25 and 50 up to a maximum of 75, and we recommend at least 40 per product. Because most of the tests make use of complete balanced block designs and each subject evaluates each product, a panel could actually have as few as 24 subjects. It might be difficult to establish a statistically significant difference in a test with so few subjects, but it is still possible to identify trends and to provide direction to the requestor. As the panel size increases to a maximum of approximately 75 subjects, especially for tests involving more than two products, the statistical significance, credibility, and the importance of the results increase substantially. These economies are possible because of the test controls and the use of subjects who have prior test experience. Sidel and Stone (1976) provided a guide for selecting panel size based on the expected degree of difference in hedonic scale ratings (for the 9-point hedonic scale) and the size of the standard deviation. For example, a statistically significant difference at $\alpha = 0.05$ is possible for a 0.50-point scale difference when the standard deviation is not more than 1.50 with a panel of 40 subjects. A word of caution is necessary: The relationship among panel size, magnitude of difference, variability (subjects, products, and experiment), and statistical significance is presented as a guide. Different types of products will have their own unique requirements. For example, a manufactured product may show less variability than a natural product, which is more strongly influenced by the quality of the harvest, and so forth. In this case, the expectation of a statistically significant difference may not be realized without a sufficiently larger number of subjects (e.g. increasing the panel size to 60 or 70). Although smaller panels will yield significant differences as variability decreases, order effects and other factors also can influence results and reduce their usefulness.

As listed in Table 7.1, a laboratory panel should evaluate no more than five or six products in a single session, and responses should be limited to acceptance–preference. Limiting the number of products to no more than five or six is based, in part, on practical consideration—for employees, their time away from their regular work. Of course, nonemployees are available for longer periods of time and could easily evaluate more products. However, this depends on the kinds of products: Whether sensory fatigue will be a problem, and how much preparation and handling is required. This does not mean subjects are not capable of evaluating more products. As Kamen *et al.* (1969) observed, consumers can evaluate as many as 12 products, provided a controlled time interval between products is allowed and only acceptance is measured.

For both laboratory and central location tests, the recommended number of products is limited to six or fewer. More products could be evaluated; however, this approach would require an extended session or subjects returning for a second or third session either later that day or on subsequent days. An alternative approach for tests of larger numbers of products is the use of incomplete designs (e.g. see Gacula and Singh (1984) and Chapter 4 of this text). If an incomplete design is used, we suggest the number of incomplete cells per subject be kept to a minimum of one-third or less (e.g. 4 of 6, but not 4 of 7 or 4 of 8). Having subjects return to evaluate all products yields a more powerful database. One other issue that is often overlooked when using incomplete designs is the need to increase the number of subjects, and that often is a limiting factor in most test programs. In our experience, having several sessions is more preferable than using an incomplete design. This way, all subjects are provided with a similar context of products. This does not mean that incomplete designs should be avoided or are inappropriate. For some products (e.g. a skin-care product or an overly spicy product), it may be the only way to evaluate them without making the evaluation process unworkable. Disadvantages of the laboratory test are the potential familiarity with the product when using employees, the limited information (acceptance) derived from the test, and the limited exposure of the consumer to the product. This latter point is discussed in more detail in the next section.

7.4.2 **Central location testing**

The central location test (CLT) is one of the most frequently used of consumer tests (the laboratory sensory test is a type of CLT) and especially by marketing research. The CLT is usually conducted in a field site specifically designed for this testing, a shopping mall, or another similar type of location—for example, a school cafeteria, a store, or an office building that is accessible to large numbers of people (the public). For this test, either consumers will be pre-recruited or, if the site is a mall, an intercept procedure will be used. Pre-recruiting is generally more costly because of the time required to contact prospective candidates and because the compensation for the effort needed to get them to the site will exceed the cost of the intercept procedure and any compensation given to consumers recruited at a mall. In fact, for many products, consumers recruited at a mall are given only indirect compensation, such as discount coupons for stores located within the shopping center. However, other factors must be taken into account when making a decision as to where a test will be done and how the consumers will be recruited. If a product has a relatively low incidence of usage, then the mall intercept method will be more time-consuming than a pre-recruited test. For example, a product with an expected incidence of 10% of the population will require a minimum of 1000 contacts to find 100 product users, and not all of those who qualify will want to participate (which could be as much as 30%). Thus, the mall intercept method of testing could require considerably more time to complete when the incidence of participation is very low. If a planned mall test involves extensive product preparation and handling, then the benefit of the lower cost of the mall intercept will be easily lost as a result of the additional time required to complete the test and the additional amounts of product required. An obvious solution is the use of several malls; however, this also introduces its own set of problems, including that the malls will be a potentially confounding source of variability. On the other hand, pre-recruiting and scheduling of subjects in a CLT at specific times allows for testing to be completed within 1 or 2 days. This is especially important for products with shorter shelf life or ones that require more complex product preparation.

In addition to the general public, consumers may be solicited from religious and social clubs, schools, and other organizations. Groups such as these provide homogenous or convenience groups of subjects as needed, and this is an easy way to recruit. However, the reader is reminded that these people are representative of a narrow segment of the population, many of them know one another, their responses may be skewed, and the data may not validate when larger scale testing is conducted.

For the CLT, there are usually approximately 100 or more responses (consumers) per product; in situations in which segmentation is anticipated, even more responses might be necessary. This increase in the number of consumers (versus the laboratory test) is necessary to offset the expected increase in variability attributable to the novelty of the situation, inexperience of the consumer, and limitations in the test environment.

The number of products should be limited to approximately five or six, and possibly fewer in order to minimize test time, an important issue in the mall intercept protocol. In a mall test, participants are liable to leave in the middle of a test if the task is lengthy, unpleasant, or too complicated. However, test results have considerable face validity and credibility where the subjects are "real consumers" representing the target population, and large numbers of such tests have been conducted. Thus, increasing the number of subjects to 100 or more is both an advantage and a disadvantage.

Limited product exposure is sometimes considered a disadvantage of the laboratory and CLT methods versus the home-use test, in which the consumer will have more and longer exposure to a product. Inasmuch as the laboratory and CLT methods are used primarily for screening product alternatives and for more precisely defining the consumer segment in relation to the specific products, this should be a nonissue. Questions about the longer term sensory effects (e.g. flavor wear-out) should be addressed in an extended-use study rather than as a routine multiproduct screening method. We recall a company insisting that the subjects consume an entire container of liquid in order to evaluate the effect; however, the company requested that several units be evaluated in a single session. Not surprisingly, most participants refused to complete the test and spent much time visiting the restrooms. Although the need for consumption information is useful, the CLT venue is inappropriate.

Another unfortunate practice that has increased in popularity as testing budgets have been reduced is the "piggyback" CLT test, in which consumers recruited to a test site participate in evaluation of multiple products, often from different categories. Although the products may be independent, it is likely they have a confounding influence on the subject and the results. Sensory interaction and fatigue, response carryover from one product set to the next, and motivational change during the course of testing are but a few factors increasing the risk of contaminating results on subsequent products in the test session. We do not recommend the practice.

7.4.3 **Special types of central location tests**

In addition to the CLT described in the previous section, product acceptance information can be obtained in two additional ways. As shown in Table 7.2, a mobile serving cart or mobile laboratory can be used to bring products to consumers and still fit within the structure of a CLT.

Without question, the laboratory environment provides the greatest control for a test. However, this control can be compromised to some extent in situations in which access to sufficient numbers of people to complete a test is necessary. Sufficient product and supplies are placed on a cart, which is brought to the employees' workstations. An office area will be quieter and a more desirable area than a manufacturing location; however, there will still be some distractions. Conversations, traffic flow, visual contact with other employees, telephones, personal computers, and so forth will distract the subjects, and the sensory staff will need to take these factors into account when designing a test to be done at the subject's workstation. As noted in Table 7.2, there should be as many as 50 responses per product and a maximum of five products in a test. As might be expected, the

Table 7.2 Characteristics of Special Cases of the Central Location Test

	Mobile Serving Cart	Mobile Laboratory
Consumer type	Company employee[a]	Public (general or selected)
Responses per product	22–50	40–60
Product number	2–5	2–5
Test type	Preference, acceptance (not quality)	Same
Advantages	Cost	No employees
	Rapid data feedback	Controlled conditions
	"Test-wise" subjects	
Disadvantages	Limited control	Limited information
	Limited information	No lengthy or distasteful tasks
	Familiarity with product	Limited instructions

[a]See text and footnote to Table 7.1 for additional explanation.

relative advantages and disadvantages represent a combination of those listed for both the laboratory and the CLT.

Not all products lend themselves to this type of test protocol. Nonfood items are more likely candidates; however, the use of hospital-style serving carts with appropriate heat controls will permit a wider range of food product applications. Nevertheless, the sensory tasks must be simple and require little time to complete. The potential for distraction (and increased variability) is quite high. One cautionary note should be mentioned. The economics of the mobile serving cart may have considerable management appeal. The serving cart is a temporary alternative, but it is not a substitute for a permanent facility with all the necessary sensory controls. This practice is much less common today than in the past.

The mobile laboratory represents one additional means by which the controls of the laboratory can be brought to the consumer. A motor home or trailer can be built to contain product preparation and holding facilities, booths, proper lighting and ventilation, and so forth. The unit can then be driven to shopping malls, fairs, or wherever there are large gatherings of people, and the testing can be done in a way that is similar to that done in the laboratory environment. This approach is most useful when the respondent population is unique or when one is interested in sampling school-aged children at the school, for example. The recommended number of products should be limited to not more than five. One could test more than five; however, sensory fatigue has to be considered as well as the number of questions on the scorecard. If there are more than two or three questions, the response time will increase and this could result in greater than expected variability. Increasing the number of subjects to not less than 50 is also recommended. Because the subjects will be inexperienced, the testing process adds some level of anxiety; however, the presence of the sensory staff should facilitate the testing activities.

7.4.4 **In-home-use tests**

The sensory in-home-use test (IHUT) is also a valuable resource, distinct from home-use testing by marketing research. Sensory research in the home environment can provide valuable information on product preparation, family attitudes, packaging, and so forth to guide development prior to larger scale home-use or volumetric testing. This information is not obtained in the sensory laboratory setting, but the research follows sensory protocols. For some nonfood products, laboratory evaluation is difficult (e.g. in testing disposable diapers), and home-use testing becomes a feasible and necessary alternative. Similarly, if one is testing running shoes, shower gels, golf clubs, or recipe kits, this usually is best accomplished before, during, and after usage in a more typical environment rather than in a laboratory environment.

Often, the first exposure that sensory evaluation will have to a product in its proposed container is when it is going into home-use testing. Because products can fail due to problems with packaging, containers, preparation instructions, or product usage, it is reasonable to have a home-use capability within the sensory department to assess products in the early stages of product formulation or reformulation.

In the home-use test, there is no control of environmental or other test variables. Therefore, panel size should be larger than that used in the laboratory test. The number of subjects is always impacted by the specifics of the test—that is, the protocol, the experience of the subjects, and, as always, the type of product. Some products are inherently variable, and this would necessitate increasing the number of subjects so as to take this into account in the analysis. If the subjects are accustomed to sensory testing, the panel size can be reduced; however, this reduction in panel size should be empirically based rather than an arbitrary decision. We expect that "test-wise" subjects (i.e. experienced subjects who know how to test and who are comfortable in the test situation) will be less susceptible to error as a result of psychological variables associated with being a subject. Further improvements are possible by eliminating subjects who have been erratic or are otherwise so different from the norm that they may be considered representative of a small population of little commercial interest.

Home-use tests often typically involve only two products, primarily because of the duration of time needed for each product to be evaluated. Extended use in the home of 4–7 days is relatively common. If there were interest in evaluation of additional products, test time would have to be extended and the rate of return of completed data generally decreases. This is due to many factors, such as boredom, lack of necessary time or interest, and vacations. One obvious alternative is to have different groups of subjects evaluate different products in an incomplete block design, which, as noted previously, also has its own set of limitations. With advances in the use of tablets for recording responses and use of the web for transmitting information, this decrease may be less than it has been in the past.

As indicated in Table 7.1, the products are tested under actual usage conditions and the entire household's opinion is usually obtained. As with all sensory tests, the primary focus of the evaluation is on sensory acceptance and may extend to packaging, preparation, and usage occasions. Test design and instructions should be kept uncomplicated.

The home-use test provides subjects with the first opportunity to form an opinion based on consuming or using the product (food or nonfood). With foods, this test may also provide subjects with their first opportunity to actively prepare and serve the item as a normal part of the daily routine. However, it should be remembered that many of the perceptual components associated with product handling can and should be isolated and studied in the laboratory or CLT model. For example, it is reasonable to have consumers individually prepare a product and evaluate the sensory components associated with the preparation procedure before moving to the home-use environment. This identification of product use variables is important in understanding their effect on product acceptance.

The home environment allows a multitude of unique variables to have their "normal" influence during product preparation and use. In the early stages of development, these variables may represent undesirable noise in the research design, whereas in the latter stages of development, they may represent a desirable resource for learning about conditions that could lead to potential product failure.

The home-use test presents sensory evaluation with several additional issues that need resolution that are not encountered in either the laboratory or the CLT designs. These decisions include packaging containers and their labels, product preparation instructions, product placement, scorecard, and questionnaire design.

Products in home-use tests are usually packaged in containers similar in size and shape to those in which the product will be sold but with the standard graphics omitted. Often, a plain container is used. There should be at least a three-digit code, a label containing the ingredient list, preparation and usage instructions, and contact information including a telephone number in case of problems or questions.

Preparation instructions must be simple but complete in terms of how the product is to be used, along with directions for completion of the questionnaire. The importance of these instructions should not be underestimated because they are the context within which the household members make their decision. Consultation with home economists and other specialists familiar with product preparation is essential to minimize confusion and ambiguous results because of improperly worded instructions. Because there is no experimenter available to the subjects at the time of preparation and evaluation, the instructions must be unambiguous.

Product placement is a factor that is often determined by budgetary considerations but should be carefully considered because it can impact the quality of the resulting data. Providing a participant with all products at the start of the test minimizes costs; even lower costs can be achieved by mailing all products to the subjects. However, for a sensory IHUT, we have observed that significantly higher participation rates can be obtained when subjects report to a central location facility where they are handed the product, read the test instructions, and the questionnaires are reviewed with them prior to taking products home. This in-person contact greatly enhances the participation and completion rates and provides a level of control in a sensory-based home-use test.

Providing both products at the same time is usually not recommended because it enables the subjects to make direct comparisons. This can be minimized by having the subjects return the used container and obtain the next product. A second barrier is the use of the tablet for data collection. If the subject enters the wrong code, the system will not allow information to be recorded, and it can also issue a warning message to the sensory staff. With employees, it is relatively easy for them to come to a specific location to receive their product and instructions and then to return the scorecard or enter their data directly on a web survey, along with the unused portion and container after completing an evaluation and before receiving another product. If one is using local residents instead of employees, they still come to a central location to obtain product.

Mailing products to consumers across a large geographical region that targets the consumer demographics fits within the marketing research IHUT process. Sensory research objectives are focused on product, packaging, and preparation. The marketing research focus is more on purchase behavior and use variables, and sensory staff should participate only when requested by brand managers. Although the two gather product and product-related information, the use of that information is different.

Larger scale home-use tests that do not involve employees or a local panel require just as much concern about product placement. In these situations, the product placement may be done at the time an individual is qualified—for example, in a mall intercept recruiting drive. If recruiting is done by telephone, the product may be mailed; however, as noted previously, the potential for product misuse is very high. Product delivery directly to the home is costly, and when it is used, the tendency is to provide all products at the same time and to return them when the test is completed. In these test situations, the greatest opportunity for error is that subjects will open and evaluate all products at the same time. In the past, interviewers placed product and scorecard in the home and then, after the specified time, returned and repeated the process. In some situations, product was mailed or the subjects came to the location for orientation and product collection. With the increasing cost of testing, companies often had all products delivered at once. Control is best achieved by having the interviewer place the first product and scorecard following respondent qualification and, on return, retrieve both prior to placing the second. This process is followed until the final product and scorecard have been retrieved. Personal interviews, unlike the telephone or the mail, provide an opportunity to review the procedures with the respondent and determine whether the product was used properly (or at all) and, if necessary, to clarify the test procedure or product-use conditions, should there be any questions. This procedure is quite similar to that followed when using employees or a local panel. In our experience, having the consumer come to a location and be given a briefing and review of the test procedure and scorecard is the best option. It enables the experimenter to minimize misunderstandings, answer questions, and improve the quality of the information obtained. With the advent of the tablet, the at-home testing process has been revolutionized, making it an easier system for capturing responses and completing analyses quickly.

The primary focus in the sensory home-use test is measuring overall product acceptability. This may be obtained through use of one or more of the scales described in Chapter 3. To maintain continuity among laboratory, central location, and home-use tests, similar procedures, analyses, significance criteria, and the same acceptance scale should be used throughout (e.g. the 9-point hedonic or LAM scale).

In addition to overall acceptance, responses to product characteristics such as appearance, aroma, taste, and texture and selected JAR scales may be obtained. Although this information can be used to highlight general product issues, it is not recommended for use in the same way as a result from a more controlled sensory analytical test. As with any untrained panel, beyond the overall acceptance judgment, there is no assurance that the responses are reliable or valid. Subjects will complete the scorecard, but it is challenging to determine the meaning of the result. In the sensory acceptance test, the value, meaning, and interpretation of those responses are more often a source of error and conflict than a source of useful information. It is not unusual for researchers to collect pages of such information, although they admittedly do not use much of what is collected. However, once collected, there exists a tendency for researchers to use this information to guide development, particularly if the other results do not meet their expectations.

Another problem associated with asking diagnostic acceptance questions is the different "psychological set" this establishes for the subject and the consequent changes in response behavior that this may cause. Overall acceptance requires that the subject view the product as a whole without direction to what sensory aspects are considered important. It is a holistic approach to responding to the product. Conversely, diagnostic assessment requires the subject to approach the product from an analytical perspective, and it directs subjects to focus on specific sensory components. It is a rather simple task to change overall attitude behavior by focusing subjects on different product attributes. Asking diagnostic questions serves to focus attention on those specific attributes. Focusing on one set of selected attributes may result in a different acceptance response than may occur if attention were focused on a different set of attributes or if there was no focusing whatsoever. The line of questioning or set of attributes selected for a questionnaire has the potential for revealing more about the researcher's product knowledge and expectations than providing unbiased and useful consumer information for product or management decisions. However, including diagnostic acceptance questions is a common practice, and it does not guarantee data will be flawed. It simply increases the risk that the development team may focus on the wrong attributes, and for ongoing sensory programs there is no way to determine for each test when and to what degree the data have been compromised.

Because the subject's response in a purchase-and-use situation will not include the same diagnostic focusing as in the analytical sensory research test, it is better to exclude that dimension from the test in the first place and only focus on information that can be useful in the development process. With only product acceptability to attend to, the consumer has little doubt that we are interested in his or her overall attitude (like or dislike) for the product. Without the potential frustrations, tedium, and misguidance associated with asking consumers attribute questions, we can be more confident about the acceptance data obtained.

If there is interest in learning more than overall acceptance from the home-use test, it is reasonable to obtain limited information about product performance.

For example, if a product was prepared and intended to serve four, did it in fact do so. Or, if the product was designed as a multipurpose product (e.g. a cleaner for glass and tile), how acceptable was it in its various usage categories. A product that was supposed to take 10 minutes to prepare may have taken 25 minutes or more; such information can easily be obtained and should be in the home-use test. This information may be obtained in addition to acceptance because it does not require an analytical sensory test, and as such, it should not create the same challenges in interpretation as those associated with sensory attribute questions.

Although a monadic sequential placement procedure may be used in conjunction with the hedonic scale, many investigators find it too difficult to disassociate themselves from a paired preference question and will persist in adding it as a final overall question. If so, the question may be phrased as follows: "Thinking about the product you tested last week and the one you tested this week, which one do you like best?" Usually this question is included to satisfy individuals in the company who mistakenly believe that the paired-preference test is the most sensitive or only valid way to measure product acceptance. Using a paired-preference question in the home-use test introduces a memory variable if the subject has to remember what was tested first, how well it was liked, and so on. Therefore, we do not consider the paired-preference question as a recommended sensory technique where a monadic sequential usage procedure has been used.

A more serious problem with using a paired-preference question in conjunction with monadic sequential scaling of acceptance is created when the two measurement techniques produce contradictory results. Which technique (and result) is then to be believed? Actually, both can be correct, but because they are different measurement techniques, they may produce different results. Because the paired question allows only a count of responses for or against a product, it is possible to get results similar to those shown in Table 7.3. If we assume that a respondent who scored one product higher than another would also prefer that product in a forced-choice situation, we may view the data in Table 7.3 as demonstrating a significant preference for product A. However, only 12 of the 40 respondents scored product B higher than product A (mean score of 6.6 vs 4.9).

If only means, percentage preferring, and statistical significance are reported (as is usually the case), one has a reporting dilemma. However, if only paired-preference data are obtained, the magnitude of difference recorded for 30% of the panel is not obtained. This can be a critical omission. For scaled data, the number of subjects scoring one product higher than another is easily derived and clearly indicates all the important information: (1) mean differences and level of significance, (2) number of subjects scoring one product higher than another, and (3) the magnitude of difference for subgroups in the population. Regarding the latter detail, a magnitude of difference of 8 scale points is considerably larger than a magnitude of difference of 1 scale point.

Equipped with the information from the scaled data, a lower risk business decision is possible than when only paired-preference data are available. Because paired-preference data can be derived from scaled data, their inclusion in the

Table 7.3 Example of the Potential Conflict between Scaled and Paired-Preference Data

| No. of Respondents | Hedonic Scale Rating | | Magnitude of Difference in Hedonic Scores (in Scale Units) |
	Sample A	Sample B	
12[a]	1	9	8
8	5	4	1
4	6	5	1
8	7	6	1
8	8	7	1
Mean score	4.9[b]	6.6	
Number preferring each product	28[c]	12	

[a]*In this example, 40 respondents participated, of whom 12 assigned "dislike extremely" to product A and "like extremely" to product B, and so forth. For more information on the 9-point hedonic scale, see Section 7.2.2.*
[b]*Denotes a significant difference (p = 0.002). B is scored higher than A.*
[c]*Denotes a significant difference (p = 0.008). More respondents scored A higher than B.*

home-use test is unnecessary. The home-use test is usually reserved for the latter stage of sensory evaluation testing; it has a high degree of face validity, and it provides an opportunity to measure performance and acceptance by household members under normal usage conditions; however, it may be more costly than other sensory tests, it requires more time to complete, and there are no environmental controls.

7.4.5 Other types of acceptance tests

In addition to the previously discussed approaches to consumer testing, there are several other options available to the sensory scientist. One of these, as discussed previously, is to bring local residents into a sensory laboratory. This approach has been increasing in use because it is convenient for the sensory staff in terms of scheduling of tests, the speed with which requests can be satisfied, and it minimizes reliance on employees. However, this use of local residents has its own set of requirements that include recruiting, scheduling, budgeting, accessibility, confidentiality, and security. A system must be established to contact and schedule subjects along with a budget and means for compensating them after each test or block of tests. Accessibility to the test site must minimize the risk of subjects wandering into restricted areas, and so forth. Provided the sensory facilities are accessible, this can be one way of extending sensory resources, and especially those needed for acceptance testing. Participants are highly motivated, will be very prompt, and willing to provide considerable product information. Compensation and employment contracts with local residents, if any, are issues generally handled through the company's human resources department.

Another possibility is for a company to develop and maintain an off-premises test facility. This eliminates many of the problems associated with bringing local residents into company premises; however, it will be more costly than bringing people to the sensory facility. It may also be possible to contract with a research fielding service or temporary agency supplier, provided they have the necessary resources.

In each case, sensory evaluation will need to assess its current acceptance test resources and anticipated volume of work before determining whether any of these options are feasible. There is no question that sensory acceptance–preference testing must be done to assist in the various product development phases. The questions to be answered relate to where it will be done, who will be the subjects, and what will be the relative costs required to provide valid and reliable data.

7.5 **Special issues**

7.5.1 **Sensory science versus marketing research/consumer insights**

Acceptance testing, as in discrimination and descriptive testing, has its share of special challenges and practices that have had and will continue to have an impact both in terms of the way sensory tests are organized and fielded and in terms of how the results are used by the business. Section 7.1 contained a brief discussion about the relationship between the smaller scale sensory acceptance model and the larger scale marketing research model.

Interaction with the market research department is one of the most significant areas that sensory scientists will encounter in their planning and conducting of research, especially affective research. When discussing market research, we are also including departments that go by other names, such as "market intelligence," "consumer insights," "consumer intelligence," and "innovation." Market research has an entirely different approach to the way issues of product development are approached. Sensory scientists must take responsibility to contribute to the education of market researchers and, in turn, be educated by market research. There is no question as to the value of sensory affective testing to the company's common goal of successful product development. Sensory is focused on products and works closely with the development teams, whereas marketing research helps manage the portfolio and works closely with marketing to promote the brands. When the two groups converge and work together, they create a synergy that allows for an in-depth exploration of products and consumers. One way to create synergy can be through information transmitted in team research projects or through seminars specifically designed to share information. Historically, the early evaluation of products was the responsibility of market research in many companies. Also, many vendor research projects involving R&D are funded by market research, so it is particularly important to have a good working relationship with this department.

There are several important areas in which the sensory science component is of real value to market research. Obviously, one of the critical contributions of sensory to marketing research is to assist in the development and selection of the "best" product(s) to be used in larger scale and, therefore, more expensive market research evaluations. This includes the delivery of well-liked products relative to current and/or competitive products and, in most instances, a descriptive evaluation of the products to be tested. Information may also be available from R&D with regard to the characteristics of the consuming population that can be used in market research screening. After market research results are analyzed, there can be questions about the unexpected performance of product supplied by production. In this case, sensory staff can review available product evaluation records, and time permitting, it can undertake a descriptive or acceptance test to identify the nature of the problem, if it is product. This kind of information helps to define the role of sensory versus the role of marketing research.

In this chapter, we have made clear the role of the sensory acceptance model and its responsibilities in the overall scheme of product evaluation. Nonetheless, situations will arise in which product information is questioned or tests are proposed that do not fit within the conventional scheme of sensory evaluation, but the results will have a direct impact. These include the use of the discrimination model to screen consumers for a preference test, the use of large numbers of consumers to achieve a statistically valid sample, the use of multipage scorecards, and the use of open-ended questions. It is our position that an awareness of these practices will be helpful to sensory evaluation and to marketing and market research. Once again, the reader is reminded that in most instances, the issue is not so much that one approach is totally correct and the other is totally wrong as it is an issue of risk and the interpretation of results. Now that we have it, what do we do with it?

7.5.2 The difference–preference test

In the discussion on discrimination testing (see Chapter 2), the practice of asking each subject to express a preference after the discrimination response was discussed and its weaknesses were identified. As a test procedure, it is not recommended. Unfortunately, this same concept has also been applied to consumer testing with an equally risky outcome. The description of its use by Morrison (1981), citing work by Moskowitz *et al.* (1980) and earlier discussion by Day (1974), pointed to the continuing interest in this use of the discrimination model with consumers (Rousseau *et al.*, 2002). We have already pointed out the basic flaw in this approach, namely that subjects require a considerable number of trials, usually approximately 30, before a discriminator (and nondiscriminator) can be identified. Day's discussion about the weakness of this discrimination–preference model is equally clear, and we are in agreement that this is an inappropriate procedure for measuring preference. Because the two tasks are quite separate—one is analytical and the other is attitudinal—it is not realistic to expect the naive consumer to handle both without considerable confusion and most likely a confounding of the preference judgment (remember that odd-sample bias will operate here).

7.5.3 **The curse of *N***

One of the features of the sensory acceptance model that is most often cited as a weakness is the relatively small number of consumers (usually 50–75) compared with the larger numbers used in the marketing research test (usually more than 100 and sometimes as many as 500 or more). We refer to this as the curse of *N*. Note that this same "curse" can be applied to any sensory test as well; that is, the more subjects in a test, the greater the believability of the results. This idea appears to originate from the mistaken notion that validity of results is stronger the more consumers participate in a test. As already noted, a statistically significant result is easily obtained as the number of judgments increases; however, this significance may not be practical. Because senior managers are more comfortable making business investment decisions when the information is based on large numbers of responses, the emphasis on the number of participants can overwhelm other factors, including the specific recruiting qualifications, the cost, and the risks associated with the decisions made from the results. There is no question that having a larger number of consumers in a test enhances the likelihood of finding product differences and of determining the preferred product. However, there are practical, product, and statistical considerations that must also be taken into consideration before one simply increases the number of consumers in a test. Time and cost are practical problems that preclude obtaining responses from every current and potential consumer. This difficulty is overcome, in part, through the application of statistical sampling techniques that permit one to use a subset of the population, and this allows one to generalize the results to a larger population.

Many companies have developed guidelines for consumer testing as a means of standardizing testing protocols and to ensure that management decisions based on results are derived from a common information source. These guidelines were formulated from the aforementioned statistical procedures, taking into account the potential for Type 1 and Type 2 decision errors, the statistical power of the test, and so on (for more discussion on this topic, see Cohen, 1977). Unfortunately, over time, these guidelines have undergone a metamorphosis and become rules, such that no product decisions are made by management unless a specified number of consumers have evaluated a product (and usually using a specified methodology). Although it is not the purpose of this discussion or of this book to review marketing research practices, this particular issue directly impacts the usefulness and challenges the validity of the sensory acceptance model and, therefore, warrants some attention.

The first comment to be made is that the sensory acceptance model is intended to screen products, to identify any products that are significantly disliked and any that closely match or exceed a specified target product. This process is completed with selection of a product (or products) that will continue on to larger scale testing. The sensory acceptance model is not designed to obtain purchase intent information nor can it be used for issues such as segmentation or related demographic tabulations. In effect, results from the sensory acceptance test serve as an early warning system as to a product's acceptance without the embellishments of concept, brand, package, label, pricing, and so forth. In view of the ease with which the test can be fielded and the results obtained, it should be a highly valued resource. Empirically, we observe that results from the sensory acceptance test translate well in the larger scale marketing research test; it is a good indicator of the future test. It is not a replacement but a valid and effective screening system.

The question of practical versus statistical significance must also be addressed. For example, how much practical importance can be attached to a statistically significant difference of 0.1–0.2 scale units with the 9-point hedonic scale derived from 300 consumers? It does not take very many responses, especially if there are some extreme scores, to shift a result to significance or nonsignificance. Before accepting such a result, it is essential to examine the response patterns in detail, preferably in the same way that one does with results from other sensory analytical tests (see Chapter 3).

Another concern in all large-scale tests, most often ignored, is product variability. With tests involving large numbers of consumers (e.g. several hundred), there is no question that production variation has a confounding effect on preference judgments. In some situations, this variation can be as great as the differences between test products, leading to the cautionary statement that results from a test with the largest number of consumers are not automatically assumed to be more valid than those of tests involving fewer consumers. Also, it is essential to examine results in detail and not just in the aggregate, as noted previously.

Clearly, then, the idea that the more test participants, the better, carries with it risks that are different from those associated with tests involving fewer participants. For sensory evaluation, this issue of risk is less important because the sensory acceptance test focuses on product (multiple products preferably); it is not to be used in place of larger scale marketing research tests.

7.5.4 The scorecard as a short story

Having consumers respond to numerous product questions and complete a long questionnaire including open-ended questions are traditional practices in marketing research tests, but they should not be part of the sensory acceptance test. The sensory test measures product liking, identifying which product (or products) will be recommended for larger scale evaluation and typically use as many as six products. A multipage scorecard will require re-testing and significantly more time to complete, and this increases the potential for sensory fatigue. The typical multipage scorecard usually includes a series of descriptive questions, preference measures, purchase intent, and a request to state the reasons for the product's preference (or lack thereof).

Asking the naive and inexperienced consumer to respond to specific descriptive attributes is very risky unless there is good reason (data) to believe that the consumer will understand the particular words. It also has the function of focusing subjects on attributes that they may or may not have thought important to their affective response. As discussed in Chapter 6, sensory descriptive attributes must be defined and understood, if the responses are to be meaningful. Often, these attributes are included as another quick disaster check of the product. The problem with use of them by naive consumers occurs when the results are used as the basis for specific product changes. Remember that with a large N, it does not take much of a difference in score to be significant. Obviously, the sensory analytical test (e.g. descriptive analysis) will have sufficient data of its own to provide for a more realistic and practical assessment of any proposed product changes.

Limiting the sensory acceptance test to measuring acceptance will reduce the risk of obtaining biased, unreliable, or invalid results. The scorecard must be a "null" instrument, having no influence on the subject's preference behavior. Including a battery of diagnostic and attribute questions increases the opportunity for various context effects to develop and subsequently affect the preference response. As reported by Mela (1989), the social psychology literature contains many examples of the influence of the contextual effects of questionnaire design on opinion survey responses. Contrary to that literature and what is known about sensory fatigue and various testing errors, Christensen *et al.* (1988) and Mela (1989) reported only negligible effects on preference scores when attribute questions are included. However, Mela's conclusions were based on 29 subjects who participated in six different test conditions, and the stimuli were experimental samples uniquely different from what can be expected in the marketplace. Christensen *et al.*'s study appears to have not been published.

Perhaps the most inappropriate question for the sensory acceptance test is to ask the subject why a product was liked and/or why it was disliked. As noted in Chapter 3, this is not a recommended practice for sensory evaluation. The question will be answered by some and ignored by others, unless prompted for a response. The information is difficult to interpret, costly to decode, and is generally operator dependent. Although it may have some value in the early stages of concept and prototype evaluation, it does not belong in a sensory acceptance test. As Payne (1965) noted many years ago, open-ended questions are not worthwhile sources of information, considering the difficulties entailed in their collection and interpretation.

These procedures may have a role in market research, in the study of consumer attitudes, and so forth; however, they are not appropriate in the study of perception and the sensory properties of products.

7.5.5 **The many ways to ask the preference question**

The reader will observe that emphasis has been placed on keeping the acceptance scorecard simple—that is, focusing on the key question, "All things considered, how well do you like this product?" In the sensory acceptance test, this rather simple task is often subverted by requests to include numerous additional questions focused on measures of liking for various product attributes such as appearance, aroma, and flavor or a number of JAR scales. The requester of such information mistakenly assumes that this will identify the precise product characteristic requiring change when the product is less liked versus another. Regrettably, people's behavior does not often work in such a simplistic way, as noted elsewhere in this book. The perceptual process is complex, interactive, and influenced by the context in which it is functioning.

As a result, responses to the attributes tend to be highly correlated with the initial response of overall liking. Review of literally hundreds of these tests confirms the minimalist nature of the results and the general lack of usefulness of the information. In fact, if the specific attribute were significantly less liked, this could not result in reformulation efforts because the attribute is generic, not specific. In other words, if a product scored significantly less liked for flavor, is it the quality, the amount, and/or the interaction of the two? A second issue is whether the respondent understands the meaning of the attribute being measured. One can argue that everyone should know what flavor means; however, descriptive analysis subjects spend between 6 and 8 hours learning how to describe perceptions and minimize confusion about the meaning of words, so the potential for confusion is very high for the untrained participant in a preference test. A similar argument can be made if the task is to measure the strength of the characteristic rather than affect. For the sensory acceptance model, measuring liking should be the only task, and information about specific sensory attributes should be obtained from a descriptive analysis test.

In the larger scale market research test, these same attribute questions are usually present, often along with numerous other questions. Here, too, we believe the sensory attribute questions should not be asked for diagnostic and product reformulation purposes, for the very same reasons noted previously, especially if sensory has completed the early stage research. In some instances, the sensory attributes may be used in a related context of linking response to those obtained from a trained panel. In this case, their inclusion may serve another purpose.

As previously mentioned, asking for judgments from naive consumers to questions that have a high potential for misunderstanding is not recommended and can easily lead to reduced sensitivity and high potential for incorrect interpretation. It is often assumed that consumers come to a test with a thorough knowledge of the products, especially if they are qualified as heavy product users. This does not guarantee that they understand the tasks or have the ability to respond in consistent, repeatable, and reliable ways that are expected by the researcher to guide reformulation efforts. If the sensory department has reliable descriptive analysis methods in place, asking these types of diagnostic attributes of consumers is usually unnecessary.

7.5.6 What question do I ask first?

A frequently asked question is "Should I ask the liking question first and the attributes second or should I reverse the sequence?" Setting aside the comments about attributes in Section 7.5.5, the best answer is another question: "What is the most important question you want to answer?" If preference is the primary reason for the test, then it should be asked first because the responses that follow will be influenced by the initial judgment of liking. This is the halo effect, discussed in the testing errors section of Chapter 4. It is unlikely that a product liked very much will have one or more attributes that are significantly disliked.

7.6 **Conclusions**

In this chapter, we described a sensory evaluation model for measuring product acceptance and provided additional insight into the valuable synergy with the market research department. In the overall scheme, sensory acceptance represents the third and final phase of test resources along with discrimination and descriptive analysis tests. As a resource, it provides continuity between the controlled laboratory environment and the typical product use situation (home, restaurant, events, etc.). The types of sensory acceptance tests include the laboratory, central location (a restaurant, at play, etc.), and home use. All of these test types follow accepted sensory evaluation procedures and practices, which include the following:

1. All subjects are qualified based on their attitudes for selected products.
2. Tests generally use 75 or fewer subjects per test, with exceptions to reflect specific demographic requirements.
3. Product preparation, handling, and coding adhere to accepted laboratory practices.
4. Measure acceptance–preference; difference and descriptive attribute measures are not recommended in sensory affective testing.
5. The sensory acceptance test does not replace large-scale consumer tests.
6. The sensory acceptance test is intended to minimize testing products that do not warrant further consideration.

The sensory acceptance test is a very cost-effective resource that has a major role to play in the development of successful products. Properly used, it will have a significant impact on the growth and long-term development of the strategic use of sensory evaluation.

Strategic Applications

8.1 Introduction

This chapter explores a number of topics that are of particular importance for every sensory program. They represent opportunities for sensory professionals to demonstrate the value of their participation in projects and especially the product sensory information that they provide. These are complex projects with far-reaching consequences and, not surprisingly, they require considerably more planning time than is needed to organize a discrimination test. These include qualitative research, benchtop screening, product development, optimization, quality control, and so on. It is likely that most readers have had some experiences with these kinds of projects.

They are a challenge and an opportunity for sensory staff and for a program not only because of the complexities of the specific problem but also because of their potential economic impact for a business. They will also impact professional activities in other areas of that company and usually are enmeshed in corporate (or individual) politics. Being able to develop product descriptive information and linking it with a market strategy is not available by other means and therefore has significant value (financial and/or otherwise). Whenever a project has such an impact,

H. Stone, R. Bleibaum, H. A. Thomas: **Sensory Evaluation Practices, fourth edition.**
DOI: http://dx.doi.org/10.1016/B978-0-12-382086-0.00008-X

direct or indirect, the sensory professional must be sensitive to the needs of the managers of that project insofar as success or failure is concerned and in what ways that information may have other implications. This includes the practical questions of why this particular request, who made the request, the business objective(s), the decision time lines associated with each task, along with the information delivery dates. Less obvious, but equally important, is the need to establish criteria by which project success and failure are measured, the need to determine how test results will be used, and the need to assess the sensory knowledge of those who will use the results, especially if they might enhance, threaten, or otherwise change their positions about a project's success. This latter issue is critical to the success of any test that is performed, but it is particularly important in these more strategic projects.

The sensory professional must also consider the impact on the sensory program. There can be situations in which the sensory program will lose credibility if little attention has been given to the details of a project and especially where there is inadequate documentation leading to a test that does not provide the expected information. In this book, emphasis has been placed on developing the tools to operate a sensory testing program in its entirety: Having a pool of qualified subjects, being able to organize, field, analyze, and report results for any test type (discrimination, descriptive, and affective), and being able to provide information that is understood and actionable are all essential for success. As discussed in Chapter 4, detailed and documented knowledge about a project and the test objective(s) is absolutely essential before an appropriate test plan and design can be formulated. In some situations, this knowledge is sufficient to identify potential challenges in the proposed project and to consider alternative courses of action. However, this does not prevent sensory professionals from becoming involved in more complex problems, often having some political undertones. Some examples of these are comparing results obtained from consumer insights with those obtained from a trained panel, or comparing results from a central location test with those from a home-use test or with those from an employee acceptance test. In the case of experts versus a trained panel, there can be considerable confusion regarding the points of difference between the two functions, and some managers might question the need for both, especially if both appear to provide the same information. The sensory professional must address these concerns; however, any response must be taken only after consideration of all the issues. The first and most important step is to determine why there is interest in such comparisons. The answer to that question will usually lead to further discussion, including how the information is intended to be used before a decision can be reached regarding the next steps. Note that up to this point, no decision has been reached regarding what testing, if any, will be done.

All of this effort is intended to define what sensory evaluation does without making any judgment, for example, about what experts are doing and how the obtained information is used. In most instances, experts focus on a company's products, and rarely do experts evaluate competitive products, whereas trained panels typically evaluate company and competitive products. In most instances, experts make quality judgments (directly or indirectly) on a finite scale, and the correlations between the two panels' attribute responses will therefore not be direct, which will make for a more complex comparison. Each panel provides information with different applications, and any comparison could be counterproductive. Of course, there is no guarantee that these suggestions will be accepted; however, it is clear that sensory professionals must recognize there is a constant need to educate managers as to what sensory can and cannot provide and to be able to recognize requests that represent potential problems. Should actual testing be done, the rules by which success is measured must be established before any testing is initiated. This particular issue was discussed in Chapter 6.

The ultimate success of a sensory program in ventures such as those listed previously will depend on the care with which each problem is studied and the exchange of information that occurs before considering even the most rudimentary of tests. In general, comparisons of results from tests done by different groups in a company should be avoided because they serve no useful purpose and are a waste of resources. Often, it derives from a lack of understanding as to what kind of information each is providing. This is especially important because a sensory program's history may be new and there is little positive experience to draw from.

In the following sections, we discuss each of the topics listed at the beginning of the chapter. The first four are grouped together because they follow a logical sequence involving activities associated with product development opportunities, whereas the others are listed in no particular sequence. The primary emphasis is to provide a dialogue about each problem that may be helpful to the reader, who at some time will be confronted with similar situations, if they have not already experienced some or all of them.

8.2 **Front end of innovation**

This section details some of the applications of sensory methods to innovation and the new-product development process. Overall, qualitative methods are becoming more commonly used by sensory research teams with a focus on product development at the front end of innovation. It may also be appropriate to use observational research, traditional types of focus groups, or multiday programs such as Qualitative Sensory Immersion (Bleibaum *et al.*, 2011) to better understand aspects or attributes of a product that the consumer views as needing change in a product, package, or the product category.

The innovation process can be divided into four general areas:

- Concept generation
- Concept evaluation and development
- Product development and evaluation
- Commercialization

When generating ideas, the first part of this process is arguably the most important because it has the greatest opportunity to provide a strategic plan to achieve corporate business objectives. Most companies have dramatically improved their product development cycle time and efficiency by implementing formal programs such as Quality Function Deployment, the House of Quality by Hauser and Clausing (1988), Stage-Gate (Cooper, 1993), or PACE (McGrath, 1996). A useful summary background to these approaches is described by Farnbach and Kuesten (2012) especially as it relates to decisions that need to be made at the early stages to minimize project termination before a product gets to the market and much time has been wasted pursuing an idea not consistent with management's business plan.

Attention is increasingly focused on front-end activities that precede the formal and structured product development process in order to increase the value, amount, and probability of high-profit concepts entering product development and subsequent commercialization. This is a welcome change from more traditional approaches in which key functional groups such as sensory evaluation were requested to do testing without being aware of the background.

There are two primary types of concept-generation research. The first is identification research, in which the emphasis is on identifying unmet needs in the marketplace. The second is concept identification, a research technique that is used to determine concepts that might fill an identified need. There are various ways to identify unmet needs: Some are qualitative (focus groups, observational research, ethnographic research, etc.), and some, such as segmentation research, are quantitative. In addition, marketplace trends and social media can be analyzed to generate ideas.

Perceptual maps are frequently used to generate new product ideas or to determine unmet needs. Within a category, all products may fulfill the same basic need, whereas a new product in a separate space may fulfill an unmet consumer need. Observational or ethnographic research may also be used to identify unmet needs. Trained ethnographers or anthropologists go into the field, in consumers' homes or places where the product is typically used. These trained researchers may spend hours or days observing consumer behavior, searching for problems, issues, and new product ideas.

In the early stages of new product development, one of the best ways to understand spoken and unspoken customer needs is to observe the consumer "in the field," or where the product is or could be typically used. There is clear evidence that consumers are not good at articulating their wants and needs. What is obvious to a consumer may not be clear to the product developer, marketer, or manufacturer.

From these research processes conducted early in the product development cycle, a variety of new product ideas may be generated. At this point, concepts may be identified that form a basis for proceeding to the next step of concept identification or the next phase of research.

There are various ways to identify concepts. Some are neither user based nor follow from an identification of unmet needs. For example, a technological breakthrough can suggest concepts, or competitors may introduce a new product that represents either a threat to or an opportunity for an organization.

During the new-product development process, there is usually a point at which a concept is formed, but there is no tangible usable product that can be tested.

The concept should be defined well enough so that it is communicable. There may be simply a verbal description, or there may be a rough idea for a name, a package, or an advertising approach.

Usually, concept testing will explore several versions of a concept or several product concepts responsive to a user's needs. Most businesses spend a great deal of time and resources on the upfront activities to develop ideas for a new product, explore the appropriate positioning for product concepts, and test these concepts to determine which one(s) should proceed to protocept (physical product that is typically made on a small scale to represent the concept idea) and prototype development. Generally, concepts must achieve specific corporate or category benchmarks or "hurdles" to move forward.

Exploratory research is usually conducted by marketing research and/or consumer insights and is closely associated with the subsequent packaging and advertising research. Research involved in the early stages may also include focus groups, omnibus studies, attitude and usage questionnaires, and actual product usage and purchase research designed to better understand consumer needs and to determine potential for the new product. This could also include an example of "descriptive analysis," in which consumers are asked to review a concept and then to describe the sensory characteristics that they would expect in that product and its perceived usage (when, where, and with whom).

Once a concept has been developed and approved for the business, the product developers start their work. It is at this stage when the sensory evaluation testing comes into the process. However, there are benefits to involving sensory evaluation and the product technical teams early in the front end of the innovation process.

Innovative new product development requires prolonged interaction between innovators and users on problems that are important for both parties. However, caution must be taken to ensure the changes are for a broad enough audience. Customization may not lead to innovation or increased usage. The goal of researchers or innovators is to recognize needs that consumers may not articulate. Observing behaviors can often provide more insight into unmet needs than direct discussions that are typical of focus groups (hence the growing popularity of ethnography and immersion techniques). During the innovation stage, it is important to reflect on all of the consumers' potential problems and needs. In addition, any innovative technology should enhance consumers' experience rather than impede or diminish it.

In product development, there is a trend toward segmentation (Gruenwald, 1992)—either market segments or sensory preference segments which use information to design products for specific target markets and away from the middle majority. This is where most of the innovation is occurring today. There are also newly developed business units based on a combination of superior technology and marketing.

Although we tend to think of segmentation as relatively new, Daniel Yankelovich alluded to this in 1964: "While demographic segmentation is a traditional way of analyzing markets, it is only one of many analytic methods. Segmentation analysis allows marketing executives to consider buyer attitudes, motivations, values, patterns of usage, aesthetic preferences, and degree of susceptibility."

Marketing segmentation is a critical step in developing a competitive strategy in today's business environment. Effective segmentation requires understanding consumer values, usage, and attitudes. Segmentation methods should lead to insights about consumers, competitors, and new product opportunities. For sensory science, preference segmentation is of primary interest. That is, segmentation is based on consumer responses to various products in a given array. The primary reason to explore segmentation is to better understand the target consumers such that products can be designed to meet their expectations for key attributes. This involves a combination of descriptive analysis and affective testing to fully understand which attributes drive liking for various consumer preference segments.

When there are "gaps" in the marketplace, and the corporate business strategy is on new product innovation, a project team involving staff from marketing, marketing research/consumer insights, product development, and sensory evaluation can be critical for long-term project success. Typically, this also involves working with a research supplier, such as an advertising agency, marketing research, or sensory and consumer testing firm. Often, suppliers are used to develop unique approaches using proprietary research techniques. The best suppliers have a wide range of experience and knowledge in a large variety of products and categories—in many instances, beyond that which one would learn inside one's own corporate environment.

Early stage descriptive analysis will identify attributes and combinations of attributes that represent a product that can be applied to a variety of situations when exploring "gaps" in the marketplace. By studying a preference map, for example, researchers can learn how consumers place products by usage occasion, primary and secondary benefits, features, emotions, and other product, packaging, and marketing aspects to best determine areas for new product opportunities.

Once the project team has been assembled, it must clearly define the business and research objectives which may include the following:

- Define and target consumer needs to be fulfilled.
- Describe which products consumers currently use to satisfy those needs; this may include substitute products, such as ones consumers would use on similar occasions, situations, etc., or a directly competitive set.
- Screen substitute products and/or the competitive set to develop the sensory and marketing array.
- Develop and screen prototypes and concepts that may meet consumer needs and that reflect a broad range of sensory impact and marketing/usage experiences.
- Regularly check with management to ensure that the product fits within the company's market strategy.
- Identify potential weaknesses of the product relative to the competition; that is, are there technological secrets that could limit competitive options?

Usually, the project team identifies the target market and consumer needs, which is done by reviewing existing research, usage and attitude studies, business plans, etc., but it can be most beneficial to involve the consumer early in this process. The primary goal is to identify any unmet needs or marketing opportunities for products that are closely aligned with the corporate strategy and which ones satisfy underlying needs of the consumer.

There could be an array of short messages or concept statements that deliver a specific idea regarding consumer needs. Statements represent the core message of consumer need and product fulfillment. These statements could be used to develop a more formal marketing concept following initial consumer affective testing.

Focus groups generally bring together 6–10 consumers per group to discuss various issues and/or to react to concepts or products. Behind a one-way mirror or via a web-based connection, carefully selected observers monitor the behavior of the participants. A professional moderator runs the focus group, following a general script that includes 5–10 key question areas to ask or behaviors to probe. It usually, but not always, takes three or four groups to collect the information. Each group should last a maximum of 2 hours, and in some cases multiday groups may be required.

Stimulus material may be real material (e.g. products, advertisement, and promotional material), rough material (e.g. concept boards, prompt or word boards, photomatics, narrative videos, storyboards, mock promotions, mock editorial, blogs, and mock websites), or open-ended or ambiguous material (e.g. words and picture sorts, collage boards, and mock interviews). The analysis and report writing from the focus groups involves gathering all the materials and documents collected, reading the moderator's transcript and report, viewing or listening to the recorded sessions, and reading each transcript. Key learning is derived from the exercise, which is then used in the decision-making process. Some companies offer simultaneous translation of each focus group such that the research team documents its key learning immediately following the groups.

Individual or in-depth interviews are one-to-one discussions that allow the interviewer to probe the responses of the interviewee in more depth than in focus groups. From 10 to 40 interviews are typical, depending on the nature of the problem being studied. Individual interviews are well suited for the following situations:

- When the topic of discussion is confidential, personal, or potentially embarrassing
- To study behaviors where socially acceptable norms exist and where pressure to conform in a group discussion would influence responses
- When detailed individual information is required concerning complex behavioral or decision-making processes
- When interviewing professional people about their work
- To obtain feedback on a lengthy piece of written or taped material

Projective techniques have consumers engage in various qualitative exercises, individually or in groups. Respondents are not asked to talk about themselves but, rather, about other people or imaginary situations, objects, events, etc. Each respondent must own and describe his or her "projections" to the moderator and others present. Analysis of the data involves data reduction, data display, and drawing and verifying conclusions.

In the context of consumer testing, ethnography is defined as the observation of the consumer in his or her natural environment (or a contrived environment) by professional ethnographers or observers, who may or may not be disguised, with or without the assistance of specialized equipment such as video cameras.

Observation is a useful approach when

- the phenomenon under investigation is easily observable;
- the phenomenon under investigation is a social process or mass activity;
- the consumers under investigation are unable or unwilling to communicate directly with the researcher; and
- the phenomenon under investigation occurs at a subconscious level.

Observation gets at unarticulated needs of the consumers and is a great innovation tool. The analysis, interpretation, and use of the data are complex, however, because the data are qualitative.

For the front end of innovation activities, concepts and new product ideas are generated using a wide variety of methods. Sensory professionals can become strategic partners in these front-end activities to help provide focus and structure to the research team in developing new products. It is important to keep in mind that front-end research is qualitative and should be verified at key stages. Reviewing results from a series of qualitative sessions should yield information that should be rephrased and tested with another and larger population. External validation is accomplished in today's wired world, provides confidence in the project, and adds a level of comfort when management decisions are needed.

8.3 Product development

Product development is an important function in every company. By product development, we refer to such activities as the formulation of a new product (new to that company or entirely new in that market), reformulation of an existing product, use of new technology, a new ingredient, or some other activity that directly impacts a product to a degree that it can be promoted to the consumer (or target market) as new or some similar designation. Interest in new products is always very high because of its contributions to profits, the opportunity to achieve growth in the marketplace, as well as the halo effects associated with the brand achieving success in the marketplace impacting other products associated with that brand. Despite this high level of interest and investment, much of it is never fully recovered because of the high failure rate, and this situation has not changed very much in the past several decades (Friedman, 1990; Fuller, 1994; Gorman, 1990). As a result, an enormous amount of attention has been given to correcting this situation. It has attracted all sorts of attention in the form of people with ideas, training programs, books, symposia, etc. See Aaker (1996) for a purely marketing focus; Baker *et al.* (1988), Fuller (1994), and Graf and Saguy (1991) provide a more technical product orientation; and Lyon *et al.* (1992) offer a practical primer on using sensory testing in product development. See also the two-part article by Manoski and Gantwerker (2002a,b) and a related article by Silver (2003) for some practical observations regarding the challenges of product development and some insights about product success and failure. Publications such as *Food Technology*, *Food Product Design*, *Prepared Foods*, and *New Products Magazine* devote a significant portion of every issue to this topic. In the case of the latter, its focus is exclusively new products, acknowledging its importance to the growth of the food and other consumer products industries. Annual conferences focus on the topic, bringing together individuals who are suppliers of services as well as those who describe their successes and, as noted previously, giving rise to an entire industry of information sources about identifying unmet consumers. New methods and techniques are described, such as the use of ethnographics (discussed previously) that are intended to directly or indirectly enhance the rate of success by getting closer to the consumer to identify unmet needs, product opportunities unique to a brand, etc. Older methods such as laddering are being used as well. Companies have also made significant investments in personnel by hiring chefs as a means of exploiting their creative efforts. Whether these investments will improve the rate of success remains to be seen. It is also not clear that existing methods are inappropriate, only that what is new receives more interest and attention—the implication being that because it is new, it must be better. Although it is not the intent to provide an exhaustive review and analysis for success or the lack thereof, it is useful to comment on some of these practices as they impact the role that sensory evaluation plays in the product development process. Perhaps more important, some companies have taken a more strategic approach to how the entire process is structured. Highsmith (2009) and Farnbach and Kuesten (2012) provide useful information about the process along with some case studies.

It should be clear, at the outset, that sensory evaluation has a very important role in product development, beginning with involvement in the early planning stages (as discussed previously) and progressing to the more typical role of evaluating products during the formulation–reformulation stages and into full-scale production. Involvement during planning is essential to identify the resources required, including a description of the target consumer, and to estimate appropriate timing for implementation of steps such as initial product evaluation and expected dates of test markets. Although a product development project may represent a line extension or a modification of existing technology and no problems are anticipated, it is nevertheless important that sensory professionals are aware of the project plan and the proposed timetables for the key decisions. Often, an existing database for the particular product category can be accessed and be of value to technology, before any actual work is initiated. For example, a proposed line extension might include formulations previously tested, and the data (from these earlier tests) could provide an additional perspective to this planned change.

Ideas for new products are derived from a variety of sources, including employees, letters from consumers, market intelligence, new technologies, as well as more formal approaches, many of which are described in the literature (Aaker, 1995; Bone, 1987; Engel *et al.*, 1995; Fuller, 1994; Henry and Walker, 1994; Meyer,

1984; Prince, 1970). These ideas are then evaluated, usually through use of the focus group technique; however, the focus group itself may also be used for generating new product ideas. Information about a particular concept is obtained from selected consumers, usually in groups of approximately 10–12. In some situations, the discussions are done on a one-on-one basis. The choice of technique depends on the status of the project, the nature of the topic being discussed, etc. The advantage of group sessions is associated with the interactions among attendees that are not possible in the one-on-one environment. In the group environment, a professional moderator is responsible for managing the process, starting from a general introduction and then directing the participants' conversation to a particular topic, eventually to discuss the specific idea, type of product, or specific features that might be incorporated into a product. Sessions last approximately 1 or 2 hours; the information is recorded, and a detailed report is prepared. This information is carefully reviewed to identify positive and negative attributes and is used as an aid to better characterize the particular concept. This procedure may be repeated with other groups within the same market region and in other regions, depending on the outcome of the initial focus groups and on various market and business considerations. It is also reasonable to expect that information developed from initial groups will be used with subsequent groups. This serves as a kind of cross-validation of the information developed.

The focus group technique is extremely popular with marketing and market research professionals, and in recent years it has also become popular in the technical community. Although the technique is primarily qualitative, it is popular because it is a relatively easy window into the minds of consumers who are representative of the intended customer. Focus groups are easy to organize and are relatively low cost. As a result, they are often used in extreme; that is, a particular project could initiate 10 or more on a single topic and would cost approximately the same as a product test involving 150–200 consumers. In addition, the sessions are recorded and transmitted to other locations worldwide. Individual experimenters can participate regardless of their location. Developments in the use of focus groups can be found in various marketing research publications (e.g. see *Quirk's Marketing Research Review* 16(11) (2002) as well as Fern (1982), Fuller (1994), and Goldman and McDonald (1987)).

In recent years, techniques such as ethnographics have been added to the qualitative tools available to the creative process, to identify product gaps/opportunities, etc. The technique allows a trained researcher to "move in with a family" and/or allows cameras into a home to fully capture how a family functions and specifically when, where, and how it uses a specific product. The process is especially attractive, in part, because it fulfills the marketing team's desires to get close to the consumer. With in-home cameras, one can directly observe the consumer (the entire family) use/prepare and consume a particular product. Not surprisingly, the process has achieved some successes, but as with any qualitative tool, it is subject to different interpretation based on the choice of experimenter or observer. Most practitioners acknowledge that interpretation is critical, but equally critical is the need for independent validation. Together with traditional focus groups and other qualitative techniques, the process of developing and refining ideas into product concepts is an essential first step. Alternatively, there are those who develop a product, do no formal qualitative research, and are successful, proving that proceeding "by the book" does not guarantee success. However, following the aforementioned procedures should, at the least, identify those ideas and/or concepts that are not well received sufficiently early in the process to minimize the expense of moving ahead with a concept that will not be successful.

Quantification of information obtained from focus groups should be encouraged. We have converted information from the focus group into a series of closed-ended questions that are scored by a larger group of consumers. This procedure provides an independent assessment of the qualitative information.

A different perspective is achieved when a food company employs a chef to create products that reflect that individual's creative efforts. Historically, companies employed technical groups (the D of R&D) to create new products. With restructuring and mergers, much of the development effort was outsourced. Today, companies have changed, enhancing the development effort with chefs. However, the basic challenge remains—that is, satisfying consumer expectations with positive sensory experiences. It remains to be seen whether the use of alternative approaches to product development, such as the chef as technology's creative arm, will be sustained. After all, the successful product must provide a positive sensory experience that is consistent with expectations and associations engendered first by seeing and/or smelling the product. This experience is confirmed (or not) by taste, mouthfeel, and aftertaste. The "tasting" experience reinforces the sensory and imagery expectation created by the product's visual and olfactory characteristics.

At this stage of a project, various types of research are initiated to assess concept/product feasibility. This research, which is sometimes collectively referred to as front-end research, should provide five types of information: a concept that scores well with target consumers, a measure of concept viability (validation), a list of consumer-defined attributes (most likely benefits and uses), target population demographics, and identification of an initial control or competitive product. Although some of this information is qualitative, especially the consumer-defined attributes, it is very important for sensory project staff to be aware of the information or at least to know who to contact for the information. Knowledge regarding what type of consumer should participate and what products should be included also needs to be defined because this will have a major impact on results. In situations in which the consumer is not the actual target, a request for an updated definition of the target is needed. What is important here is to emphasize communications among all parties, especially in the early qualitative stages.

The focus group in all its versions, as already noted, is an essential tool for the product development process; however, it can be overused and/or the results can be subject to misinterpretation. This happens most often when an inexperienced (or experienced but biased) individual is observing the group and assigning too much importance to some of the comments from one (or two) participants. Unfortunately, this selective listening, a trait common to individuals who hear only what they want to hear rather than the entire discussion, can have disastrous consequences for a project. Development efforts can move in the wrong direction, etc. Some individuals will field 10 or more groups and collate the information but present it as quantitative without ever establishing that it is reproducible. Although it is not a sensory responsibility to question this approach, it is important for the sensory professional to be aware of information sources and be able to place them within the context of the project. In most instances, however, once agreement is reached concerning concept viability, formulation efforts are initiated and sensory evaluation should have a more active role in developing quantitative product information.

As we observed in a previous publication (Sidel *et al.*, 1975),

> *The objectives of marketing and sensory testing during product development are to determine the relationship between a product concept, a control product (when available), and the experimental product for the purpose of aiding product development efforts to converge on these points. (p. 109)*

This is the desired goal; however, there are many pitfalls in this process, and the most logical of plans are frequently changed for reasons that can be wise, and sometimes not so wise. Sensory evaluation must provide needed services, maintaining an independent but cooperative position as concepts appear to change, priorities and timetables shift, and key personnel are transferred.

As soon as is practical, testing should be initiated to establish a sensory database for the product or product category. If control product is available (identified from the concept and focus groups), then the question of what to test is relatively easy—control product and other similar products. In developing an initial database, some care must be exercised in product selection, especially with regard to formulations that may not yet be ready for evaluation. This testing may be intended to further refine the consumer language, to understand the ways in which the products differ, and/or to provide a more precise sensory description of the concept. As noted elsewhere, product source is very important and should be documented, including preparation location and date. This information becomes increasingly important when the concept and/or the target population shifts (discussed later).

From a strategic standpoint, the most useful sensory methodology will be descriptive analysis (e.g. Quantitative Descriptive Analysis (QDA)) for several reasons. First, it provides a quantitative map of the products, delineating all the characteristics and their differences. Second, the method enables one to test multiple products. Third, it enables one to compare results from one test to another. As many as 20 or more products can be evaluated, and the evaluations can be done in a laboratory environment or in typical use situations. Mention was made previously about product selection, and particularly selecting formulations that may not be ready for testing. Once a concept has been described, formulation efforts should result in a large number of prototypes. Some of these prototypes may be very different from others, reflecting different interpretations of the concept (or of the target product) or the creative efforts of technologists in association with chefs. One of the challenges that the sensory staff faces at this stage is the difficulty of convincing developers to prepare sufficient numbers of products for evaluation without regard for whether the products are considered "acceptable." There is a reluctance of developers to provide more than two or three samples for testing. This often derives from a mistaken belief that the test results might be used to judge their progress or a desire to bypass any sensory analysis and go directly to an acceptance test (thus saving time and money). In most instances, this results in failure and time is lost. The value of the information obtained from a multiproduct test is far greater than that obtained from a test of two or three products. Multiproduct tests have the potential for developing causal relationships that have even greater value by identifying those formulation changes that impact attributes in negative ways (e.g. increasing the strength of characteristics such as artificial, mealiness, etc. that are known to reduce preferences) or are not perceived as significantly changing the strength of specific product characteristics. This latter situation is particularly important because technologists and chefs may assume, usually incorrectly, that any formulation change (ingredient type and/or amount) regardless of the magnitude of the change will result in a perceived change. Without empirical evidence derived from a design study, such assumptions are only that. In fact, one needs to make a change of at least 10% for it to be detected. Nonetheless, the creeping process seems to be deeply embedded in the subconscious, which will hinder progress.

The sensory professional must make a concerted effort to reassure the development specialist of the usefulness of having a range of products and not be overly concerned about whether they are good, bad, or indifferent. At the same time, the sensory professional may find it necessary to modify test plans to incorporate a more diverse array of products than originally intended. As discussed in Chapter 4 on errors in testing, inclusion of one or two products that are very different from others will very likely cause contrast and convergence effects. Inclusion of such products may be necessary, and this would require a change in the test design—for example, blocking the serving order such that these products would be served last. In effect, the sensory professional must be flexible in terms of adapting a methodology when presented with atypical products. The decision as to which products will be included in a test should be based on bench screening and extensive discussion with the project specialists and any other project team members willing to participate. It is wrong to test only those products that represent best effort. Unfortunately, this latter practice makes it very difficult, if not impossible, to understand the effects of ingredient and process variables, and particularly those variables that affect desired attributes in a negative way. What should be kept in mind is the need to include a sufficient number of products to yield an array of differences (and similarities) that will be helpful in subsequent formulation and processing efforts. In addition, the project staff should be reminded that results of these initial tests are a part of the database and not intended as the basis for changing the concept or assessing progress of development efforts. Too often, decisions are reached about the progress of development based on tests designed primarily for other purposes (the database), and in this particular instance, such decisions could have significant consequences for the project staff. Some individuals may decide that product development emphasis should be changed or will recommend termination of the project. After a few tests have been completed, there will be sufficient information to allow for an assessment of progress relative to the program objectives.

A brief comment is warranted about the problem of the changing product concept. Product concepts change for any number of reasons. For example, new market information may reveal that the original concept was not sufficiently well defined, additional testing with consumers may dictate a change in positioning, or competition may have changed their product. Technology may change, or realignment of marketing personnel and of responsibility may cause changes in project direction. These changes are often made with no change in budget or timetable, and that clearly impacts a variety of activities, including planned sensory tests. This is an opportunity for sensory evaluation, given the responsiveness that is possible—for example, to field a series of descriptive tests in a week or less and provide actionable information. The existing product sensory information will be especially helpful in this situation. If the product range (in the initial tests) is sufficiently broad, then many concept changes will probably continue to fall within the product category. However, not all changes can be anticipated, and sensory evaluation cannot test all product possibilities. The product development process is never completely predictable, and sensory evaluation should accept this lack of predictability as an aspect of the work, especially during the initial planning and work stages.

Up to this point, some of the more obvious actions taken during the early stages of a new product venture have been emphasized. As these activities are formalized, the program should develop into a more orderly process. As shown in Table 8.1, events in the development process have a logical sequence (assuming there are no unanticipated diversions). However, the development process is neither logical nor orderly, and the ability to respond quickly, to take a flexible approach, will enhance the sensory group's credibility with project managers and management in general. For purposes of this discussion, we have focused on three functions: market research, product development, and sensory testing. However, it is likely that product development will involve the efforts of many other groups, such as purchasing, advertising, production, and quality control, but that their involvement will not always be continuous. The sequence of events, as presented in Table 8.1, is based on the assumption that there is an identified control product representing the concept. Two other scenarios are possible: if no control product is available and if the concept is so novel as to have no direct counterpart in the marketplace. The effect of these scenarios on the development sequence will be considered later in this discussion.

The first three steps in Table 8.1 may be considered as the source for the initial database; extensive communication about a concept will be required, and bench screening of various prototypes before and possibly after sensory testing will be necessary. Once the concept has been refined (steps 4a and 4b), it is likely that the development staff will have refined their skills and will have several prototypes available for comparison with the control product (step 5). The most promising products are screened using the discrimination model (step 6a) or using descriptive analysis, if the differences are obvious. The descriptive model is probably

Table 8.1 A List of the Sequence of Events in the Product Development Process When Control Product Is Available	
1. Market research	a. Identify product category.
	b. Develop and test concept.
	c. Identify control product.
2. Product development	Formulate prototypes based on 1a–1c.
3. Sensory evaluation	
Descriptive analysis	Develop database with control, competitive, and prototype (or protocept) products.
4. Market research	a. Refine concept with focus groups.
	b. Develop market and use strategy.
5. Product development	Formulate protocytes based on three to match control.
6. Sensory evaluation	
Discrimination	a. Determine which prototypes match the control product.
	b. For product that matches, verify with additional formulation.
Affective	c. Test developed and control products in laboratory and at central location sites.
	d. Test developed and control products in home use.
Descriptive analysis	e. Define current developed product and other formulation for use in manufacturing and quality control specifications, and identify any sensory properties responsible for unsatisfactory acceptance ratings.
7. Product development	a. Shift from pilot plant to test production.
	b. Initiate cost reduction.
8. Sensory evaluation	
Discrimination	a. Evaluate production product and impact of cost reduction.
Descriptive analysis	b. Evaluate any products that fail discrimination test.
9. Market research	a. Large-scale consumer tests in selected markets.
	b. Evaluate advertising, package, and pricing.
10. Product development	Initiate line extensions.
11. Sensory evaluation	
Descriptive analysis	a. Evaluate line extensions.
	b. Evaluate competitive products.
Affective	c. Test line extensions and competitive products.

Source: This table was developed from a previously reported article by Sidel et al. (1975).

most useful at this stage because an operational panel will be able to evaluate six to eight products and results available within 3 days or less. If the discrimination model is used, then any of the prototypes that match the control will be verified using another batch of product (step 6b). Obviously, confidence will be increased if developers can formulate the product a second time that is perceived as not different from the control.

This leads to the next step, controlled or preliminary affective testing (step 6c), which could involve an employee panel and/or selected consumers brought to a central location. The primary goal for this activity is to establish a baseline for product acceptance without benefit of any imagery. If the prototype product meets acceptance expectations, then home-use testing may be initiated (step 6d). By "meets acceptance expectations," we mean that the product achieves a selected liking-preference score (as a minimum) considered as necessary before the product will be considered for greater investment, a larger scale test, or some other type of qualifying challenge. One of the responsibilities of sensory is to determine in advance of these tests if minimum scores are considered necessary or if some other criteria will be used by management to assess a product's market potential. Once results from the home-use test are confirmed, sensory must evaluate the product in a descriptive test (step 6e) to quantify the product profile. This information should be used by quality control and manufacturing, but it may also be used by others as product development efforts change, etc.

Assuming the product passes all requirements, the project moves to step 7, and the development effort switches from reliance on pilot plant samples to use of regular production facilities. At this stage, purchasing, production, and marketing will focus on ingredient cost and availability, estimated production volume, etc. It is at this stage that ingredient changes are likely to occur, leading to reformulation and retesting. Once again, the discrimination model is most appropriate (step 8a); however, inherent variability in raw materials may necessitate going directly to descriptive analysis.

At this stage, the program will receive considerably more attention from the project management team or possibly senior company management. There is an even greater financial commitment to provide product for large-scale marketing research tests, for production to determine the ease with which the product can be manufactured, packaging issues (type and label contents), distribution, advertising, etc. All of this comes with an increased cost and risk. If sensory data are used in support of label or other advertising claims, then a thorough and detailed review of these data is necessary. It is also possible that the competition may introduce products, may change their advertising, or may undertake some similar action designed to offset the uniqueness of the new product. This competitive action may force a reassessment of the product, and additional formulation work may be initiated. Descriptive analysis (e.g. QDA) would be appropriate for these products.

The results from these descriptive tests will be very important if project management is concerned about the competition and is contemplating reformulation. It is relatively easy to chase competition and to lose sight of the original goal. The descriptive analysis results serve as a reminder and a focal point for any formulation work. However, this reminder will be disregarded unless the sensory professional makes the effort to communicate the information in a way that is easily understood. At the conclusion of step 9, the project enters a new stage and may be transferred to specialists that move projects into full-scale production. For sensory evaluation, most of the work will have been completed and pertinent information communicated to legal, quality control, and product management.

Typically, project staff attention now shifts to the formulation of line extensions (steps 10 and 11). Here, too, the database (descriptive analysis and acceptance) serves as a frame of reference in the screening of submissions before more extensive testing is initiated. Sensory and development specialists will be coordinating their work and functioning at optimal efficiency. It is possible that some (or all) of the line extensions will bypass many of the aforementioned testing steps; a strong correlation between internal and external results serves as the basis for reducing the total number of required tests, especially some of the larger scale consumer tests. Once the product and line extensions have been introduced and are in full production, sensory evaluation should summarize all test results in a single document. Documentation is important in that it serves as a record of the range of products and product variables that were evaluated and the relation of the results to other tests (home-use, central location, and test market) and provides other pertinent information.

As noted previously, there can be other scenarios: No control product matches the concept sufficiently well enough to represent a basis for proceeding, or no formulation matches the concept well enough. In the former situation, in which no control product matches the concept, the initial steps are identical with those in the first example—developing a database, refining the concept, and product formulation (steps 1–4). In some instances, the concept can be stated in several ways, and one can measure which formulation best matches which concept. With the control expected to be similar but not identical with the concept, formulation work and bench screening may actually be more difficult. For example, it may be difficult to determine how similar the products should be to the concept (this is discussed further in Section 8.4 on optimization) and which variables should be emphasized.

It should be obvious that this scenario and the subsequent scenario are the most challenging from the development and sensory testing perspectives. The products to be included in the initial tests are not so easily delineated. If the products are not expected to match the concept, the discrimination model is inappropriate at this stage of the project (step 6a).

Acceptance testing (step 6c) will be helpful and will provide direction to the formulation efforts (especially when combined with the QDA data) and an estimate of the degree of liking for the products. The latter condition is important because the product will not match the concept. If none of the products score well in the test—for example, achieving a 7.0 (±1.0) on the 9-point hedonic scale—then some additional descriptive analysis may be necessary.

In some instances, one can have subjects develop their own ideal product using their existing descriptive scorecard. Subjects score the characteristics based on their "ideal" product. The resulting data may help identify those experimental products closest to the "ideal," and this could aid the developers in focusing their work. However, some additional comments are appropriate here. We observe that a subject's "ideal" often represents a target to aim for but subjects may not "expect" or believe it can be reached. In addition, asking subjects for their ideal as a direct question does not necessarily mean that it is also important (enough so as to have a direct effect on preference behavior and purchase intent). The discussion on optimization (Section 8.4) provides a description of an alternative and potentially more successful approach to identify what is necessary for a product to be ideal.

If results of the acceptance test (step 6c) meet the score specification, then the sequence of events should follow those described previously, with the descriptive data from step 6e used as the new target. Steps 7–11 will follow in sequence, much as occurred when control product was available from the start of the cycle.

When a product counterpart to a concept is not available in the marketplace, then it will be necessary for a protocept to be developed. A protocept is a product that represents the concept prepared with ingredients that would be available in a kitchen (if it were a food), using best ingredients. This should be prepared by a chef familiar with the project and able to provide options for further consideration. This protocept would not likely be manufactured; however, its role is very important in representing the characteristics of the concept. The protocept would be tested by market research to ensure that this best possible example of the concept is viable and it is as well liked as the concept. If it is not, then it will be necessary to repeat the process; however, preparing alternatives increases the likelihood that a successful match will be identified within a single test versus a series of tests that are time-consuming and expensive. Only if the protocept is viable should product development proceed to formulate various prototypes (step 2) and sensory evaluation proceed (step 3). It will probably be necessary to include the protocept in the descriptive analysis to provide for a record of this type of ideal product. Protocepts that are formulated "by hand" can be difficult to duplicate, and small differences should not be cause for major concern. The descriptive data provide a basis for assessing the impact of these differences.

Continued refinement of the concept takes place (step 4) at the same time that product development proceeds with formulation of prototypes (step 5). For sensory evaluation, as before, the descriptive methodology is the most useful technique (steps 3 and 6e). As the project proceeds, some problems can occur, primarily those associated with the protocept. If it is used for too long, the subjects may consider it a control product and diverting them from their primary goal. Because it could never be matched, a substantial amount of time and effort could be wasted on this quest. Again, descriptive analysis will help to identify the prototype that most closely resembles the protocept, which can then be eliminated. The prototype will be evaluated in a series of affective tests (steps 6c and 6d). If a favorable response pattern is demonstrated, the project proceeds through the remaining steps (7–11) in much the same manner as the previous scenarios.

Throughout this development process, sensory and market research must maintain a dialogue along with the rest of the project team, comparing test results, being very sure about the source of the products (protocept, prototype, and production) relative to the results, and deciding how to best use available resources. Additional acceptance testing will be required to ensure that the product (in several versions) continues to meet the acceptability requirements. It is also possible that some modifications will be made to the concept to accommodate control product.

Finally, some comments are warranted about some newer developments such as optimization techniques and their usefulness in product development. These techniques have many benefits and can, in fact, enable a project team to reduce the development process time and provide a direct link between product, attitude, and imagery. It provides a technical basis for formulating a product to a specified target of consumers, but in the process it provides an extensive database about product formulations, what sensory characteristics are most important, etc. Because of the multivariate nature of the testing, the program requires more planning time; however, the actual testing is more efficient.

In the beginning of this discussion about product development, mention was made of the need for information about key decision points to be communicated among a team. Topics such as the criteria by which progress is judged, how sensory information is communicated, and the extent of involvement of all team members in all meetings need to be discussed and agreed to at the start. Establishing a project team with direct communication links is especially important in any product development effort. Often, such access to information is restricted unnecessarily, and this leads to problems after tests are finished and reported. Projects become misdirected and failure risk is much higher than anticipated. In many projects, the role that sensory plays is not as prominent as it should be primarily because of a lack of understanding of sensory's contribution. Again, the reader is referred to Highsmith (2009) and Farnbach and Kuesten (2012).

This brief discussion about sensory's role in product development has focused on principles and practices, in the broadest sense, recognizing that the process of developing a product is difficult, very costly, and quite unpredictable. It is difficult because business plans and strategies will change, and sensory resources must be capable of responding rapidly to these changes. Of course, having a product database will enhance sensory's ability to be responsive and to receive recognition as a contributor to the product development process. Although much has been written about the product development process and much more will continue to be written, it remains to be seen how effective sensory evaluation can be in this process. Clearly, it will depend on the skill of the individual sensory professional and having the resources available when needed. Examining all of the information written about product development, it is clear that the role sensory has is never a prominent one, which could explain the poor success rate for new products.

8.4 **Product optimization**

Developing new products or improving existing products is a costly and time-consuming process. Companies continually seek new approaches for achieving greater success in the marketplace. Brand managers and their staff want to know the relative importance of product sensory characteristics that impact preference and purchase intent and support the development effort. Optimization has become a system of increasing importance as competition increases and markets mature. By optimization, we mean a procedure for developing the best possible product in its class. This procedure implies that an opinion of best possible is provided, and in sensory evaluation this means responses for the most liked/preferred product (McBride, 1990; Schutz, 1983; Sidel and Stone, 1983; Sidel *et al.*, 1994). The methods are based on use of multivariate designs and made practical by the availability of computer hardware and software. Optimization information has considerable appeal because of its immediate and practical applications for the marketing specialist seeking a competitive advantage. For the product specialist, the information provides a focus for formulation efforts; for quality control, it provides a basis for identifying those specific product variables that require monitoring before, during, and after processing, and so on. With unlimited development possibilities, it is understandable that optimization has considerable appeal. However, as with any new approach, optimization warrants careful study before it can be fully embraced as a worthwhile sensory evaluation resource. In this discussion, our attention is focused on different approaches to sensory optimization, with particular consideration given to the development of guidelines for the planning and design of these tests and some very practical aspects of the testing itself. Other types of optimization programs are possible—for example, ingredient costs, processing, and nutritive value. In most instances, the particular models to be discussed are applicable. In this discussion, we focus on the sensory model but with consideration of some newer options, including consumer attitudes and imagery, that further enhance the usefulness of the information. For an example of a typical application, see Robichaud *et al.* (2007).

The use of optimization models received its greatest impetus from the work of Box and associates (for more detail on the method, see Box, 1954; Box *et al.*, 1978), who developed an approach to multivariate designs described as response surface methodology (RSM). This statistical design concept was especially valuable in engineering and manufacturing in general, where it was possible to identify and control input and output parameters and to mathematically identify the combination of input variables necessary to achieve an optimal result. For example, it could be the maximal yield from a chemical or food process, achieved with the minimum input of energy within specified cost constraints. The concept of response surface was well received, leading to a relatively wide range of industrial and research applications including its use in quality control (for more discussion on this application, see Taguchi, 1987). The use of response surface in sensory evaluation has become considerably more commonplace since its earliest applications (Henika, 1982; Mullet, 1988; Resurreccion, 1988; Schutz, 1988a).

In addition to and independent of response surface designs, developments in the behavioral sciences were providing additional statistical models that were especially appropriate for sensory optimization. Multiple regression/correlation (MR/C) techniques are sophisticated and useful statistical procedures that enable the investigator to understand the ways by which variables are related to each other, as well as to develop mathematical expressions for predicting future events based on those variables without knowing, in advance, which variables are most important. These techniques are analytical in the sense that they are able to sort information, to categorize details that can be categorized, and to identify aspects that may require further inquiry. For sensory evaluation and the behavioral sciences in general, the MR/C techniques are especially useful because they have, as noted by Cohen and Cohen (1983), "capacity to mirror, with high fidelity, the complexity of the relationships that characterize the behavioral sciences." In sensory optimization, we want to identify those product variables that are important to acceptance; however, we usually do not have *a priori* knowledge about which variables should be manipulated to have the greatest effect on preference, and therefore we look to these statistical techniques to assist us in sorting the possibilities. From this sorting, a mathematical model will evolve that can be tested and refined, leading to a predictive tool that has widespread product application. The interested reader will find useful information on the topic in Cohen and Cohen (1983), Draper and Smith (1981), Gunst and Mason (1980), and Manly (1986). Both RSM and MR/C use similar types of computations in deriving the various models (both are different types of multivariate designs).

However, there are some differences between the two approaches, and we believe that these differences are important insofar as sensory evaluation is concerned. As already noted, one objective of the optimization process is to identify those sensory and other independent variables that influence preference choices (or any other dependent variable, e.g. cost of ingredients) and the degree of importance for each of those variables. If important variables are not represented in a test, then the various computations will not identify them. This is an especially important point; that is, these designs are interpolative, not extrapolative. In response surface designs, the important input variables are known in advance and systematically varied, and their output is measured. For sensory evaluation, it is usually not possible, *a priori*, to specify the important sensory variables, and therefore the selection of products for the project becomes very important. It is discussed later in this section. The inexperienced investigator should be aware that a result is always obtained from a multivariate analysis regardless of whether product selection or test variable preparation was correct. This also applies to data that are highly variable and exhibit little or no face validity. The analyses will iterate until a statistical significant result is obtained, and in the process it may eliminate some data. There is no argument about the use of tests for outliers or with the investigator removing data for a product because it is distorting the model; however, this should be done with full knowledge of the consequences (for comments on these issues, see Manly, 1986). In that regard, it is worth noting the use of a technique referred to as partial least squares (PLS) regression as a third approach, at least insofar as concerns the data analysis (Martens and Martens, 1986). The methodology was developed for econometric applications but has been applied to other fields in which "the emphasis is on predicting responses and not necessarily on trying to understand the underlying relationship between variables" (Tobias, 1999). PLS procedures are particularly appropriate when knowledge about the independent variables and their relationship with dependent variables is unclear and when there is an expected high degree of multicollinearity among the independent variables. In most instances, one wants to minimize the problems associated with multicollinearity. Also, relying on a default system without fully appreciating the consequences can lead to solutions that are interesting but may not answer the question being considered. Inasmuch as the results from a multivariate analysis are usually the basis for further work (e.g. product change) and multivariate analysis is not an end in itself, one should apply it with care (an argument for all multivariate analyses). Another modeling approach that has generated interest is the use of neural networks (Bomio, 1998; Ferrier and Block, 2001). The methodology derives from an effort to incorporate large data sets for which there are no obvious relationships except they are derived for a particular product. For situations in which little or no sensory analyses have been done, and/or there is little understanding of the relationships between ingredients and sensory responses, this approach may have relevance. However, the literature on the use of neural networks has not shown it to be as useful as originally proposed. This may be related to the finite number of variables in an optimization project and the knowledge that the ingredients in a product directly impact liking. All these techniques represent potential opportunities that warrant consideration; however, each must be viewed in the context of its usefulness in providing specific direction for requestors and not simply because it is new. In our experience, MR/C followed by RSM represents the best use of regression to identify the combination of sensory and other measures that will yield a best liked product.

As already noted, for MR/C the important variables must be represented in the set of products, whereas RSM requires manipulation of the important variables (known in advance). If the variables only need to be represented, then we can include other products of interest, especially competitive products. This approach has the additional advantage of ensuring that the important sensory variables are included in the set of products. We assume that if our products have little or none of the important sensory variables, then certainly they will be contained in other competitive products available in the marketplace. Implicit in this discussion is the expectation that many, if not all, of the products will be different from each other (different because of sensory qualities and not just by formulation or by chemical analysis). If this were not the case, then one might have far fewer products (and a much smaller database), which significantly reduces the effectiveness of the resulting model; predictions would be based on a relatively limited database. Pretesting of products will minimize the problem, but this could mean elimination of some products and the loss of the ordered matrix (which is especially important in RSM designs). As with many other sensory test plans and procedures, attaining the ideal design can be confounded by practical considerations. To address the issue of having so many products as would be encountered in a full factorial design, the use of partial factorials and related types of designs is appropriate. Such designs provide the means by which the variables of interest can be evaluated. To assist the investigator, several computer programs are available for these applications.

In an optimization study, a logical sequence of events is usually followed. As shown in Figure 8.1, the process begins with the planning stage, which includes selecting the product category, obtaining products, deciding which optimization methodology will be used, defining the consumer population, and preparation of

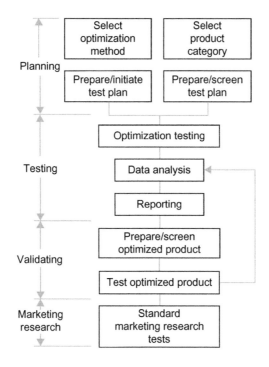

FIGURE 8.1

Block diagram of the major components of an optimization scheme.

Source: Reprinted from Sidel et al. (1983). Food Technol. Vol. 37, No. 11. Copyright © by Institute of Food Technologists.

the actual test plan. It is important to obtain products that fully reflect the marketplace and whatever is available through formulation. Based on the specific optimization method chosen, the products will be screened and selected for testing, usually done in a bench screening environment by the project team members. The methods used for the analyses are interpolative, not extrapolative, so it is essential for the products to represent the sensory space for that product category. If one is using an RSM approach, additional formulation may be necessary to ensure a suitable matrix of products. Once testing is initiated, the project proceeds in a more systematic way. Detailed descriptions of the statistical procedures can be found in the texts cited previously. These are omitted here because there are many alternatives (as described in the literature), and the specific steps are left to the individual investigator.

So far, the emphasis in this discussion has been on the concept of optimization and some of the differences between RSM and MR/C designs. It is useful if we also consider the data derived from the testing. The sensory information is the descriptive data derived from a QDA analysis, whereas the consumer's degree of liking for the products is obtained from the target population. Other data obtained should include chemical and physical analyses, formulation variables, and other kinds of consumer information (e.g. purchase intent and imagery). All this information forms the database for the subsequent analyses. The first step in this process is to examine the results in a univariate format to establish that the results are consistent with expectations regarding product sensory differences and consumer preferences. Although one might assume this to be an obvious first step, it is often overlooked, in part, because of use of default systems or because the only interest is to obtain a significant regression equation. Unfortunately, some default systems also remove responses that are mathematically outliers, further "adjusting" results with no appreciation for the behavioral consequences. The sensory and all other product analytical data are subjected to appropriate data reduction techniques (Kline, 1994). These results are plotted and examined to verify that they reflect the product differences, once again applying the "intraocular" shock test of asking if the results make sense. Attributes and analytical measures are then selected to represent the various factors, and one proceeds with the relevant regression analyses. Where there are preference segments, discussed later in this section, the results are related to the preference segments to identify which characteristics define the segments. For example, if there were two preference segments, one would want to examine the relationship between preferences and specific sensory attributes, as shown in Figure 8.2. Obviously, the presence of the segments leads to development of two

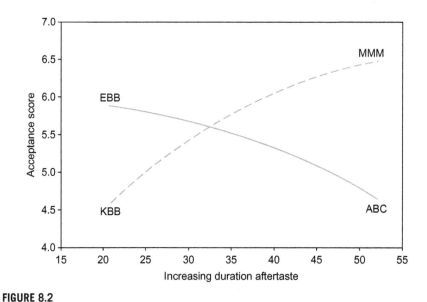

FIGURE 8.2

A plot showing a sensory attribute that differentiates two preference segments.

regression equations, etc. A linear multiple regression equation would take the following form:

$$Y = k + \text{descriptor A } (W_a) + \text{descriptor B } (W_b) - \text{descriptor C } (W_c)$$
$$+ \text{chemical measure 1 } (W_1) + \text{physical measure 2 } (W_2)$$

where Y is the acceptance preference value (dependent variable); the constant k is the equation intercept; descriptors A–C are the most important variables derived from the analysis; $W_{a,b,c,1,2}$ are the weightings for those variables; and sign $(+, -)$ indicates scale direction for that variable.

The technologist then formulates products based on his or her knowledge of ingredient and process effects on the specific attributes, and the products are tested relative to the target. Once there is confidence that significant progress has been achieved, a validation test would be initiated with target consumers. Acceptance testing would be an appropriate test of the model. As a measure of test-to-test reliability, two or three products from the initial study would be included.

Optimization studies would be incomplete if they identified only the important variables and their contributions to product acceptability. From these data, it is also possible to predict the combination of variables that will yield optimum acceptance. The prediction will consist of mathematical and written statements based on assumptions developed from the test data and will provide for more than a single optimal product; that is, optimal acceptance may be possible based on different combinations of the important independent variables.

For sensory optimization, proper product selection and the type of sensory information that is obtained are essential. The importance of these requirements cannot be underestimated. Product selection is difficult because the criteria by which products are selected will include sensory properties, technology, formulation differences, and marketplace considerations. It is important that the full range of anticipated sensory differences is represented in the test products; this could mean the inclusion of products that might not be economically viable but that have a particular sensory characteristic not readily perceived in other test products. Specific formulations might be included because of interest in particular process variables or ingredients. The inclusion of competitive products further ensures that a full range of sensory properties will be represented. In some instances, companies will undertake an optimization project but exclude competitive products without realizing that it can identify the means for optimizing one's product only to discover that it is significantly inferior to other products in the market. Satisfying the sensory requirements involves bench screening and possibly some testing; however, our experience suggests that the screening is usually sufficient to yield an acceptable array of products, usually 20–30. Bench screening will require several sessions with sufficient time to recover from the sensory challenge of sampling that many products. This is also necessary to ensure that the project staff gains a full appreciation for the range of product differences and similarities in the market relative to the project objective. In some categories, it is not unusual to start with as many as 100 or more products. This number should not be surprising if one considers the large number of formulations that are possible just with ingredient changes. As noted previously, the intention is to select an array of products that encompasses the sensory world of the product category independent of whether the products are liked. Products in regional and local markets could be included to ensure that a wide range of differences are represented.

The sensory data are developed using descriptive analysis, and the consumer provides the dependent judgments, preference, purchase intent, etc. Separating the two kinds of information is critical to minimizing the halo effects and the very high potential for multicollinearity when consumers are also providing the descriptive information. It is remarkable that this reliance on the consumer for all the judgments continues with the efficiencies and availability of descriptive analysis capabilities. Similarly, the trained panel's liking responses are biased and should not be obtained.

Descriptive analysis is the most appropriate sensory tool for optimization because there is no *a priori* knowledge concerning the important sensory characteristics. Therefore, all of the sensory properties of the products should be described, and then various factor analytic techniques and regression analyses can be applied to establish the relationships between the various sensory attributes. These results can then be used in combination with the consumer responses to identify the important variables. The literature on sensory optimization provides numerous examples of design approaches (e.g. Henika, 1982; Horsfield and Taylor, 1976; Moskowitz, 1972; Mullet, 1988; Robichaud *et al.*, 2007; Schutz, 1983; Schutz *et al.*, 1972; Sidel *et al.*, 1994). Most of these investigators incorporated sensory evaluation into their tests; however, few used a quantitative descriptive analysis for the product sensory database. With application of quantitative descriptive procedures such as QDA and a clear delineation between the analytical and the affective responses, sensory optimization has become more reasonable and a potentially valuable resource (facilitated by the access to and power of PCs to analyze results).

By coupling the different sensory capabilities within the MR/C design, there is a much greater likelihood for identifying those product sensory variables that influence acceptance in the marketplace and that combination of variables that will lead to optimal acceptance (Robichaud *et al.*, 2007; Sidel and Stone, 1983; Sidel *et al.*, 1994; Stone and Sidel, 1981; Stone *et al.*, 1991). In our experience, sensory optimizations are successful when beginning with an MR/C design approach and using the RSM design after identifying the important variables. This approach capitalizes on the power of both designs.

We would be remiss if we did not comment on the complexity of optimization from a testing as well as a conceptual viewpoint. The elements of testing were previously described, as were the challenges of design and product selection. Conceptually, optimization may appear to be an attempt to control the development process or possibly the creative efforts of specialists. It is foolish to conclude that a mathematical model will mean the end of product creativity. There is no substitute for the intellectual efforts of the specialist capable of interpreting and incorporating sensory responses into finished products that possess that combination of unique properties not possible by other means. Creativity will also be an integral part of the process not only for the product specialist but also for the sensory professional. The purpose of the optimization model is to provide a more precise delineation of the approach that has the highest probability for success with the target population(s).

Optimization procedures continue to be of significant value and interest in sensory evaluation and are now a staple in the consumer products industry. Studies have revealed new information unobtainable with smaller and less diverse product sets. One finding is that consumer preferences are often heterogeneous, and the heterogeneity is related more to sensory difference among products than to demographic differences among populations. We have observed this type of result on a worldwide basis, in very diverse countries, regions, and cultures (Bleibaum *et al.*, 2011). In unbranded evaluations, we frequently detect different "preference groups" or "preference clusters" imbedded within the aggregate test population. After data are obtained from the consumers, one can apply clustering procedures to the responses to establish whether such segments exist. Figure 8.3 is an example of an output in which three clusters or segments are identified. Because the testing was done on a blind basis, the basis for the clustering is not explained by brand usage or typical demographic criteria and more often explained by other emotional and related criteria. This information, by itself, is interesting because it can help

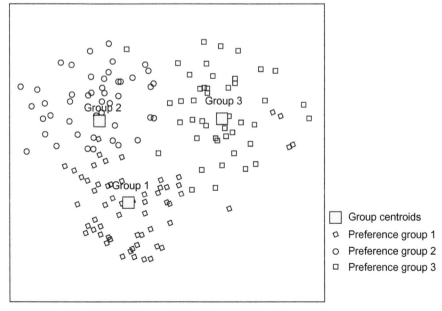

FIGURE 8.3

Plot of the output from a cluster analysis identifying three segments and their corresponding predicted optimum product.

to explain why products do well in one test and less so in another, simply because of the type of person in the test. However, the real value rests in identifying product opportunities in the marketplace driven by specific sensory characteristics and related imagery. For example, some groups are differentiated by their preference for low-intensity (flavor and aroma) products; others prefer attributes in one modality to the exclusion of those in another, whereas others prefer products that have unique attributes or intensities (see Figure 8.2). We have observed many patterns, and the task is to identify attributes important to each preference group. Once this is accomplished, a meaningful development and business strategy is possible or the converse—that is, identifying which product has the appropriate combination of attributes that best fit a specific market strategy. This may require unique products optimized for the different groups or a "bridge" product that represents a compromise between disparate preference groups. Deriving a bridge product from known preference differences is preferred to using averages from aggregate data. The latter discounts important differences and results in products with little or no unique appeal. Some attributes increase acceptance with one preference group while decreasing acceptance with another group. When group size is equal, such attributes have no influence on aggregate acceptance, and this leads to an erroneous conclusion that the attribute is unimportant or that the development changes were too small and not recognized by consumers. A worse-case scenario is a test that includes significantly more consumers from one preference group than occurred in a previous test. In this case, aggregate results will represent preferences of the dominant group, which may not agree with previous results or changes in the test product. Knowledge about preference groups and the attribute intensities they prefer removes much of the mystery of why some consumer tests seem to go awry and results do not support previous research or the current hypothesis.

A basic optimization procedure limits consumer response to product liking and models those responses to descriptive and/or analytical measures for the purpose of determining which attribute combination will result in a best liked, or optimized, product. The context provided by the products set is sufficient for deriving a successful optimization model, and this is all that may be needed. However, purchase behavior is influenced by additional factors, many of which can be, and need to be, assessed in the optimization process. The optimized product can be defined in relation to an "ideal,"—to specific uses defined by a situation and user group, packaging, pricing, and brand imagery. Products with diverse uses (e.g. an ingredient and spread) and used by different groups (e.g. children and adults) require an approach that addresses those issues. Usage and Attitude (U&A) data assist understanding consumer attitudes and behavior, and where these data are unavailable or outdated, the optimization project must include many of the same elements. Assessing and integrating these elements provides better understanding of the wants, needs, and purchase interest of the consumer. It also improves potential for developing the right product for the right population and identifying product-related benefits important to that population. The following paragraphs describe some of the enhanced measures we have used in our own optimization research.

At this juncture, we reinforce the importance of the team approach in optimization research, as additional elements from development and marketing are included in this research.

The initial phase for this research requires gathering available information. One needs to begin with usage and attitude information about the target consumer. This information may be available from U&A data, or it may need to be developed for the research. To supplement U&A data, we recruit target consumers and mail (or otherwise provide) them a proprietary situation usage survey (Item-by-Use (IBU)) that they complete in advance of the qualitative or quantitative testing. Consumers selected for the qualitative sessions are assigned "homework" involving one or more creative tasks (e.g. prepare a collage) to focus them on the product category and ease them into sharing their views and opinions at two consecutive days of qualitative sessions. Information from the survey and group sessions is analyzed and is the basis for subsequent questionnaires for the quantitative phase of consumer testing.

Consumers recruited to the test location are asked to complete a questionnaire about their "ideal" product, its attributes and uses, and a series of questions derived from the IBU consumer survey and discussion groups described previously. Products are then evaluated, one at a time, in a standard central location test format. Consumers complete a hedonic question for a product followed by another proprietary questionnaire (Sample-by-Use (SBU)). The SBU contains questions related to the perceived intensity for sensory attributes (also measured in the preliminary "ideal" questionnaire), situation appropriateness (initially measured by the IBU), and a conjoint linked proprietary questionnaire (SensMark for combining sensory and marketing information) assessing the degree to which the products sensory cues are judged to deliver on marketing-related issues such as brand, perceived value, and perceived benefit. Proprietary exit questionnaires include demographic information, lifestyle and attitude questions, anticipated category and brand purchase behavior for the immediate future, and a post-test "ideal" assessment. Information from the product evaluation and exit questionnaires provides a user profile and, where possible, a discriminate function linked to preference group membership. This information provides preference group-specific screening criteria, as well as identifies whether, or how, preference groups differ in their attitudes, lifestyle, and purchase interests. It would be naive to assume that preference group members have homogeneous attitudes, values, and behavior. "Knowing the consumer" is raised to a new level of understanding with these enhancements to optimization research.

Consumer focus on the category in optimization research offers an opportunity to measure response to a variety of other issues. Packaging container sizes, shape, structure, graphics, and so on are readily accommodated in the research, as are alternate concepts and communication options. Quantitative survey information and qualitative follow-up groups are used to further explore topics of interest for groups exhibiting homogeneous attitudes and preferences. Our results have convincingly demonstrated the importance for sensory professionals to work closely with professionals in other product-related business units (e.g. marketing) in a systematic and integrated approach to product optimization. Robichaud and co-workers (2007) provide an example of the integration of different kinds of information to identify what combination of variables will appeal the most to specific market segments in the wine industry. Rarely will isolation or working in an information vacuum produce the success needed to conduct this research and to compete in the marketplace of today and the future.

8.5 Sensory, physical, and chemical relationships

There are numerous reasons for establishing relationships between one or more sensory measures and one or more physical and chemical measures. One can reduce reliance on panel tests for some products or where the test is repetitive in nature, identify the strength of a sensory measure that has been demonstrated to impact preference negatively, and develop causal relationships between the physical and chemical measures and preferences. Routine evaluations might include applications in a manufacturing facility for purposes of assessing raw materials and/ or finished products. In addition to reducing dependence on panels, use of instruments or other physical measures, especially those of the imitative type, has provided insight into the sensory processes involved in product evaluation (Bartlett *et al.*, 1997; Bleibaum *et al.*, 2002; Noble, 1975). By imitative, we refer to the definition by Noble "in which the property is assessed by a device which imitates the way in which humans perceive the sensory property." We suggest a more specific definition; by imitative, we mean the way in which the property is assessed by a device that correlates with the way in which humans respond to that property. For example, extensive studies on the mechanical properties of food, such as chewing, have led to the development of instruments that more accurately record the various kinds of movements perceived as chewing as well as provide a mathematical explanation for these movements. Although practical applications remain limited, it is reasonable to expect this will change as users become more adept at identifying the relationships and simplifying the process.

The literature provides ample evidence for the high level of interest as well as some insights into the problems of achieving a direct causal relationship between specific chemical and physical measures and product quality (Noble, 1975; Trant *et al.*, 1981), but the evidence to date suggests that either the information is not easily translated into a practical system or the cost is too high. The development of electronic noses and tongues has renewed interest in this topic (Bartlett *et al.*, 1997; Bleibaum *et al.*, 2002). There is no question that knowledge of relationships between sensory and physical and chemical measures will have a significant economic impact in addition to improving product quality measurement and the associated improvement in product specifications. The use of an instrument or a system of instruments will be of a decided advantage in a number of situations—for example, in the evaluation of heat in pepper oils and similar difficult or unpleasant tasks, or in situations in which a limited number of people are available and frequent and repetitive evaluations cannot be sustained. In practice today, however, it is the exception to find instrument systems functioning as a complete or even partial replacement for panel tests on any wide scale or to find panels being used as a complete replacement for instruments. This should not be interpreted as meaning that instrument–sensory relationships are a research curiosity only, or that they are too difficult or impractical to apply within the framework of a manufacturing environment. A more likely explanation may be the lack of understanding about the various measures that are obtained, what is being measured, how it can fit within existing operations, and, most important for management, the cost. Of the many types of physical and chemical measures, determining which ones correlate on a causal basis and directly linking these measures to specific product sensory characteristics is an important first step. If the goal is to incorporate the results into an ongoing evaluation of a raw material or as part of a process, then management involvement and commitment become critical.

Another issue that can be a potential problem is the idea that the instrument measure(s) will yield a numerical value that is directly equated with preference. As Noble (1975) and Trant *et al.* (1981) and others cautioned, considerable risk is entailed in relating a specific instrument or a specific sensory measure with the hedonic response—that is, the preference for a product. The hedonic response may be expressed as a parabolic function, whereas the other measures could be linear, curvilinear, or sigmoidal, as depicted in Figure 8.4. Product preference may be changed as a result of a combination of ingredient, process, and package changes; however, such changes are not likely to be explained by a single physical, chemical, or sensory measure. As additional changes (in a product) are made, the preference

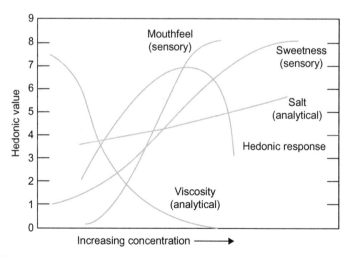

FIGURE 8.4

Theoretical representation of sensory and analytical measures for a product combined with the hedonic response pattern, which is depicted as an inverted "U."

for a product may not change but the analysis indicates a change. This does not mean that physical and chemical measures cannot be related to specific sensory measures or that some combination of physical, chemical, and sensory measures cannot be used to predict product acceptance; it is simply to emphasize that the problem is more complex and considerable risk is entailed in relying on a single measure to serve as a basis for determining that a product will be more or less liked. It must also be kept in mind that there are other marketing-related variables that will influence preference, and often these will have a not insignificant effect compared to the product itself. Some of this was covered in Sections 8.3 and 8.4.

For purposes of this discussion, we return to the issue of identifying the relationships among selected physical and chemical and sensory measures. An easy example to demonstrate is the relationship between perceived bitterness in beer and the concentration of isohumulones. Testing an array of beers with bitterness units (related to the amount of isohumulones) ranging from approximately 10 to 30 yields a correlation $\geq +0.75$. One can also show a similar correlation with liking for beer; however, it would be a mistake to assume that bitterness intensity (perceived or chemical) alone can explain preferences. In fact, beers with differences as large as 10 units can be equally liked, which is further confirmation of the multivariate nature of the relationship. Examination of a correlation matrix will identify other relationships among the attributes measured. These univariate relationships provide a basis for further computations and reinforce the value of such experiments. For example, results could identify whether specific analyses are redundant and thus allow for potential cost savings with no loss of product information.

In the past three decades and after hundreds of projects, we have always identified reasonably robust relationships among the various measures; that is, R^2 values greater than 0.85. At the outset of any such research, it is necessary to establish the specific objective(s) as to which instrument or instruments are under consideration (sensory or physical/chemical) and, especially, how the information will be used. For example, if the intent is to develop an instrument in place of a sensory test as a basis for raw material purchasing, then the approach will be different than using the information in a quality control process. Raw material sampling will likely be done less frequently than in-process or finished product sampling, so the type and frequency of testing would be different. Before any actual data are obtained, attention must be given to the test plan and to selecting the appropriate products and methods. This planning takes into account the sensitivity of the various instruments and any other relevant background information. For example, a costly and time-consuming physical or chemical measure would not be very practical, and such information should be taken into account in the planning process. In addition, the instrument may be too sensitive for the intended application, and its degree of sensitivity would have to be adjusted to reflect what it can measure and control. Other issues that warrant attention are the criteria by which a product (raw material or finished product) is rejected, if that is a possible outcome of the test, availability of other resources in support of the test, the decision criteria, and the responsibilities for those decisions. Awareness of this information will be important in minimizing conflict about acceptable product or whatever other criteria are used and the consequences. Other information, such as estimated frequency of testing and magnitude of expected differences (derived from preliminary tests), is used to determine the scope of the effort, which in turn enables management to estimate cost-effectiveness of the entire program.

If these relationships are not yet established to everyone's satisfaction, then an investment in research will be necessary. As emphasized previously in this discussion, sensory measures are usually multidimensional, whereas the physical and chemical ones are typically unidimensional. A particular physical measure will be represented perceptually by several sensory characteristics; so it should not be surprising that several physical and chemical measures may be required to establish a potential causal relationship. For simple systems, it may be easier to demonstrate a univariate relationship between instrument and sensory measures (i.e. the physical element and the sensory element are more alike), but as the stimulus system becomes more complex, this relationship may not hold. For example, changing the concentration of sucrose in water may be readily perceived as a change in the strength of sweetness and the relationship with a particular measure such as viscosity of the solution could be quite high; for example, greater than 0.75, this relationship would be deceptive, if the appearance or some other product characteristics also changed.

Once the decision is made to explore these relationships, it is necessary to identify the products and the methods. The sensory measure is easiest to identify—use descriptive analysis as the core sensory database. The physical and chemical measures are more difficult; however, we have found it best to start with current analytical methods now being used; there may be few or there may be many. We have encountered situations in which fewer than 10 are used, but in most companies the number of physical and chemical tests can be in the 50s, and in some instances we have dealt with several hundred measures. Inasmuch as there is no *a priori* knowledge as to which measures are relevant, it is better to err on the side of more than too few. In fact, one side benefit of such an approach is the ability to identify physical and chemical measures that are providing the same kinds of information but are obtained by methods that are more or less costly. There is no question that some redundancy is not undesirable as a kind of insurance should a particular method become inoperable; however, it would be useful to know where there are cost-saving opportunities.

For the information to have relevance, as many products as possible should be evaluated, ideally as many as 20 or more. Analyses of these types, involving MR/C, require a sufficient array of product differences to yield meaningful relationships, as previously emphasized in Sections 8.3 and 8.4. Therefore, the products selected should represent a typical array of differences expected without regard for whether the products will be well liked. Discussion with technologists will help identify possible products for inclusion. The emphasis on including a product should favor one's own brand. In effect, it may be necessary to formulate products to stretch some of the variables; otherwise, the results could yield relatively small differences, making the correlations tenuous in that the standard error of the equation could be large. This concept is critical to obtaining useful information. If product differences are too small, then any useful correlations can be compromised. As mentioned previously, the interpolation is much easier when the variable differences are large. Once data are obtained, each database is subjected to typical univariate analyses to ensure that the results "make sense" based on what is known about the products. This is then followed by data reduction procedures such as principal components analysis and factor analysis to identify any underlying relationships among the measures and at the same time identify the redundancies. In most instances, we expect a reduction of approximately 80% from variables to factors.

As shown in Table 8.2, the factor analysis (accomplished with a varimax rotation) output from a combined physical, chemical, and sensory data file yielded a 92.89% resolution; that is, the percentage accounting for the differences that were measured was explained by seven factors (20 sensory measures and 19 physical and chemical measures of 14 products). The individual analyses yielded 83.9 and 87.27%, respectively, each of which was interesting; however, the combined analysis provided a focus to possible connections between the different measures and those for which no connection was identified by examining the output shown. This kind of information can be mapped and examined for clues as to the relationships depicted. As shown in Figure 8.5, the map is particularly useful when reviewed with the analytical chemist whose knowledge of the product's chemistry can explain some of the relationships. The next step would involve selecting a sensory measure such as sour flavor and identifying the combination of physical and chemical measures that predict the obtained values. The result, shown in Table 8.3, identifies the combination of measures that best explains sour flavor. This process can be repeated with other measures of specific interest. Based on this information, one could focus various analyses on those measures that are most important relative to the production process and/or consumer behavior. In later sections of this chapter, we discuss use of these methods for related applications, identifying those attributes that influence consumer purchase behavior, the relationship between the important sensory measures and imagery, etc. Multivariate analyses are being used much more frequently due to the speed and power of the PC, but with as any of these tools, the opportunity for misreading of results is high (and much more common based on a quick perusal of today's literature). As one would expect, one test regardless of the magnitude is not likely to be an end point. More likely, it will define some relationships and provide a focus for the next steps.

There is no question that meaningful instrument–sensory relationships can be identified, and should be. Both physical/chemical and sensory capabilities are sophisticated and readily available. Electronic noses and tongues are a part of this; although they have been used primarily as a research tool (Bleibaum *et al.*, 2002). Few companies have taken advantage of these sensing systems, in part because of cost for maintenance on a real-time basis. As with any of the relationships that are developed, there is a financial benefit; identifying which physical and chemical tests provide the most information enables one to decide which measures should

Table 8.2 The Factor Analysis Output for the Combined Sensory, Physical, and Chemical Data

	Factor 1	Factor 2	Factor 3	Factor 4	Factor 5	Factor 6	Factor 7	
Bitter Fl	0.95							
Astringent Aft	0.94							
Bitter Aft	0.94							
Overall Aft	0.93							
Astringency	0.92							
Hot Fl	0.89							
Hot Aft	0.87							
Overall Fl	0.85							
Specific gravity	0.77							
Apparent extract (AE)	0.77							
Apparent attention (AA)	−0.74							
Fruity Fl	−0.71							
Original extract (OE)	0.70							
Fructose	0.53							
Phenlyl alanine		0.96						
Isoleucine		0.96						
Leucine		0.95						
Total amino acid		0.94						
Glysine		0.92						
Chloride		0.88						
Phosphate		0.80						
Glucose		0.78						
pH		0.75						
Total polyphenol	0.57	0.66						
Overall Ar		0.64	0.58					
Total organic acid		0.61			0.57			
iso-Amyl acetate			0.84					
Fruity Ar	−0.52		0.65					
Sour Ar		0.60	0.65					
Sweet Ar			0.63					
Sweet Aft				0.93				
Sweet Fl				0.88				
Fruity Aft	−0.56			0.69				
Clarity Ap					−0.84			
Color Ap	0.55				0.73			
Color	0.57				0.72			
Sour Fl						0.82		
Sour Aft						0.80		
SO_4							0.93	
								Total
Percent variance explained	32.20	24.92	8.73	8.48	7.34	6.94	4.28	92.89

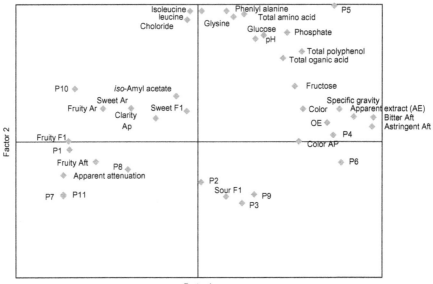

FIGURE 8.5

A map of the first two factors depicting the sensory and physical and chemical measures. In this map, the two factors account for 57.12% of the differences measured.

Table 8.3 The Regression Analysis Output Identifying the Physical and Chemical Variables That Best Predict the Sensory Sour Flavor

Model	B	Beta	Zero Order	Partial	%Contribution
Chloride	−0.050	−0.844	−0.438	−0.769	37
Total organic acid	0.014	0.877	0.248	0.743	36
Apparent attenuation	0.163	0.489	0.114	0.577	28
Constant	10.932				
R	0.80				
R square	0.64				
Adjusted R square	0.57				
Standard error of the estimate	1.65				

continue to be used versus those that should be no longer used. Part of the problem may be that these kinds of projects usually originate within R&D and often do not leave. They should find a receptive home in many sections of a company, especially in those business units in which availability of subjects and time are limited, such as quality control. Of course, the success of these programs is highly dependent on the individuals involved, the thoroughness with which the system has been studied, and a management commitment to the program. The reader will have already realized that this research is similar to that used in optimization, relying on the same kinds of designs and analyses but with a somewhat different product set and a different end point/application.

8.6 **Stability testing**

In recent years, determining product stability and establishing shelf life have become increasingly more important, especially for foods, beverages, and other biologically active products. This interest can be attributed to the impact of new processing methods, new products, increased use of product dating (i.e. the assigning of an expiration date for product sale or use by the consumer), and so on. For products not requiring an expiration date, establishing shelf life could be driven by competitive and business issues. In either situation, there are compelling reasons for knowing what product changes occur over time and their impact in the marketplace. For the dated product (sell by, use by, etc.), there is the cost of its retrieval from the marketplace and the attitude of consumers toward a product (avoidance) that is close to or at the end of its shelf life. For a nondated product, there can be shifts in consumer purchasing habits or a competitive advertising campaign that makes consumers more aware of sensory differences as a function of time. Quite apart from these issues, technological developments (processing and ingredients) and new packaging materials will also affect product stability. The financial consequences of a business decision based in part on an inappropriate shelf life can be significant and, therefore, establishing shelf life is very important.

Companies have always measured product stability and, in general, were aware of the need to know when a product's shelf life ended; however, the aforementioned developments have placed an entirely new perspective on the problem. There has been a substantial increase in stability testing but not always with an appreciation for the complexities of the problem in terms of test design and analysis, methodology, the relationship between product change and product acceptance in the marketplace, and the economic impact of increasing or decreasing shelf life. As a result, sensory resources are often overwhelmed with requests for testing to establish a product's shelf life. A simple example of this situation is the request for products to be tested monthly for 18 months when fewer than half that number of tests would have been appropriate and no significant changes are expected for at least the first 8 or 9 months. All sensory programs will be challenged by the need to schedule tests for six or seven product categories in addition to other activities. For refrigerated products, the testing can be especially challenging, possibly requiring tests twice a day for consecutive days. Obviously, a process has to be developed that balances the frequency with which products are tested versus the availability of resources, but without sacrificing needed information. Test frequency is discussed in greater detail later.

At this point, it is useful to consider some basic issues of stability testing as a means for determining a product's shelf life:

1. Over time, all products undergo change, and this change is usually a deleterious one; chemical and biological changes are initiated leading eventually to either product safety concerns or product sensory concerns.
2. Measuring stability does not, by itself, establish a product's shelf life, consumption date, or some similar designation. Responsibility for that decision should be identified.
3. Test planning must take into account that products change as the testing proceeds, necessitating reliance on specific kinds of test designs (split-plot, etc.).
4. Because testing is usually done over time, control product tested at the start cannot be placed in suspended animation and assumed it has not changed.
5. Stability testing should be initiated only after criteria have been specified for starting and ending a test. What is time zero, day of manufacture, or some other date?
6. There must be sufficient product from a single lot to complete the entire study— enough for all sensory, physical, chemical, and consumer testing.
7. Product sourcing must be specified and representative of what is currently available in the market.
8. The basis for stopping a test before the design is completed must be specified.

When a request for shelf life determination is submitted, it is important for sensory staff to meet with the requester and discuss all the details about the project, including those listed previously; that is, the context for the request. Additional questions include the following: Is the request based on a new technology or one that is contemplated? Does it relate to new packaging material or a competitive claim? What is the current shelf life? What stability information is already available? If the product is currently being reformulated, this too should be known. Establishing stability when the formulation work is still in progress can be a waste of resources unless an objective is to identify which formulation best meets a desired degree of stability.

In addition to these criteria, other important considerations include the frequency with which products are removed from storage, the kinds of tests that are most likely to measure expected changes, and the products themselves. Typically, a technologist will request frequent testing, possibly at monthly intervals (e.g. monthly for 18 months) throughout the anticipated life of the product, and with weekly withdrawals if accelerated storage is included. On the other hand, the sensory approach is to minimize the number of withdrawals, to minimize the demands on testing resources (facilities, staff, and subjects). Ideally, one should be establishing baseline information at zero time and at one or possibly two additional times and should cluster the remaining tests around the anticipated end of shelf life. It is not realistic to precisely state the number of withdrawals because it is entirely dependent on the nature of the product and background information about it, the stability of the ingredients, and the overall project objective. Some products are not released from storage until a specific time period has elapsed; this time lapse may be necessary for the ingredients to blend, to satisfy a regulation, and so on. Thus, the zero time for a product may not be the date of manufacture. Therefore, it is important that this circumstance be considered (and documented) when preparing the test design and the planned withdrawal/test dates. Frequent testing places an undue demand on sensory resources; however, it may have a potential value by identifying changes early in the life of the product. Alternatively, if product does not reach retail until 1 month after production and 95% of product is purchased and consumed within 6 months, then testing frequency would focus on initial, 1-month, and a cluster of tests just before, at, and after the 6-month period. Testing at 2, 3, and 4 months would not accomplish very much relative to the sensory investment. One can easily see the testing challenge for a product with a 12-month or longer expected stability/shelf life. The converse of this situation is the refrigerated product whose shelf life is counted in days. The challenge for sensory is devising a protocol in which typical testing will be modified to provide the information without sacrificing its quality. For a refrigerated product, the test will likely be limited to a few weeks or less. Designing the appropriate test is possible when all the information is available from the product specialists and the brand manager.

Additional points of discussion are the availability of sufficient product for all proposed tests, product source, and the matter of having a control or reference product. Amounts of product can be determined only after shelf life, frequency of withdrawal, and types of tests (number of subjects per test, etc.) are estimated. It is especially important to have these amounts specified as soon as possible so that the responsible individual can determine if enough product is available. Often, an insufficient amount of the product is set aside before the test is planned, leaving the sensory staff with limited options. A related issue is to be sure that the amounts needed are in the typical retail-sized units, not one large container. The latter will not be representative product. When insufficient product is available, some compromise will be necessary. The most frequently selected option is to reduce the sensory database, usually by decreasing the number of responses per test, using less satisfactory test methods, providing less product per serving, or having fewer withdrawals. Although a certain amount of flexibility can be incorporated in most test plans, these compromises are not equivalent. For example, changing test methods would be a poor compromise. Reducing the number of judgments or numbers of subjects risks not achieving expected significant differences, whereas a reduction in serving portion might be a better option. To minimize this particular problem, the experienced sensory professional will develop guidelines for stability testing that will include recommended amounts of product needed for specific tests and amounts needed for a typical stability test. This may not solve the problem of the forgetful technologist, but it will minimize the frequency with which tests are fielded with insufficient data.

Product source is also an important issue because it is critical to the decision about product stability. Any test plan must always include the question, "How representative is this production lot?" If the lot is atypical, then there is a high degree of risk that the shelf life could be under- or overestimated. If product is available from several manufacturing plants, a business decision must be made as to which plant or plants are to be included in the test. If product from more than one plant is being considered for the test, then the analyses will be confronted with an additional degree of complexity. Although it is possible to design a test that incorporates products from different plants (considered in Chapter 4 under split-plot designs and reconsidered in this discussion), there are other considerations. For example, the number of evaluations will increase; another consideration is the likelihood that the plant results will be compared and used for other purposes.

Another issue is the inclusion of a reference (or control) in the test. It is typical to expect the technologists to ask about use of a control product. In foods and beverages, considerable effort is expended on using some kind of technology to place a control in a reduced-temperature (chilled or frozen) environment and/or use of an inert gas to maintain product in a suspended state. This assumes it will not change or has undergone minimal change. However, all products change regardless of storage conditions, and therefore it is quite unrealistic to treat the product as static. In effect, it will become another variable. It is not uncommon to learn that this product is usually presented to the subjects as the control or reference and that will change subject response behavior (the context in which the judgment is made), a topic discussed later in this section. The sensory professional has to address this issue by not designing the test with an identified reference; it must be treated as another variable. The control is changing but not at the same rate as the other products in the test. Perhaps the simplest and recommended solution to this problem is to exclude any reference product held under conditions that are not related to the problem—for example, product held at −20°F in an atmosphere of nitrogen. If a product is normally held at ambient conditions, then the experimental conditions would most likely encompass the norm but certainly would not include a condition as extreme as −20°F in an atmosphere of nitrogen. For those unfamiliar with contemporary measurement techniques and use of split-plot designs, the lack of a reference creates insecurity and doubt as to the merits of testing without a "reference." Requests are often made to have the subjects score the products relative to a reference (e.g. measuring degree of difference). Despite these concerns, designated references continued to be used along with a degree-of-difference methodology, and we offer some comments about them here.

The following are two popular approaches for constructing a difference-from-reference scale:

1. Represent the reference as being at the center of a scale for each attribute.
2. Represent the reference as having a different scale value for each attribute.

In both conditions, the subject examines the reference (but does not evaluate it) and then scores a test product by indicating its distance from the reference. This procedure has a number of inherent problems. Frequently, a separate reference is included with each test product, thereby doubling the number of products for subjects to examine. Also, the experimenter is assuming that the reference is invariant between experiments, subjects, and products; however, experimenters and subjects frequently report that references vary. Finally, specifying that a reference is at the scale center for each attribute is not credible with testers who perceive the reference to be different. When the perceived attribute intensity for a reference is either high or low, displaying it at the center of the scale will compress one end of the scale while expanding the other. Compression and expansion occur whenever the reference is displayed at one location and perceived to belong elsewhere. Adjusting judgments to compensate for the discrepancy between where the reference is depicted and where it is actually perceived defeats the primary purpose for specifying reference location. In the process, subject sensitivity is compromised as subjects adjust their responses to reflect their changing perceptions.

Difference-from-reference methods are popular with proponents of absolute scoring and invariant subjects and products (e.g. the concept of "the" gold standard). There appears to be a belief, albeit mistaken, that a method such as difference-from-reference would be more sensitive than monadic scoring; however, this has not been demonstrated for reasons already noted. It will be more useful, reliable, and valid to evaluate the reference as another coded sample along with the test products and statistically analyze for differences. The measurement concepts described in Chapters 3 and 6 should be referenced.

Once product and test design requirements have been satisfied, it is necessary to select methodology. No single test method will be entirely satisfactory. Discrimination would determine whether products are perceived as different but not whether they are liked. Descriptive analysis would identify specific product differences but not whether the products are liked. Affective testing would provide measures of liking but not the basis for the liking.

Including all three categories of methods at each withdrawal could place a strong demand on sensory test resources, especially subject availability, so some type of testing compromise would seem reasonable. A recommended test approach would be to establish an initial database at time zero using descriptive analysis and preference testing. This plan would be repeated at the estimated 50, 80 or 90, 100, and 110% of current or estimated shelf life. If, however, the acceptance measures at 80% were not significantly different from the previous withdrawal, the descriptive analysis might not be done. This would be a sensory decision. As a minimum, there would be results from four of the withdrawals in which all products (still in the test) were tested with both methods.

This approach requires more decision-making responsibility by the sensory staff versus following the original plan regardless of the results to date. This is a realistic approach with regard to the testing program; however, those that requested the testing will need to know why the design has been changed. Unnecessary tests are eliminated, and the test capability has more flexibility, which is especially important. For example, if a product at the 50% (of estimated shelf life) test shows more change than expected, a decision could be made to move the next test to 60 or 70% of shelf life to determine if the rate of change is accelerating.

An alternative plan makes use of all three types of sensory test methods. Initially (time zero), discrimination, descriptive, and acceptance information is obtained for all products. At the first withdrawal, there is a discrimination test of the control versus the stored products. The discrimination test is recommended only if no statistically significant differences were found at the initial evaluation.

If discrimination results showed no significant differences, no further testing need be done at this time. If differences are obtained, an acceptance test of the control and different products would be appropriate to determine whether there has been a decrease in acceptability. If there was a significant difference in acceptance, the study may be terminated. The decision to terminate the study would be based on the magnitude of the decrease in acceptance and any other available information—for example, if chemical analyses indicated significant changes had occurred. If there were no significant differences, the study would continue. At this time, a descriptive test to identify the perceived differences between the control and experimental products is optional. If the products represent an established brand and if a substantial data file exists, then the descriptive test at this first withdrawal might not be worthwhile. As results are obtained, they need to be shared with the requesting group so that any decisions that are reached are open and agreed to by all parties.

This procedure would be used for the remaining withdrawals, but with some modifications. If on two successive withdrawals there were no significant differences between the control and test samples, acceptance and/or descriptive tests would be implemented on the next withdrawal. This minimizes the risk associated with a control that is changing at a rate similar to that of the test products.

If significant differences were obtained from the acceptance test at the start of the project, then the test should be stopped and a new sample obtained. Similarly, if product differences are obtained from the descriptive test, a decision to proceed or to begin over (obtain a new sample) must be made depending on whether or not the differences occurred within an expected range for the product studied. Criteria for determining whether differences are "expected" or "reasonable" may be established according to the principles described in the next section on quality control. Obviously, this is an important issue—that is, the criteria by which products are considered to be within specification. The issue of restarting the test should be communicated in advance of testing with key project staff. Failure to plan for this situation will seriously undermine the value of the program and of this particular approach to stability testing. One also needs to have a pool of qualified subjects who can be easily contacted for scheduling and rescheduling.

Finally, the criteria for project termination should require that the difference or lowered acceptance be demonstrated on two consecutive withdrawals, as a verification of the product change. Although there can be great interest in completing a test, in some instances the products have changed so much as to be disliked extremely and/or subjects refuse to continue the test.

A number of test designs and analyses have been described in the literature (Gacula, 1975; Gacula and Kubala, 1975; Lawless and Heymann, 2010). In most instances, these designs and particularly those described by Gacula emphasize a statistical approach with only limited consideration of the sensory input. For example, the same subjects are not likely to be available for every test, and therefore subjects are a source of variability within each test and from one test to another. Whereas the analysis of each test within a withdrawal is quite straightforward, the analysis across time is more complicated. Product has also undergone change and is another source of variability within and across test variables. The most useful design is the split-plot analysis (general linear hypothesis), which was described in Chapter 4. In the preparation of any design, the variables need to be specified in detail. For example, if there are three packaging materials and three storage conditions (temperature and humidity), then there will be a nine-product matrix, including the control conditions. From a single production lot, sufficient amounts will be allocated for the nine-product matrix and the actual test is initiated. The reader will find it helpful to review that section of Chapter 4 when planning a stability test.

There are some aspects of stability testing not already discussed that warrant comment. For example, it has been suggested that the control product after the initial test be a sample from current production on that day when products are removed from storage and tested. This approach is suggested as a way to address the issue of the control product changing over time (regardless of how stored). It is stated/argued that because the product is typical of production and therefore representative, it should be a reasonable alternative control sample. However, it would be a weak choice because it will be a source of variability not part of the original sample lot. This could exaggerate product differences and lead to false conclusions about differences (Type 1 error). An equally challenging problem is the actual choice of the "control" from production—that is, who chooses it, what criteria are used, and so on. This should not be an issue; the control should be treated as a variable in the test and not as some specially designated sample that serves to confuse subjects, leading to increased variability and decreased sensitivity. Again, the reader is directed to Chapter 4 and the section on split-plot designs.

In some instances, stability testing is planned around products of different ages obtained from the marketplace. This procedure obviates the need for testing over time; all products are available for evaluation at the same time. This approach will establish the condition of products in the marketplace; however, it cannot substitute for a purposefully designed stability test. Products will be from different production lots, different distribution systems, etc. If it is intended to function as the basis for designing an appropriate stability test, then it has a value. Often, such a test serves to highlight the extent of differences that exist among products in the marketplace—a topic discussed in Section 8.8.

Finally, there will be situations in which budgets and time constraints preclude the appropriate design and one is left with test plans described in the previous paragraphs. In that situation, it will be important for the sensory staff to include additional trials so as to have a large enough database for any comparisons with product tested previously.

Once a stability test is completed and the results are analyzed, the information becomes part of the shelf life determination. As mentioned previously, a product's shelf life is based on several criteria, including safety, sensory, consumer attitudes and perceptions, and business inputs. If a product meets all safety, sensory, and consumer preference requirements, then it becomes a business issue as to what sell-by or consume-by date will be printed on the label. For example, consumers may "believe" that a particular product is fresh and have high purchase intent as long as the sell-by date is 3 months or less. If the actual stability results indicate that a 6-month shelf life is feasible, that does not mean it should be. Similarly, the packaging material may afford a much greater degree of stability than is required, representing a potential cost savings. Often, decisions are made regarding stability testing without an appreciation for the potential benefits from a planned approach. Historically, decisions about product shelf life were based primarily on packaging materials and safety (and the latter is certainly paramount); however, much more can be gained from measuring stability with a design approach, taking into account ingredients, technology, packaging materials, and various consumer and sensory measures. This approach will more than recover the investment made to organize and field it.

8.7 Quality control

Quality is an integral part of every company's business strategy. Substantial investments are made in advertising to promote product quality to the consumer; however, in recent times, the main focus has been directed toward safety, nutritive value, and having a connection with organic and natural (whatever that means). Nonetheless, companies continue to make investments in equipment and staff to address quality from the raw materials through to retail. It is interesting to note that this investment in quality remains a challenge, as noted by Daniels (personal communication, 2000), because as many as 25% of products in retail exceed company specifications, excluding physical damage. In effect, there is considerable variation among products available to the consumer at retail. The importance of this variation as measured by preference and purchase intent needs continued attention. How much should be invested in changing a process and/or raw materials that result in a product with less variation but no change in purchase intent or the converse? In fact, when these questions are asked, the frequent response is one of surprise and/or disbelief. As already noted, in recent times, business executives have given more attention and investment to food safety, food security, and new products, especially those with an organic connection. As a result, the focus on quality has been overshadowed but not forgotten.

The word quality is used in so many different ways that it is often difficult to understand what is meant by the word, especially when used by those in R&D versus marketing or production (Carpenter *et al.*, 2000; Clute, 2008; Sidel and Stone, 2006; Stone *et al.*, 1991). All companies operate quality control (QC) activities at production facilities, with support when needed from their R&D operations. This activity is an integral part of manufacturing to ensure that what comes off the production line is consistent and meets the product specification. Considerable literature on quality, its definition, and how to maintain it is also available; however, only limited information has been available specifically on sensory evaluation's role in QC. Several publications on QC for the food industry provide a synopsis of applications for sensory evaluation (Gridgeman, 1984; Hubbard, 2003; Lyon *et al.*, 1992). A discussion about coffee quality and chocolate quality is summarized by Mermelstein (2012a,b). What is most interesting is that the various individuals cited made note of the importance of sensory evaluation as being part of the process and under consideration. Any mention of sensory evaluation in a discussion about QC is encouraging. However, as already noted, more needs to be done to address the issues of what is sensory quality, how it is measured, and how sensory information and resources can be incorporated into the product quality decision-making process. The discussion here focuses on various approaches one can use that the authors have found to be particularly helpful when considering sensory participation in product QC. The discussion includes both global issues and specifics about the sensory resources and their applications. We begin with an older but useful working explanation of QC and its function in a business provided by Hawthorn (in Herschdoerfer, 1984):

> The aim of quality control is to achieve as consistent a standard of quality in the product being produced as is compatible with the market for which the product is designed, and the price at which it will sell.

This particular explanation incorporates many of the variables that are integral to the QC process and, more broadly, quality itself. For example, it uses words such as consistent, product compatible with the market, and price. In effect, quality and quality control should start with the market and work back to product and ultimately to the raw materials. In a symposium titled "Sensory Quality in Foods and Beverages" (Sidel *et al.*, 1983), we noted that the word "quality" had different meanings and had evolved from an expert-dictated concept to a more production-oriented system and gradually to a more consumer-influenced concept. The word quality became a "buzz" word used by marketing executives and in advertising about a company's products. A discussion on this issue can be found in Stone *et al.* (1991). Traditionally, "quality" implied "degree of excellence" (Kramer and Twigg, 1970; Sidel *et al.*, 1983; Wolfe, 1979). Used in this way, the term implied a universal, objective, absolute, and context-free criteria for establishing and judging the quality of a product. Examples of quality standards are represented by such procedures as the oil quality scale, butter and ice cream scorecards, and government standards for a wide range of agricultural crops such as olive oil and meat. Alternatively, consumer acceptance (and purchase behavior), which is a driving force in the marketplace, implies none of these standards. It is segmented, subjective, variable, and context dependent. The notion that an expert's opinion of what constitutes an excellent or high-quality product will invariably reflect consumer acceptance should no longer be an issue. Sufficient documentation to the contrary is directly available (McBride and Hall, 1979) or can be easily deduced by reading articles about product quality (Smith *et al.*, 1983; Stone *et al.*, 1991). In this discussion, it is important to recognize that use of quality measures as a part of the manufacturing process is part of the equation, and the other part is the quality of the product as perceived by the consumer.

As noted in the third edition and continuing today, most QC sensory programs continue to struggle with being an integral part of the manufacturing process. This lack of success can be attributed to a number of reasons; however, it is clear that it is not related to sensory's ability to provide information or to establish a system that is accepted by plant managers. It appears to be much more related to organizational issues, plant consolidations, partially processed products moving from one region of the world to another, reluctance to move away from reliance on experts and/or existing practices, cost constraints, and inadequate understanding as to how sensory information will be integrated with other product quality measures. Considering the amount of attention given to product quality in the technical and trade press, it is an opportunity that should not be ignored. The idea of bringing sensory evaluation into this system makes sense because a part of the quality monitoring involves evaluating/tasting product.

We believe that a successful integration of sensory resources with existing QC activities begins with an agreement that sensory quality has value and that plant management will support the research and provide a budget (usually from corporate) and a time line. There is no question that sensory can provide data on raw materials and products within and across shifts and can provide this in a timely manner; however, the more basic issue is how to integrate the information into the QC process. After all, QC activities are time driven because product cannot be held back waiting for a typical sensory test. To be successful, sensory staff has to develop product information and a plan to integrate it with the other quality measures. Without this information and support of plant management, no sensory QC program can expect any success.

Once there is agreement to proceed, a discussion needs to ensue in which product quality is defined, and whether it will include manufacturing quality or is entirely based on consumer quality criteria. As Hawthorn noted, QC is based on an integration of information that must include the consumer—that is, meeting the expectations of the consumer balanced by what can be manufactured at a price that will sell and yield a profit. These elements are at the core of what should be a more comprehensive QC program. Many of the activities of current QC programs will not change because of their focus on the production process; however, some parts, especially the evaluation part, will change, taking into account the time constraints of existing systems. Existing sensory quality monitoring systems are out-of-date because they rely on insensitive scales, experienced but not qualified subjects, and no regular external audits to determine that some kind of corrective action needs to be taken. Such results should not be surprising. As already noted, resistance to change is driven by the observation that all production is sold, so why make changes? The reports by Mermelstein (2012a,b) suggest that measuring product quality continues to be a problem, so some change is necessary. The challenge for sensory staff is to establish how best to integrate consumer responses into the quality monitoring process and QC in a meaningful way given the time constraints and the number of products that are evaluated.

Establishing a consumer-oriented sensory quality program is realized with consumers providing their assessment of product quality information and integrating it with production and raw material quality information. As previously noted, companies developed quality standards based on what was being manufactured and built monitoring systems around it, with input from technology and other product specialists. It has been driven by manufacturing, a realistic focus as long as the market was expanding greater than population growth, competition could be minimized, and one was confident that acceptance by the consumer was consistent with the market strategy (image, price, etc.). Over time, however, this situation has changed, almost dramatically considering the competitive nature of the marketplace. It would not be surprising, for example, to learn that some companies use grading systems unchanged for more than three decades without any verification that their systems are consistent with consumer behavior. Products are incorrectly downgraded or the grade of a raw material is higher than what is needed. Auditing of a QC program revealed just such a situation. Approximately 20% of existing product was downgraded and sold at a significant discount when its scores were not significantly different from the highest grade. It could have been sold in other markets. Equally important was the knowledge that the company was purchasing the key raw material at a premium price when the consumer's response was the same regardless of the source of this raw material. This particular project served as a good example of using sensory product quality information and incorporating it into the grading system, thus bringing it more in line with consumer quality expectations as well as adding to the bottom line without diminishing the image of the brand. Not surprisingly, the QC unit was initially resistant to accept this information: unwilling to change and considering it a threat to its function. Ultimately, the group recognized the benefits and changes were implemented. It is a reminder that change is difficult, especially when the basis for the change has the appearance of challenging long-held beliefs.

Sensory input to QC activities should be a part of every QC system provided there is agreement from manufacturing and QC itself because they will be responsible for approving any changes to the current system. Once there is agreement, a project team including sensory, production, and QC should be organized and meet to identify all the information needs that do not already exist, including the current evaluation process, frequency of testing, and documentation on the extent of variation as seen by QC. Discussion will also be needed on the issue deviations from a standard, a process commonly used as part of the QC process that leads to withholding a product from normal distribution channels.

The procedure used to establish a standard for production is important because it will have a direct influence on determining the degree of deviation that will be allowed and the measure of that deviation. If expert opinion were used to establish the standard, most likely the expert would establish the extent of deviation and the amount of change before quality would be lowered significantly. The expert, as noted previously in this chapter, must rely on some type of grading systems to measure deviation from a standard (Amerine et al., 1959; Baten, 1946; Downs et al., 1954; Mosher et al., 1950). The weaknesses of grading systems have been covered elsewhere in this book (see Chapter 3) and in the literature by Pangborn and Dunkley (1964), O'Mahony (1979), and Sidel et al. (1981, 1983).

When a sensory approach is used to establish a standard, it enables development of a regression model that can be used to determine the effect on acceptance of any number of product variations. Once a model is developed, the effect of changes can be predicted as a paper-and-pencil exploration, or it may involve testing actual samples representing a wide range of possible manufactured variables or deviations and this serves as a validation of the model. This leads to development of a scoring system that identifies which characteristics have exceeded a standard (± range) and leaves to management the decision as to what action will be taken. Obviously, there can be considerable pressure when a product is on hold; for example, production, traffic, warehousing, and so on will be disrupted. On the other hand, permitting product that exceeds the established specification to leave a plant and enter the retail system is more damaging and the effects will be more long-lasting, although these latter effects are not directly observed, at least not in the immediate future. Over time, however, their effects can be very significant. Companies attempt to minimize this problem through a separate program, often referred to as Quality Assurance. This program monitors longer term shifts through testing product obtained from retail outlets. However, any results obtained from retail are influenced by the distribution system and related handling issues. The key to minimizing the long-term changes is to reduce the variations at the source. Before describing a means for developing and integrating a real-time sensory quality system, it is useful to reflect on how the information should be used and the main pitfalls to avoid. As noted previously, measuring the strengths of important sensory quality characteristics is relatively easy, as is determining whether the strengths exceed the acceptable range; however, it is far more challenging to identify the actions that should be recommended. In most situations, quality measures were (and continue to be) made using a scale that included acceptance/rejection and most test participants learned to avoid the latter and thus avoid providing explanations to plant managers. As discussed later in this section, sensory measures should not include accept/reject judgments directly or indirectly, leaving those decisions to plant management. This minimizes conflict and enables the process to function solely as a measuring system. Manufacturing emphasizes output and will pressure any function that impedes that flow, except of course product safety. By shifting the decision to management, the sensory information represents what is perceived without determining that it is or is not approving release of a production lot. It is a management responsibility to address the issue of risk using a combination of sensory, physical, chemical, and business information to release a specific lot.

As noted more than four decades ago (Stewart, 1971) and still correct today, a sensory quality monitoring system must be rapid, reliable, consistent, valid, and accepted by plant managers. It will require compromises from sensory as well as from the current QC process and agreement from manufacturing as to their participation in the decision-making process. For sensory, it means that typical tests involving 20+ subjects are unrealistic and therefore traditional statistical tests, even those intended as small panel statistics, will not be used. It also means there will be some retraining of current QC test subjects.

Before describing a sensory system, some additional comments are warranted about developing the standard, sometimes referred to as the "gold standard." As previously mentioned, the sensory quality model is based on regression analyses to identify those sensory attributes that have the largest effects on the key quality measures along with their associated variance measures. Once this information has been developed, the project team will need to communicate their findings with management as a way of positioning the standard as actually a family of products that constitutes the standard. To fully accept and appreciate this concept, a set of descriptive profiles is developed from products representing the tolerance limits and from products within and outside those limits. An example of a profile is shown in Figure 8.6. Product accept/reject decisions are based on deviations from the standard. The process of identifying the important sensory quality attributes and establishing their ranges and converting these into an operational scorecard is a multistage process. The initial task is to identify an array of products representing within and outside the range that would be acceptable based on current criteria. Products beyond what would be considered acceptable are also included. The second task is to obtain a sensory analysis, consumer quality measures, and physical and chemical analyses for all the products. Through a series of regression analyses, the most important sensory, physical, and chemical attributes are identified. Results from such analyses are summarized in Table 8.4. Once the attributes are identified, they are plotted versus the obtained acceptance scores and the limits established from the consumer judgments that

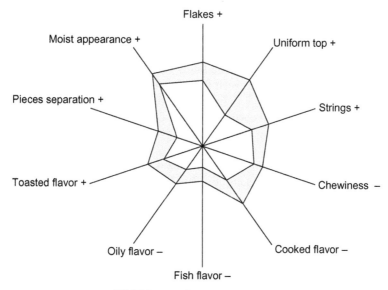

☐ QDA range for specification attributes

FIGURE 8.6

An example of a plot of the important sensory quality characteristics along with the range within which a product can vary but still meet the specification.

Table 8.4 Summary from a Regression Analysis Identifying the Most Important Analytical and Sensory Attributes and Their Relative Contributions to Acceptance[a]

Analytical	B	Beta	r	Beta × r	% Relative Contribution
Color	−0.009	−0.388	−0.739	0.287	21.40
Reduced sugars	−5.508	−0.236	−0.282	0.067	5.00
Sensory					
Spicy aroma	−0.043	−0.280	−0.559	0.156	11.63
Sweet flavor	0.075	0.265	0.241	0.064	4.77
Onion flavor	−0.085	−0.398	−0.885	0.352	26.25
Dairy flavor	0.108	0.322	0.129	0.415	30.95

[a]*Based on the variances and the slopes of the individual lines, the best estimates for the ranges are determined, leading to the scoring system shown in Figure 8.7.*

indicate a significant change in preferences and/or purchase intent. Figure 8.7 (identified as the product quality index) shows the scorecard (Figure 8.7a), the scorecard with the limits (Figure 8.7b), and the template (Figure 8.7c). During training, subjects use the scorecard (Figure 8.7a) to familiarize themselves with the attributes and practice scoring with products representing within and outside the range. The system is dynamic and representative of what is manufactured, and the limits can be changed based on manufacturing issues and consumer considerations. Not surprisingly, there will be adjustments; for example, if the limits are too narrow for an attribute of less importance, the limits can be extended or the converse can be applied. It is critical to make such decisions based on both manufacturing and marketplace information and not to base them solely on what can be manufactured or to create limits that are too costly to achieve or result in too high a rate of rejection. In effect, one seeks to develop a dynamic system that reflects both internal and external considerations.

In addition to the delineation of the standard products, there will also be reliance on specific methods and subjects. This will require the adaptation of standard analytical sensory test methods (mostly discrimination and descriptive) to fit within the manufacturing environment. There is no universal test method for this application.

Monitoring product quality requires specific sensory test resources and the availability of qualified subjects and an individual responsible for organizing and administering the tests. The resources described here have been used successfully in manufacturing (Nakayama and Wessman, 1979) and include other developments (Gridgeman, 1984; Hubbard, 1996; Stone et al., 1991). For most evaluations, discrimination and descriptive tests will be used; however, their use will be modified to reflect the nature of the problem, the limited availability of subjects, and the limited time in which to reach decisions.

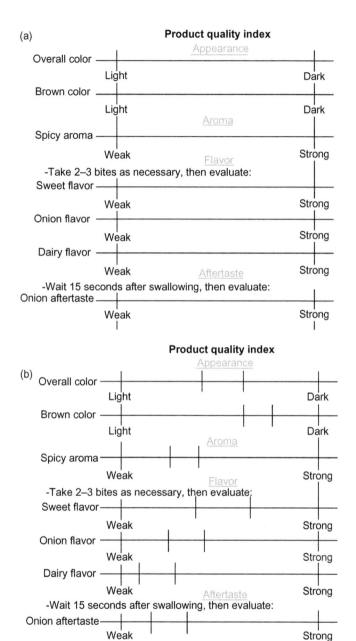

FIGURE 8.7

(a) The product quality index scorecard developed from the regression analyses and plots versus acceptance. (b) The index with the ranges identified. (c) The template applied to the index. A rapid scan of the template identifies which attribute scores are outside the range.

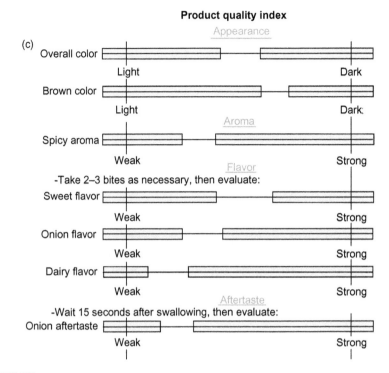

FIGURE 8.7

(Continued)

1. An area for testing should be designated at each plant. It is unlikely that there will be an area as extensive as that available at R&D, but lighting and related environmental controls, booths, and accessibility are necessary. Typically, there would be three or four booths, a small preparation area, and a panel discussion room (this room will most likely be used as a general-purpose room by others). This accommodation would allow for a range of testing activities, with R&D providing the necessary support for more extensive testing. For further detail on facilities, the reader is referred to that section of Chapter 2.

2. All subjects must be screened for testing. Guidelines for screening are provided in Chapter 5. The importance of having qualified subjects has been emphasized throughout this book, and it is no less important here. It is interesting to observe the positive impact that screening (and training) has on plant personnel. Some production personnel have, for years, evaluated products as part of their daily work routine but have never been given instructions as to how to use their senses and whether the information they provided had any impact. It may also reveal that some individuals are not as sensitive as believed or, in some instances, they are quite insensitive to specific characteristics. The screening procedures need to be well-defined such that all subjects follow the same protocol. Usually, the

R&D sensory staff develops this protocol (with suitable modifications as experience is gained).

3. Reference materials (products and/or ingredients) for discrimination and descriptive tests may be identified and their preparation and use documented (if appropriate). In our experience, only a limited number of references will be or should be necessary. For subject training, they can be helpful whether for the descriptive analysis or if the multiple standard discrimination test is used. Over time, discrimination and/or descriptive evaluations will be used to replenish the references. If the evaluations include ingredients, product during processing, in addition to finished product, then references may be needed for each of these test activities. Use of references should always be approached with care because they can communicate a sensory message that is not relevant to the task (see Chapter 6).

4. Once subjects are screened, they proceed with descriptive training to recognize product attributes that are within and are outside of a specification. Training for multiple-standard discrimination tests involves first familiarizing subjects with the reference range and the evaluation procedure. Next, subjects evaluate sets of products and their performance is monitored. Qualifying subjects follow the same general plan as described previously for discrimination testing.

   For descriptive analysis, subjects are trained according to the guidelines described in Chapter 6. Because the products will have been evaluated as part of the specification, the subjects are trained with the existing language, definitions, and any reference materials. However, it is not necessary for these subjects to be trained for an evaluation of all of a product's attributes. In fact, the results from the descriptive analysis of the products used to establish the quality sZtandard will have identified the most important attributes, and it is these attributes that are the focus of training. A typical scorecard used at plant will be limited to a single page and not more than 8–10 attributes; otherwise, it will be too cumbersome to use and give an impression of complexity that is inappropriate.

5. The frequency of the evaluations and when done will be specified, as part of the initial development of the program, as will the testing protocol and related issues. Such documentation is important so as to ensure that the same procedures are used all the time.

6. The criteria for product acceptance/rejection must be documented and agreed to by plant management. From a strategic standpoint, it is necessary to remove from a subject direct responsibility for placing product on hold or for rejecting a specific lot. Subjects should focus solely on providing responses (to specific attributes), and the test administrator reads the results and advises QC and/or the plant manager as to the results. Traditionally, QC programs used a procedure that trained the individual subject to a specific score below which (or the converse) product was acceptable and above which it was not. In addition, the subject also had to explain each judgment that resulted in product rejection. Clearly, this approach discouraged dissent, and individual subjects quickly learned to

"adjust" their responses to conform, which was not in the best interest of consistent product quality. However, it did provide false evidence of the consistency of the products. It is therefore very important for the success of a sensory QC program to ensure that subjects are not directly responsible for the product decisions. In addition to not being directly responsible, the use of a second tier of testing will further enhance the program's success. For example, the QC manager, in concert with the appropriate plant managers, can evaluate all products that are just outside the range and make the final decision (using all available information, of which sensory will be one part).

As previously mentioned, the in-plant QC program will rely primarily on two methods: discrimination and descriptive analysis. For the former, the multiple-standard discrimination method is recommended. Product that is in and out of specification must be available during and after subject screening and training. For each test, such standards are also included. This procedure requires each subject to determine whether products are within the specified range. These tests are simple; that is, they are go/no-go tests, and no other information is required. As few as three subjects will evaluate a product. If two of the three subjects state that the product is outside the specification, the product either is not accepted or is put on "hold" for further evaluation (see the previous note about second-tier testing). Reference products should be available for the subjects at these tests. The use of very few subjects and a methodology that relies on the go/no-go decision at this stage is especially appropriate. Most manufacturing sites have limited numbers of people available for product evaluation, and still fewer to participate on a daily basis, willing to evaluate multiple sets of products.

On a daily basis, finished product will also be evaluated using a modified version of the standard descriptive test. The modifications, described previously, will include fewer respondents (two or three) and fewer descriptive attributes (6–10). Results from the testing that established the product quality standard will have identified the attributes that are most important as well as the range of scores around each computed mean. The subjects are trained using the original protocol but only for these important attributes. Each subject scores a product once, and the test administrator places the template (see Figure 8.7) on each scorecard or enters data electronically and quickly identifies the attributes outside the range. If two of the subjects score an important attribute outside the range, it is identified in the reporting to the QC manager. Here, too, some responses beyond the acceptable range are expected, and it could require two (of three) subjects marking outside the range for the three or four most important attributes to result in a hold. Note that there is no statistical treatment of the responses (there are too few and no replication), and the intent is to provide a rapid response that is not dependent on a specific numerical value. If a particular lot is put on hold and plant management cannot reach a decision, then samples could be re-evaluated by a second group of subjects or R&D could be requested to make a more complete analysis. The requirements for hold and reject will be product specific; there are no universal rules other than to use the most important sensory quality attributes, qualified subjects, and have the active support of plant management. Regardless of the path that is followed, the idea is to keep the evaluations in balance with the product and the production process.

It has been suggested that magnitude-of-difference scoring might be an appropriate methodology. In this test, the subject provides a measure of the difference from a standard. We see no particular advantage to this method; the magnitude-of-difference measure will have its own problems (the type of analysis and having a standard that is reasonably constant), and it is still a go/no-go decision. For this particular application, the emphasis must be on a dichotomous outcome that is easy to understand. Experience has shown that the go/no-go minimizes the potential for conflict and/or the misuse of scores to promote a particular position.

7. As with any sensory analytical test, subject performance is also monitored. This assessment should include use of reference products and those used during training. Subjects should be aware of the performance monitoring and have the opportunity to review their records. Every effort must be made to keep subjects motivated and to help them maintain their sensory skills. Subjects who exhibit inconsistencies in their performance will have to be dropped (sometimes temporarily and then recertified). The extent of these inconsistencies will depend on the product, and especially on the magnitude of difference representing the specification.

8. A formal record-keeping and reporting procedure is also essential. Because little published literature on sensory evaluation procedures in production exists, it is important that a company have complete documentation of its systems.

The foregoing considerations should be viewed as guidelines for developing a QC sensory testing system.

There are three stages in the manufacturing process at which product quality can be monitored. The first stage covers the continuing go/no-go decisions that are made as raw materials are received, the second stage occurs during their storage, and the third stage occurs during one or more stages of manufacture. Sensory monitoring during these periods serves to minimize release of out-of-specification product. Tests at these stages should be intuitively obvious, but surprisingly little of this testing is done formally. If such testing is done, relatively few records are kept, and many of the procedures have been so drastically modified as to be almost unrecognizable compared with written company protocols. Although it is possible to accept tests performed in a less adequate facility, the test procedures must adhere to accepted sensory practices. Considering the importance of product quality at this stage of its life, it is surprising that so few companies are aware of the weaknesses of their current procedures and the relative ease with which these procedures can be improved. It is hoped that the guidelines and descriptions in this discussion will provide the necessary stimulus to improve the monitoring of product quality.

The remainder of this section concerns the stages of manufacture where sensory tests can be readily made.

Ingredients are received in bulk from suppliers or from company-operated facilities. On receipt, they are checked to ensure that they meet the specification. The evaluation may be extensive, as in the case of analytical testing, or may be as uncomplicated as requiring a sniff or taste by an inspector. It should be understood that not all ingredients require formal sensory examination. In a formal monitoring program, the company will have developed a list of ingredients that require sensory examination and will have established a frequency for that examination. For example, evaluation may be required for every new lot in the case of some ingredients and every tenth lot, once a month, and so forth for other ingredients.

Once frequency of examination has been established, the sensory test method should be specified. Here, we suggest the use of an individual who has been screened, qualified, and trained to discriminate product that meets the specification. The individual may record judgments such as accept, reject, or hold. A hold requires additional and similarly qualified individuals to evaluate the material using the same test procedure. Each company must decide for itself which ingredients it will accept; however, with three trained and qualified subjects, any reject or more than 50% hold responses should be given serious consideration by management. Management may acknowledge the recommendations to reject or hold an ingredient but still approve its use based on other business reasons.

Some ingredients enter the processing system immediately on receipt and are evaluated at that time. Other ingredients may be stored for several months or longer, and a periodic assessment of these should be performed to avoid potential problems. The procedures will be similar to those used upon the ingredients' receipt. Any ingredient stored prior to use in production should be evaluated at the time of its planned introduction.

Product manufacture can range from batch to continuous operations and can involve numerous points at which ingredients are added and processes occur. Mixing, blending, fermenting, and packaging are examples of some of the types of steps. At critical points, product quality must be monitored. Consideration must be given to those points at which the potential for variability is high and will result in a noticeable defect in the finished product. Success at this stage of monitoring product quality requires a reasonable and cooperative effort on the part of sensory evaluation, production, and QC. Each process is unique, and an approach that is suitable for one product probably will not be appropriate for another. It will be necessary to evaluate products removed at different stages of the process and to determine whether the information is meaningful and has a relationship to finished product quality.

The discrimination model is the most useful sensory approach for evaluation of products during processing. The subjects (one to three) would make their evaluations relative to the multireference model, as described previously, as the criteria for acceptance, hold, or rejection of product. Descriptive analysis may be helpful, at least during the investigative stages, to identify at which step in the process the most change has occurred. However, the most useful methodology will be discrimination because of its rapid input to the decision-making process.

For finished product evaluation, the immediate decision making will be based on the discrimination model with multiple references representing acceptable product or using the modified descriptive procedure. Subjects will be trained to identify products that exceed the range of acceptable differences. Criteria for accepting, rejecting, or holding product will be similar to those applied to ingredients and to products during processing. Descriptive analysis should be used for selected products on a regular basis. The purpose would be twofold: to identify any trends away from the reference and to provide a day-to-day independent check on evaluation. The descriptive method would make use of the existing database, would have a limited number of dimensions, and so forth. The descriptive data are plotted against the reference standard(s) to determine whether the finished product is within acceptable limits.

It is assumed that the primary sensory evaluation group will conduct in-depth QDA analyses of the products to monitor long-term drift. The statistical methods for monitoring long-term descriptive drift are similar to the split-plot analyses discussed under product stability and experimental design (see Chapter 4).

As noted previously, much of the success of a sensory QC program depends on having management's commitment to the process, having identified the important sensory quality attributes, and having available products representing within and outside of the standard product for training and testing purposes. Our approach relies on a combination of consumer acceptance and descriptive analysis to establish the sensory limits for a product (the family of gold standards). Another approach relies on management and a project team to select an example(s) of their company's product as a standard, and that product is then submitted to sensory analysis. The resulting description and its limits then become the documented standard against which other descriptive profiles are compared. The approved finished product should be stored and replenished as needed for training or refreshing sensory memories. These products should be included in selected tests.

Establishing and replenishing references for ingredients and product during processing is considerably more difficult. The research required to establish these references may be prohibitive, and the effort of the project team or expert judgment may be required. Once references are established and a reasonable replenishment cycle is agreed, their use in the sensory monitoring of product quality can proceed along the lines described previously. It is possible that reference products and ingredients can be obtained and stored only at considerable expense. In this situation, more reliance will be placed on sensory resources at the company technical center to provide complete descriptive analysis of products and, in the process, identify references for use in training and monitoring the sensory skills of plant personnel. In today's business environment of heightened food safety and food security, some in-process evaluations will be better served through the use of more sophisticated sensing devices such as electronic noses and tongues (Bartlett *et al.*, 1997; Bleibaum *et al.*, 2002).

The evolution of data capture systems such as tablets and software makes the process much simpler from a data capture perspective as well as for communicating to managers. Now, the different kinds of information—physical, chemical, and sensory—can be easily collated and read anytime and anywhere. As advances are made in sensing devices and relationships among the different measures, it is reasonable to expect some reduction in use of repetitive sensory tests. This will be welcomed by all subjects in all food and beverage plants everywhere.

The approach to QC described here emphasizes the impact of the marketplace on helping to select standards, use of discrimination and descriptive testing at the plants, and a management-approved plan of the decision-making process. Reliance on a single test method is not recommended, whether there is a single plant or there are 10 plants. Each system in each plant will have its own unique features, but all will have sufficient commonality to provide for comparisons where needed. Statistical designs and analyses combined with contemporary quantitative sensory methodology will make QC systems valid and successful. Finally, this approach has proven to be very informative as to the various measurements currently being used. We have identified significant investments in QC procedures that were redundant with other measures, as well as having no bearing on finished product quality. Although some of them may be valuable from a manufacturing control perspective, many were thought to be important to the consumer only to find they had no impact.

From this discussion, the reader should develop an appreciation for the complexity of the use of sensory evaluation in QC. However, the need for sensory evaluation is so great that QC cannot be ignored. By identifying the basic elements of a QC program, the sensory professional can proceed with greater confidence than heretofore realized.

8.8 Market audits

Consumers have expectations regarding the safety and quality of the products they purchase. In today's environment, the major focus is on safety; however, one finds that safety and quality are intertwined. Companies have made substantial investments to assure customers that their products are of highest quality and are safe. This investment includes developing auditing systems to monitor product safety and quality, especially before (raw materials), during, and directly after manufacturing. Distributors, wholesalers, and retailers also have similar concerns, and they, too, have invested in their own systems to monitor safety and quality; however, they usually look to the manufacturer for assurances as to product safety and quality.

For most sensory programs, the topic of product market quality has received less attention except in situations in which there is an unusually high level of consumer complaints and/or there is reason to think that product quality has changed. For example, a consumer test indicated that current product scores were much lower than expected with no obvious explanation. Most companies have developed procedures for addressing consumer complaints, including determining where product was purchased and, if possible, retrieving units not consumed or the empty containers. This usually does not require use of sensory resources; however, when products receive unusually low scores for no obvious reasons, then it is more likely that sensory input will be needed. Often, the problem is larger than a single test, and audit is recommended. For some companies, audits are an integral part of brand maintenance usually done on an annual or biannual basis. In recent years, sensory resources have become a part of the audit process.

Retail audits provide brand managers and product developers with an evaluation of product uniformity and consumer acceptance at "point of purchase." They provide a comprehensive picture of the consumer's experience in the marketplace and reflect different stores, different regions, and so forth. These audits are a good way to obtain information on conformance to specifications after product has left the manufacturing facility. Specifications are an agreed upon set of criteria for ensuring that manufacturing procedures follow expected procedures and are consistent with a company's business strategy. The role of sensory is to measure the consistency of the products from the different regions relative to the specification taking into account each product's trip from the original manufacturing location.

For an audit to be useful, information from a broad array of products is needed, including those representing weak and strong markets, various ages, from different distribution channels and distribution facilities (retail grocer, warehouse store, small retailer, etc.), and from a variety of environmental conditions. The sampling has to be representative of where product is sold. It is typical that as many as 100+ products are obtained. Once they are available, they are bench screened to familiarize the team with the range of products and the extent of differences (and there usually are differences) reflecting plant sources, markets, and so forth and also to select a subset for testing. This screening process will likely take several sessions to be sure that products selected for evaluation reflect the range of differences in the marketplace as understood by technical and by the brand management team. Assuming product testing is needed, the goal is to have an array of approximately 20–25 products. More could be tested; however, the cost can be substantial, and there are procedures that can be used that allow for more products but not all are evaluated in their entirety—for example, just physical, chemical, and sensory analyses.

In most instances, descriptive analysis is the primary sensory tool, and it is eventually combined with consumer, physical, and chemical measures to provide a detailed picture of what the consumer experiences. Because the consumer segment can be expensive, the descriptive analysis can be used to identify any product attributes that are extreme as well as identify product redundancies and therefore not worthy of consumer testing. The results from the audit will identify those products and their path from manufacturing to the retail store where there are deviations from the specification, how representative they are, and so forth. The audit provides brand managers with a basis for determining whether any change is needed and its impact on consumer purchase behavior. Auditing programs can also be used to develop databases to monitor product performance versus competition.

8.9 Extended-use testing

Regardless of the business objective—product improvement, new product development, quality control, post-market audits, and even optimization—it is important to know how well a product performs over time—that is, how well liked is it on the third or fourth usage, etc. This could be considered as somewhat like a shelf-life test but more directed to a concern about satiety, "flavor wear-out," or, in the case of chewing gum, having a flavor last for more than a few minutes. In the past, answers to these questions were often addressed in a home-use test. The issue, however, is more complicated in today's competitive environment, in which technological advances and novel ingredients have necessitated more comprehensive analyses and information that will support the expense of the innovation and of any claimed benefit.

A product's acceptance may increase or decrease over usage occasions (time) for a number of reasons. Consumers may adapt to the product attributes, products may perform differently in a particular usage situation, monotony may occur with repeated exposure, the novelty effect of a new or unfamiliar product may wear off, the consumer may pay attention to different product attributes over time, or the consumer may change his or her opinion. All of these factors and more may influence product performance over time.

Not all products will warrant extended-use testing because it can be expensive and, if not planned, can take time. Some products and research objectives are more appropriate for extended-use testing than others. As previously mentioned, it is most beneficial for novel or unfamiliar products, products with attributes that could result in sensory adaptation, products that have a potential to be used in a number of ways, products whose use interacts with other products during normal-use situations, products new to the company, and so on.

From a sensory perspective, having consumers provide responses to the same question within a single session or returning on successive days can be done in a laboratory environment. However, these protocols are likely to yield information that represents what the subjects think the answers should be. This is not unlike time-intensity tests in which subjects are told to provide a measure of its strength at various intervals, yield a result that is consistent with the expectation that the strength of a stimulus will decrease, and lo and behold, that is the result. Such testing has to be designed in a way that does not make it easy for the subject to learn the purpose for the test. This does not mean that one cannot use the laboratory environment, but it does demand that the design reflect the potential problems of repeated usage. The in-home-use test (IHUT) evaluation is widely used for extended-use evaluations, even with the lack of control that accompanies such tests. If it is important that subjects evaluate the product with more typical or natural context aspects of product consumption, then a home-use test may be the best alternative.

The number of evaluations required to determine changes with extended-use testing depends on the product and the research objectives. If one simply wants to determine a measure of reliability for laboratory results, only one or two repeated evaluations may be required. On the other hand, if it is a novel or unfamiliar product, sufficient time must be given to the subject so that the novelty effect has at least partially worn off. If it is a product that can be used in multiple ways, sufficient time must be provided so that these usage situations can occur. The sensory scientist can either provide a specific set of usage requirements or allow the subjects enough time with the product such that a variety of usage occasions will occur. There is a trade-off to these types of IHUT evaluations; that is, one may lose the typical sensory controls that reduce variability versus evaluating the products under more normal usage conditions.

Whether different results are derived from varying lengths of evaluation can only be determined by experimentation. Ideally, one could conduct tests with the same type of subjects using the same evaluation scale, with single-sample presentation, after different periods of exposure. This is not a likely scenario because of time and cost limitations, but it would be valuable to conduct such a testing process for at least a partial representation of the targeted class of products.

Extended-use testing generally includes some measure of acceptance, but there are other types of sensory/consumer methods that could be used in this type of test, including qualitative research and analytical sensory methods such as discrimination and descriptive analysis.

To obtain a more meaningful assessment of a product's performance by repeated testing, we must devise a procedure that minimizes repetitive questions. For some foods, repeated consumption, especially over a short period of time, may result in a decrease in liking scores. This may not represent sensory adaptation but, rather, a psychological effect that can lead to incorrect assumptions about product acceptance. Thus, to attempt an extended-use test by having a subject evaluate a food product in an IHUT at every meal (when it would not ordinarily be consumed) in order to accelerate the testing is not recommended. Therefore, instructions for use occasions should allow for the more normal time between evaluation occasions. For nonfoods, good sense would apply. You would not want someone to wear the same pants or run every day in the same shirt if you wanted valid evaluation measures. The bottom line is that you must balance the efficiency of repeated product presentation against the possible development of a monotony effect, which would result in a flawed evaluation.

Measures of satiety for food and beverage products might include adaptation to the odor of nonfood products. For short-term exposure, a single tasting or sampling of a product, when consumed, may produce a degree of satiety or fullness. The larger the amount consumed, the greater the satiety. If a product were consumed over repeated occasions in a short time span, clearly there would be an increase in satiety. With repeated extended use separated by longer time intervals (e.g. 1 day), this is not an issue. A good question to ask is whether this satiety is a good condition or a poor one, and whether it interferes with the extended-use test. If we take the second aspect first, we can be reasonably sure that if one feels full consuming a product, then this will probably influence subsequent ratings of a test product and produce contrast effects if multiple products are being evaluated.

Extended-use testing should be considered an essential part of any sensory science program. It is not often required, but if there is an indication that a single or short exposure may not accurately reflect repeated usage, then an extended-usage evaluation may be useful. One of the features of a descriptive panel is the ability to evaluate product in a typical setting and to communicate results electronically as the information is recorded, and it can be done over time. Sensory staff need to be creative in how they use their resources.

8.10 Sensory and legal claims for advertising

In recent years, increased attention has been given to measuring perceived efficacy and to developing evidence in support of claims of product benefits or superiority as part of some advertising campaigns. By perceived efficacy, we refer to those product characteristics that provide a benefit that will be stated and/or advertised in some manner. Examples are a hair product containing a specific chemical or mixture of chemicals that cures dandruff, a lotion that minimizes skin dryness, and foods and beverages that promote a desired health condition based on specific nutrients. Products that are subject to defined health and safety regulations must, of course, satisfy specific regulatory requirements, such as clinical evidence substantiating the biological effectiveness of the product.

Perceived efficacy, on the other hand, addresses the issue from a behavioral standpoint; that is, the product should be perceived by the consumer as providing a particular benefit. Advertising a shampoo that cleans and leaves hair with a soft feel requires evidence that such benefits are perceived by consumers (regardless of what those words mean to those consumers). A product can have both types of efficacy; however, the presence of one does not guarantee the other. Perceived efficacy does not provide one with a basis for a biological claim; however, the reverse situation would be more likely to occur. For example, developing a lotion that is effective in preventing dry skin (as demonstrated by clinical evidence) would have a greater likelihood for market success if the user also perceived it as effective in keeping his or her skin moist or some similar benefit. Ideally, the product's sensory characteristics are consistent with the consumer's expectations for that product. Failure to appreciate these differences in efficacy has had far-reaching and primarily negative business consequences (e.g. needless reformulation or delayed introductions). For these reasons, measurement of product efficacy is a complex issue requiring the incorporation of behavioral measures in a biological study. In some instances, the two elements will be studied separately.

Measuring biological efficacy is accomplished by means of relevant physical, chemical, and biological analyses under well-defined conditions, etc. Although target consumers are the subjects in the clinical trial, their reactions to the products are usually measured by specialists. The specialists are medical and health-related people, including physicians, nurse-practitioners, biochemists, and technical personnel, who may also provide judgments of product effectiveness. Perceived efficacy, however, is strictly a consumer's perception, and this aspect may or may not be a part of a clinical trial. It is important to separate these measures to minimize any potential bias that could arise. Knowledge of the contribution of each type of efficacy to a product's success should provide a more precise focus for advertising and should aid in developing an optimal marketing strategy that emphasizes those product characteristics most readily perceived as beneficial by the consumer.

Using sensory procedures in a clinical trial is appropriate in situations in which the respondents' perceptions of product effectiveness will be a part of the product claim. Sensory evaluation also can contribute in clinical situations in which no claim is planned, but judgments of product effectiveness are made by the experimenter and/ or the subject, in addition to the usual biological measures. The perceptions of the experimenters (or others) can provide further insight into how a particular product is perceived to be working. In this situation, the need is for the sensory professional to incorporate measurement systems that are unambiguous and easy to use, such as a 9-point category scale of strength (see Chapter 3). In many situations, scales are used that may not be familiar to the sensory professional or are adaptations from more familiar scales. For example, it is very common to find 3- or 5-point scales being used with quality word anchors. Experimenters are often unaware of the insensitivity of these scales, in part because it is not their discipline and in part because they are following what was done in the past and assume that it is "standard." As discussed in detail in Chapter 3, 3- and 5-point scales are not as sensitive and are not recommended. Contributions from a sensory professional will be useful provided clinicians are aware of contemporary measurement techniques and the benefits of incorporating sensory tasks into the clinical trial.

Incorporating sensory information into a clinical trial requires a thorough knowledge of the product and the details of the trial. For example, what product biological benefits are being measured? When and how will the product be used? and at what stage (or stages) in the trial will it be possible to obtain sensory measures from the participants? The issue of how efficacy will be measured is especially critical because it will be the driving force, the major influence on the test design, and will impact the specific sensory questions asked of the respondents. Knowing what kinds of benefits and potential claims (the hypothesis) will be made before testing is started has important consequences for the sensory planning.

Planning also includes the specific sensory measures that will be obtained, the ultimate use of products, frequency of usage, and frequency of response—that is, initial and subsequent responses or only a response at the conclusion. These issues are most likely to arise in the evaluation of products such as furniture polish, room air fresheners, and skin lotions. The design must anticipate the impact of product usage over time and the extent that response changes based on initial reaction compared with changes attributable to the way the product is used or the amount of the product used. For most of these tests, the problem takes on an added dimension; that is, the product is applied to a substrate (furniture, skin, or hair), and the response is from the individual who has applied the product. It is possible to have others provide judgments about the products; however, this information would be analyzed separately. Evaluating the effectiveness of a cleaner for no-wax floors might require several different types of no-wax floors as substrates, and the respondents would evaluate all of the floors. Alternatively, each person might evaluate only his or her own floor. The specific protocol will depend on the claim that has been proposed. It is very important that whatever claim or claims are anticipated, they must be known before one can prepare a test plan.

In addition to the issue of who will provide the judgments, it is necessary to consider the nature of the response—that is, the quality or aspect to be measured. This decision will also be influenced by the claim; for example, if shine superiority is the objective, then the dimension "shine" would be measured. However, the reader is reminded that the particular dimension must be defined, or the respondent may use different criteria to assess a dimension such as "shine." As discussed in Chapter 6 on descriptive analysis, considerable risk is entailed in using terms that are not defined, especially for the untrained respondent.

Frequency, amount, and location of product application (and evaluation) must also be specified and should be based on prior test results. Each test must begin with an initial evaluation. This is especially important, even for products such as biological preparations that require repeated usage for effectiveness. Obviously, a product that is not well liked or that is considered of marginal effectiveness on initial exposure is not likely to become more likeable with time. Failure to obtain the initial response could lead to a misinterpretation of results; for example, the product could be assumed to be ineffective when, in fact, the problem might be attributable to the odor, the color, or some other related property.

Frequency of response should depend on frequency of application and the time estimated for optimal effectiveness to be realized. For personal care products, the frequency must be specified in advance and controlled to the extent that misuse is minimized. That is, one respondent may use the product five times, and another respondent may use it three times. Missing or unequal numbers of respondents are not a problem if respondents come to a test location; however, this may not be possible depending on the product and the specific claim being tested. The fact that the product is used in a noncontrolled environment or that the product is applied to different surfaces does not mean that the senses function differently or that traditional measurement techniques are not applicable. These design parameters are built into a test so that their effects can be partitioned from product effects. For most of these tests, the designs are multiple factor and require use of the multivariate analysis of variance model such as treatment×subject and split-plot, as discussed in Chapter 4.

When food and beverage companies want to make a claim and advertise it, whether to challenge a competitor or make a claim of parity (as good as), a different set of procedures and practices are superimposed on the typical sensory and/or consumer test. Depending on the type of claim that will be made, all aspects of the test can be expected to be scrutinized by the legal department of the company making the claim and competitors affected by the claim, by networks that will air those claims, and perhaps even by consumers influenced by the claim (Zelek, 1990). Like the clinical trial, the manufacturer will have large sums of money and perhaps even its credibility at risk; therefore, advertising research involves a high degree of exposure. Careful planning and attention to detail can minimize risk. It is reasonably easy to obtain data that will be useful and defensible without appearing too self-serving, provided proven sensory practices are followed and there are clear directions of the claim or claims that are being considered. In recent years, this topic has been given more attention and discussion (American Society for Testing and Materials (ASTM), 1998; Davenport, 1994; Edlestein, 1994; Passman, 1994; Read, 1994; Smithies, 1994).

Planning a test in support of a specific product claim requires ingenuity and creativity in addition to knowledge about sensory testing. First, it is necessary to determine whether the claim will imply a comparison (either to current product or to a competitor), whether it will directly challenge a competitor, or whether it will simply be a parity claim (i.e. as good as). Second, it is necessary to know whether the claim will be advertised locally, nationally, to the trade, etc. This information along with the specific statements are used to determine which product(s) will be tested, what type of sensory information will be obtained (e.g. attribute and preference), who will be tested, etc. If the claim is about a sensory attribute for a single product (i.e. not comparative), a trained descriptive panel could be sufficient or one could use a consumer "agree/disagree" Likert scale. We have successfully used QDA data where the sensory attribute(s) was defined and used consumers where more conceptual types of attributes (e.g. natural flavor) were in the claim. In the latter case, consumers may be asked whether they "agree" or "disagree" with a statement (e.g. "the product has a natural flavor").

When there is reason to believe that a comparative or challenge claim is possible, consumer testing is necessary. For national taste test claims (claims shown on the networks, broadcast, etc.), the situation is complex, and much has been written and discussed, as noted previously. For example, the National Advertising Division (NAD) of the Better Business Bureau has developed guidelines for testing with consumers (Smithies, 1994), as has ASTM International, Committee E-18 on Sensory Evaluation (E1958). For a national claim, it is recommended that a minimum of 500 consumers from the four geographic regions of the country be tested. However, this is only a guideline, and one might test more or less depending on the circumstances of the specific claim, the type of product, and the population tested. A second important element is the question of who will be tested—all consumers, customers of the products, the persons who shop, etc. Passman (1994) cited some examples in which the consumers were not sufficiently representative, thus negating the basis for a claim. This latter issue is particularly important, and more discussion about this can be found in the aforementioned references.

Once a decision is reached that a claim is warranted (and this is usually based on some prior technical knowledge, for example), a preliminary plan should be prepared that includes the potential claim statement or statements, who should be recruited, where and how the products will be obtained, what kind of a test is proposed, and the specific questions that will be asked. This information is discussed with the company's legal department before any additional effort is put to the plan. Once there is agreement that a claim has merit, a detailed test plan can be prepared. There is one note of caution regarding any plans: Documentation should be kept to a minimum. For most claims, advertisers and the networks will expect to review all the documents to satisfy themselves that the specific claim is justified. Test details should be complete and easy to read, with all parameters defined. If the claim is challenged by competition, they too will have access to all the same information, notes, etc. The competition can refer their concerns to the NAD, and this will likely result in a long and costly process, with depositions, etc. In some situations, the challenge ends in court; however, every effort is made to adjudicate the dispute via a hearing to which both parties agree. Given this background, the sensory professional should make every effort to design a study that is as simple as possible and focuses solely on substantiating the claim and nothing more. The more questions asked, the greater the likelihood that one of them will yield results that are inconsistent with the core question, and selective reporting is never acceptable.

Product source, like consumer source, also has its own set of challenges. All products in a test must be representative in terms of source and age, with no last-minute alterations. Not long ago, company A believed (technical information) its vanilla-flavored frozen dessert was preferred to that of competitor B and decided that a national claim was justified. A few weeks prior to fielding, the competitor's vanilla flavor source was changed (thought to be based on availability). A decision was made to proceed on the mistaken assumption that the original flavor would be used once the supply returned to the original source and the results would still be applicable. Unfortunately, this did not occur. In another situation, one product is obtained from retail while the other is obtained from a warehouse, which also will negate results. Often, a well-planned test is undermined because of a failure to check the smallest detail. Once a claim is challenged, numerous people will examine every aspect of the test searching for weaknesses, no matter how minor, that will preclude the information from ever being used.

The reader is reminded that each claim or potential claim has its own set of requirements beyond those noted here. There is no recipe for success when it comes to claims substantiation. Each test is unique; however, the NAD and ASTM E-18 guidelines provide a useful perspective on the current situation. Perhaps the most useful part of the guidelines is access to various case studies that enable one to read about the failures and what not to do. The reader should also note that this section does not provide recommended scorecards, scales, or methods. This is intentional. As already stated, each potential product advantage/claim has its own set of challenges. The specific claim has a major impact on what kind of test plan is most relevant, what specific question will be asked, and who will provide the responses. The sensory professional can provide the relevant test plan input, recommend the most appropriate design, and ensure that the fielding is done consistent with good sensory practices.

Advertising a product benefit and/or an advantage presents many unique elements that are not typical for sensory evaluation or market research. However, as noted by Passman (1994), the "survey must be more science and less an art." This represents an opportunity for the sensory staff to provide the science and demonstrate its value to an organization.

8.11 **Conclusion**

In this chapter, we focused the reader's attention on examples of programs that will benefit from sensory information and, equally important, the active participation of sensory staff as part of any project/program team. These programs go well beyond a request to determine whether an ingredient change can be detected or whether two products are equally liked, and so on. The benefit from participation of sensory staff and sensory information is the ability to connect the product technology, physical and chemical parameters, consumer preferences, perceived quality, purchase behavior, and imagery. All these factors have a role in ensuring product success and failure, and all would be challenged without sensory involvement.

Sensory professionals cannot expect to be successful by focusing solely on a test (or series of tests) and expect to grow as professionals. There has to be a transition from a pure service function to one that is proactive, to participate in projects that include everything from raw material procurement to the legal issues associated with label statements.

In describing specific problems, we identified ways in which sensory evaluation can contribute, and we purposely emphasized the planning and organizational aspects of the particular problems instead of presenting a step-by-step solution. Because each company and each product category has unique characteristics, we cannot provide a detailed description or recipe for every situation or product. The reader is expected to apply the information presented in this chapter to his or her specific needs. It is important to bear in mind that the principles of sensory evaluation do not change; only their applications change. Success depends on the patience, skills, and knowledge of the sensory professional.

Epilogue

9

CHAPTER OUTLINE

9.1 Introduction

The emphasis in this book has been on understanding the basic principles of the science of perception and applying them to the evaluation of various kinds of stimuli, such as foods, beverages, and personal care products. Results from any evaluation are important because they contribute to a company's product knowledge and ultimately to that company's profits. However, actionable product sensory information and the success of a sensory program do not just happen by chance. Success requires a commitment from a company's management to support a program that offers a range of services. Such a program needs to be flexible to respond rapidly to changes in the marketplace or changes in a company's business strategy.

Sensory programs require time to develop, regardless of company size. Although we originally (in the first and subsequent editions of this book) stated that as many as 3 or more years would be needed, this time line has been dramatically changed due to the increase in the number of academically trained sensory scientists, technological developments in hardware and software, and today's business consolidation and competitive environment. Companies today look for results much sooner; there is an expectation that one should be able to produce actionable information in a few weeks (or less, depending on the problem), not in several months. This means that a sensory professional will be testing products while developing resources. Depending on the type of test, using external resources for some or all of information will be necessary. Whether one is starting a new program or rebuilding an existing program, in most situations, it takes approximately 6–8 weeks to restructure capabilities, analyze available subject capabilities, and previous test results before initiating an appropriate course of action. Of course, this very much depends on the capabilities

H. Stone, R. Bleibaum, H. A. Thomas: Sensory Evaluation Practices, fourth edition.
DOI: http://dx.doi.org/10.1016/B978-0-12-382086-0.00019-4

of the sensory staff and the time needed to establish priorities and develop those resources. Program success also depends on managers understanding the benefits of sensory information and providing support during the building/rebuilding effort. One should also keep in mind that timing for adopting new methods, modifying existing ones, and qualifying subjects can be accomplished as long as there is sufficient support from those who rely on sensory information as part of their decisions. However, establishing a reputation for delivering actionable information and the respect for one's colleagues will take longer. Understanding sensory information and functioning in a competitive environment requires skills to manage information and function strategically.

A successful program requires at least one individual—the sensory professional—to assume responsibility for its technical and business development. Without this individual to develop and champion sensory evaluation, success will be difficult because it will rely on individuals unfamiliar with the science and who do not appreciate the subtleties of human behavior (Eggert, 1989; Sidel and Stone, 2006; Stone and Sidel, 1995). Regardless of who manages sensory resources, success is achieved when staff is aware of a product's business plan, when test planning derives from that plan (i.e. thinking and acting strategically), and when resources are available when needed. Historically, sensory professionals focused on laboratory testing, but there are situations in which product information obtained in a typical use situation (not as a typical home-use test) would better satisfy the project objective. An example is the evaluation of a specialty beverage such as a mixed coffee beverage prepared and usually consumed at the point of purchase. Instead of bringing the product to the subjects, bring the subjects to the shop. This avoids sample handling and travel time as intervening variables, and it provides product information in a typical use situation. Evolution of the tablet as a way of capturing and transmitting data for computing in real time makes these kinds of tests easier to complete and report. Flexibility is the key, along with the knowledge as to how to design and field the test in a nonlaboratory environment. This is one example of the sensory staff demonstrating flexibility in providing information of value to the brand. In approximately the past decade, there has been increasing evidence of success for some sensory professionals who now can serve as role models for younger professionals. Active participation in professional societies is one way to gain insight into how these individuals developed their careers. Of course, each situation will be different, and younger sensory professionals need to learn what has been successful and then determine whether it also applies in their situation.

It should be no surprise to read that success is best achieved when sensory staff thinks and acts strategically. An important related issue is to know how the information from a test will be used or know the answers to questions such as the following: What will be done with the information? How will it be used? Why it is important? Knowing that products A and F are significantly different has meaning only in the context of the purpose for the test and what action will be taken. It is challenging for younger sensory professionals to think and act beyond just reporting results from a test, but it is insight into the meaning of results, that is, how the information will be used, and communicating it to those responsible for initiating action that enhances a sensory professional's growth and value to one's company. One goal of this book is to provide sensory professionals with guidance on the development of resources and when and how to best use them. In earlier chapters, we identified the organizational requirements necessary for a successful sensory program and ways of combining them to yield a viable resource. These resources are described in detail in Chapter 2, and the interested reader may wish to review that material again. We remind the reader that sensory evaluation is one of the least understood and most misused functions in most companies. This was true two decades ago and, unfortunately, remains so today, albeit the situation has shown some improvement in more recent times. One often encounters either of two extremes: (1) Data are limited but conclusions are reached that go well beyond the limits of caution or (2) there is too much data, which yields numerous statistical but not practical significant effects.

It is interesting to observe scientists and technologists diligently adhere to the rules of science in studying their specific problems, yet these same rules are ignored when it comes to evaluation of the end point of their endeavors—a product that satisfies a marketing objective. Project teams assess progress, informally sample products, and when sensory results do not meet expectations, they ignore them out of hand. As already noted, there has been a general lack of understanding for the science of sensory evaluation. A possible explanation for this might be the ease with which one can evaluate a product; no special instrument is required, just one's senses. Technologists formulating products make use of various kinds of equipment to accomplish a specific task and then measure success using a variety of measuring devices. These latter devices are indispensable for the technologist because they reflect the effects of formulation changes, etc. In contrast, sensory test results yield numerical values obtained from human perceptions. The responses are equally important, reflecting whether a formulation change was perceived. Both types of information are important and are more meaningful when they are linked through use of appropriate analyses. As noted in Chapter 8, instruments are univariate in what they measure, whereas humans are multivariate in their responses. It is reasonable for a project team to examine products and discuss test results. However, the issue of when those individuals decide that the sensory results were not what was expected and change recommendations based on their own "evaluation" is evidence of the challenges facing sensory professionals. This example and other abuses of sensory evaluation have been described in Pangborn (1980) and Stone (1999). Although the two references are approximately 20 years apart and more than a decade old, this problem continues to occur with regularity and remains a challenge for sensory professionals.

Sensory evaluation represents a different type of challenge for managers in addition to scientists and technologists. This is more obvious when results are used for determining project accountability or as a measure of a requestor's performance. For example, when a developer works to cost-reduce a product such that the change will not be perceived, the test results can be considered as an internal check on project progress. Each time a test results in a failure to meet an objective, some resentment of sensory evaluation can and will likely develop. History cautions us about the fate of the bearer of bad news. When a sensory staff person is describing results, stating that the products were perceived as different (i.e. rejecting the null hypothesis of no difference), the requestor could misconstrue this action as a personal rejection of his or her work. If the cost savings associated with the test are substantial, there will be great interest in the project and equally great pressure to succeed. Sensory staff must be sensitive to the potential impact of a test result and think strategically about how best to communicate. Care must be taken to ensure that the results are not used as a basis for measuring a developer's competence. The problem can be difficult, if not impossible, to eliminate entirely; however, the sensory professional should have had a discussion with the requestor to be sure that the objective for the test is unambiguous and that the alternative outcomes from the test have been discussed. For example, if the results indicate that a difference was detected and the hypothesis rejected, what will be the next steps? Perhaps there is reason to believe that the difference will be perceived, so an alternative test might be more appropriate. Failure to have this kind of discussion can lead to a misunderstanding as to the meaning of the results, which is not a desired outcome. In some instances, we have observed requesters seeking alternative testing resources in the hope that it will yield a favorable result. Requestors can be very creative in describing their particular problem so as to justify an alternative test method or test capability. Although there is no way to completely prevent such a course of action, it does reinforce the importance of communications for sensory staff. Product and marketing research managers must have complete confidence in what sensory information can do for them so as to minimize risk in product marketing decisions and, most important, to not reject results from a test because they do not satisfy their personal goals.

In this book, we described numerous problems (or opportunities, depending on one's view) that are frequently encountered by sensory professionals. These problems are not confined to any specific product category; they are generic to all products. We propose alternative courses of action, presented in the context of guidelines that will enable the sensory professional to develop an appropriate course of action and lead to a successful solution. This process is highly dependent on the active participation of the sensory staff with the equally active support of management. Sensory evaluation at the professional level remains for the strong-hearted individual.

9.2 Educating the sensory professional

During the past three decades, there has been a significant increase in awareness of the science and the applications of sensory evaluation. Books, journals, and scientific and technical societies have organized symposia and technical programs on the science, and this has had a salutary effect on the practical applications of the science. Increasing public awareness of the field as a result of the enormous interest in wine and food appreciation, especially blogs, has enhanced awareness and attracted more people to the field. All this activity has stimulated the academic community to catch up with the needs of industry, government, and academia itself, creating a further demand for qualified professionals. There are now many food science departments offering courses in sensory evaluation. Eventually, this will help to alleviate the shortage of trained professionals. Various technical societies and industry associations have continued their work to develop testing guidelines. In addition, workshops on sensory evaluation are offered regularly by individuals, consulting organizations, and academic institutions. For example, the University of California has offered a certificated distance learning program for more than a decade that has attracted students from throughout the world (Bleibaum and Schutz, 2010). The American Society for Testing and Materials (ASTM) International Committee E-18 on Sensory Evaluation of Materials and Products, the Sensory and Consumer Sciences Division of the Institute of Food Technologists, the Society for Sensory Professionals, the Sensory Panel of the Food Group of the Society of Chemical Industry, and the European Chemoreception Research Organization are some of the many associations focused on sensory evaluation and offering courses in the field. Websites in multiple languages have been developed, providing an opportunity for individuals to share their interests and to seek free advice. These developments speak well for the future of sensory evaluation. ASTM Committee E-18 has been particularly active, as has been the International Organization for Standardization, with many pamphlets and special technical publications. These documents provide useful information; however, there are some aspects of these documents that warrant further consideration. After all, these guidelines and recommendations are based on the collective judgments of those having a vested interest in the outcomes. While it is important to emphasize the need for subjects to be qualified to participate in a sensory test, along with general information regarding some of the recommended procedures, it is quite another matter to specify a test method or a specific activity such as absolute thresholds, like a recipe. As mentioned at several places in this book, using threshold tests has been documented to be of little value in how well subjects will perform in the evaluation of products.

Other concerns are specifying the number of subjects and the use of a specific language for descriptive analysis when the basis for their use derives from committee decisions and not necessarily research results. Although the information is usually presented as guides, it is often misunderstood by the nonprofessional and becomes rules that hinder progress. After all, it is easy to cite these guidelines when questions arise as to the details of a test. Sensory evaluation measures the responses of people to products based on a wide range of objectives that cannot be predicted. This is a dynamic system, not static and certainly not static in the case of human behavior or the products that are tested. Sensory professionals have to make use of those resources that have proven to be useful in providing information that is reliable, has face validity, and, therefore, is actionable. Despite these concerns, these documents provide some degree of consistency and formality to the discipline, but they are not a substitute for formal academic training and rigorous evaluation by professionals as to the relevancy of a particular method for one's products. We have participated in the preparation of some of these documents and appreciate limitations; however, most sensory professionals are not familiar with these caveats. The hesitancy on the part of universities to develop an adequate curriculum has changed dramatically in recent years, but more needs to be done. It speaks to the demand for trained sensory professionals. Although many of the supporting disciplines already exist—experimental psychology, physiology, and statistics—the addition of courses on sensory evaluation has provided the student with a curriculum not available in the past. As more qualified professionals enter industry, there is a higher likelihood that companies will delegate responsibility for sensory resources to individuals with the appropriate skills.

In general, academic institutions have grown beyond considering sensory evaluation as a part of an allied discipline, such as flavor chemistry. However, there are still some places where the topic is considered as not much more than 3 hours of lectures in a course on quality control—of lesser interest than more laboratory-oriented scientific activities (e.g. cell biology, neurophysiology, and natural products chemistry) that attract greater interest and longer term financial support. The link with flavor (and aroma) chemistry is understandable inasmuch as the chemistry is accompanied by descriptions of the end products in terms of their aromas and tastes. The early literature on the chemical senses devoted considerable attention to this topic (e.g. see Moncrieff, 1951), and this trend continues (e.g. see Harper, 1982a,b). More recent publications have provided a more comprehensive description of the use of sensory evaluation in this field and its more unique components (Clarke and Bakker, 2004; Fisher and Scott, 1997). Knowledge of the aroma and taste qualities of various chemicals is essential to the chemist formulating flavors and fragrances for specific application (e.g. see Dorland and Rogers (1977) or any other recent text). Often, this technical (expert) language was and, to a considerable degree, still is equated with sensory evaluation. Not surprisingly, this causes confusion with regard to what is sensory evaluation. With increased numbers of courses offered at academic institutions, the differences between sensory and flavor chemistry are more easily recognized and the potential confusion will lessen.

In today's business environment, the sensory professional without formal academic training can be at a disadvantage in terms of (1) organizing and managing testing activities, (2) acting strategically, and (3) the professional's ability to assess published research that is directly applicable to his or her own program. It is in this latter area that more discussion is warranted. There is no question that research on the physiology of the senses and studies of binding sites, etc. is essential to understanding the perceptual process at the cellular level. Research at the cellular level has shown considerable progress in uncovering mechanisms that provide insight as to how various stimuli work. However, it would be inappropriate to allocate resources based on ongoing research without active management support or to make substantive changes in an evaluation procedure without convincing evidence. Research that focuses on the mechanisms of perception is very important, but the professional working in industry must deal with complex stimuli and limited time lines. A simple example is the request to measure how much of an ingredient change is possible before it is perceived. This relies primarily on the discrimination model. The result is very important from a business perspective. What the chemical and biological steps were or what criteria (sensory and mental processes) a subject used to reach the decision are matters that go well beyond the sensory test. As mentioned in Chapter 5, the research by O'Mahony, Ennis, and co-workers (e.g. see Bi and Ennis, 2001a; Bi *et al.*, 1997; Ennis and Bi, 1998, Ennis and Mullen, 1986; Rousseau and Ennis, 2001; Rousseau *et al.*, 1998) has failed to demonstrate any useful advantage associated with the practical issues of difference testing and the associated decision-making process This is not to imply that such information has no potential; rather, the information to date is insufficient and still speculative. It is complicated when tests are fielded with naive and unqualified subjects. The argument is made that it is justified because consumers are not qualified but this is not relevant. Companies market products to consumers of their brand, and any product changes need to be tested with those consumers and particularly the ones most likely to detect a difference and not those most unlikely to detect a difference.

Another example is magnitude estimation, a form of ratio scaling that was presented as a significantly more powerful technique for measuring strength of a perception. It was proposed as a method that could be used in place of ordinal and interval scaling, offering a wide range of advantages. For details about this methodology, see Chapter 3. What was interesting was the large number of peer-reviewed publications on ratio scaling espousing its benefits, adding to its reputation as the method of choice for the future. However, subsequent research in the evaluation of products versus simple stimuli failed to demonstrate the claimed advantages. Today, sensory testing of products that makes use of ratio scaling is rare. As noted in Chapter 3, the method requires more than the usual explanation before a subject can use the method, and no superiority could be demonstrated. Surprisingly, it took several years for most sensory professionals to become aware of this lack of advantage and to realize that there were other useful scaling methods (Anderson, 1970, 1974; Birnbaum, 1982). Although it can be argued that this "correction" reflects the strength of the science of sensory evaluation, the very fact that it has taken time can be interpreted as a weakness in the discipline. It is hoped that texts such as these will serve as a reminder to be cautious when presented with superiority claims for measuring perceptions.

The just-about-right scale is an example of a technique adopted from another discipline, and it has almost become a standard procedure used in consumer testing; however, there is little evidence that it offers any significant advantage. In many companies, it was used where no descriptive analysis capabilities existed. Today, it is used whether or not descriptive data are available. Marketing uses the responses as a guide; that is, if the percentage of responses in the just-about-right category (of three categories) equals 65%, then that product attribute is OK, no change is needed. Percentages less than 65% mean a change is needed. Companies usually have established percentages as norms for what constitutes a change versus a no change. As described in Chapter 3, the scale is a nominal scale and as such, the types of analyses are limited. However, newer approaches have provided a basis for further computations, including various methods under the heading of "penalty analysis" or "mean drop analysis." Although such procedures reflect a serious effort to obtain more information from such data, the basic method has flaws. First, a 3-point scale is insensitive. Second, there is potential confusion regarding whether the response in the too-weak, too-strong categories reflects strength or liking. Third, statistical analysis of nominal information is limited to percentages, but it is fairly common to encounter means and variance measures. Using this information for marketing purposes is reasonable; however, using it for sensory purposes to guide product development is a poor substitute for descriptive analysis.

Neural networks are another development that was thought to have a role in sensory evaluation. Neural networks came about through the idea that if one did not know what variables were important and their interrelationships, this methodology would be appropriate. The methodology incorporates fuzzy logic and essentially "learns" from previous analyses, adjusting as more data are analyzed (Bomio, 1998). In situations in which the number of variables is very large—for example, identifying the variables that contribute to adult obesity in the population—this technique could be useful. In the evaluation of foods and beverages, for example, the number of variables is finite (i.e. a product's formulation, the sensory attributes, etc.). No research could demonstrate an advantage to use of the methodology, and as a result, its use in sensory evaluation has all but disappeared.

In more recent times, the use of multivariate methods including factor analysis, multiple regression, and discriminant analysis has had a major and positive impact on sensory evaluation (Kline, 1994; Korth, 1982; Lawless and Heymann, 2010; MacFie and Thomson, 1988; Manly, 1986; Smith, 1988). Sensory professionals welcomed the availability of the methods because of their direct applications. They provide a basis for relating different kinds of product information and uncover potential underlying relationships not obvious when data are examined in a univariate way. In some environments, their use is synonymous with "cutting-edge" sensory evaluation. As noted in Chapter 8, optimization and related techniques rely on multivariate designs and contribute useful product information in relation to such dependent variables as pricing and imagery (Stone and Sidel, 2007). Use of these techniques is now so popular that they have become a standard procedure for some companies. As with any of these more sophisticated systems, a working knowledge of the methods and their limitations is required. Misuse of multivariate analyses is common because of the ease with which they can be used on a PC and the availability of software that does not come with sufficient details about the risks associated with the analyses. The programs iterate until a solution is obtained, and in many instances, the sensory staff lacks an appreciation for the limitations and recognizes only that a significant effect was achieved. In some instances, the number of products is too small (as few as six), the number of variables measured is many times greater, and the scales used are not optimal. Although the information will be displayed in map form and look informative, the results from implementation are less than successful. Some of the software will allow transformation of the data and identify and exclude mathematical outliers, and the experimenter has no way of knowing the behavioral effects of these changes. This is not to suggest that the computations for these transformations are incorrect; rather, the results do not come with warning flags, and the user could reach conclusions that are incorrect. In any complex analysis, there are numerous decision points that can have a significant impact on results. Investigators must not ignore reviewing results to ensure that they make sense before proceeding to the next steps or trying alternative analyses. As Groopman (2009) noted, "Statistical analysis is not a substitute for thinking." This caution is more thoroughly explored by May (2004) in a discussion about the benefits and the risks associated with multivariate analyses based on assumptions about which little was known. May cautioned about the misuse of the models and that users were not sufficiently well enough informed. In sensory evaluation, this is a particularly challenging situation because a large portion of any database is derived from human perceptions. As previously noted, taking any database regardless of size or how obtained and subjecting it to various computations will yield a result, but that does not mean it has any validity.

Multivariate designs enable one to study the mathematical relationships among variables not usually considered in the past—for example, competitive products, formulations, sensory characteristics, preferences, pricing, perceived quality, and product benefits. Without use of multivariate designs, such variables could not be managed within a single test and certainly not in a cost-effective way.

Newer developments in the field include emotional descriptive analysis, laddering, napping, and benchtop sorting techniques. Emotional descriptive analysis is a technique adapted from social psychological research, as are laddering and related interviewing techniques. The idea in the case of emotional descriptive analysis is to uncover product benefits not obtained through more traditional focus groups or other probing procedures intended to identify language that consumers use when discussing a product's benefits. There is no question that imagery, products, physical and chemical measures, and sensory attributes are connected. The key is to identify which attributes within each are most important to purchase behavior. Laddering is a probing technique used in one-on-one situations, in which the interviewer follows a specific line of questioning with a consumer—for example, "You said you liked product X because it tasted good. Why did you say that?" The idea is to uncover the underlying basis for the particular statement. Developing new methods and adapting methods from other disciplines is important for sensory professionals, but it is also important to understand the procedures and the practices in sufficient detail to make an informed decision as to their applicability. For example, probing consumers seems quite easy; however, professionals in the field consider it a difficult task and have certificate programs to ensure that users are qualified. Of course, like any development, the final question is whether the information can be related back to a product's sensory attributes and its formulation. Can one take action based on results? As mentioned in Chapter 1, sensory professionals must be aware of new developments and be able to assess their application to the field and avoid reliance on what may be fashionable at the time but is found to be based on questionable science.

A few comments are warranted about the increased reliance on mathematics for more than descriptive purposes. May (2004) stated, "Sadly, examples of the application of statistical 'confidence intervals' to distributions resulting from making arbitrary assumptions about essentially unknown parameters, and then endowing this with reality by passage through a computer, continue to proliferate" (p. 792). In studying human perception and specifically discrimination, ability is relevant.

Developments in the science of perception and in sensory evaluation as well as in allied sciences will continue. The sensory professional must be able to assess these developments and determine the extent to which they are applicable to his or her specific needs rather than adopting them without appreciating the consequences. Knowledge of the basic principles of sensory evaluation is essential to ensure that the risk of an incorrect or premature decision to incorporate these developments into a test program is minimized. One of the most important principles is the use of qualified subjects.

Finally, some additional comments are necessary with regard to the importance of educating managers about the science on which sensory evaluation is based. As mentioned in Chapter 2, the sensory professional must establish the basis on which a program is developed. Communication is very important. Sensory professionals must think strategically and be aware of changes in business plans and their impact on technology, which in turn impacts available sensory resources. Managers regularly receive information about new developments from external as well as internal sources; however, they cannot always assess the value without additional background knowledge. The externally derived development, because it comes from a manager, can put a sensory professional in a defensive situation and/or it raises questions about internal resources. Sensory professionals must keep current with the literature (made easy by access to the Internet) and with new developments that have potential applications.

New developments are often presented in workshops. As previously mentioned, one can offset lack of a formal education through workshop participation. For more than three decades, programs have been offered and have usually attracted a diverse audience. Workshops enable the individual to become familiar with the principles of testing, with appropriate tests for a particular problem, and with the potential value of new developments. Check the web; one can easily identify short courses and conferences for almost every week of the year.

9.3 The future

The future for sensory evaluation is bright, judging from the number of industry openings and the many companies that are expanding their sensory test capabilities. Although mergers and cutbacks have significantly changed the landscape, the need for product sensory information remains strong.

If sensory evaluation is to become more successful, it must move from a passive, service-oriented program to an active, results-oriented resource. This metamorphosis will require a conscious effort on the part of each sensory professional to anticipate needs and to be able to respond with minimal delay. Sensory professionals need to think and act according to their company's business plans and especially in relation to brand strategies. At the same time, those individuals who could be users of sensory evaluation must be made aware of its potential. Thus, the sensory professional not only must acquire new knowledge but also must educate others about the use of sensory information. Sensory evaluation has proven beneficial for many companies, and we hope that our science will continue to work its magic.

There is no question that sensory evaluation is a profitable investment. We think of it in terms of a 1-to-10 ratio; that is, for every dollar invested in sensory, it will return 10 or more. This kind of success occurs when sensory professionals actively participate in the business of sensory evaluation. The development and use of predictive models of consumer-product behavior has had and will continue to have a salutary impact, not only because of the specific and immediate value to a company but also because of effectiveness in demonstrating a higher degree of sophistication than was realized possible for sensory evaluation. For these reasons, we can be optimistic about the future for sensory evaluation in business and as a science. It is hoped that this book will continue to provide the impetus for the continued growth of the science, that it will stimulate more academic growth of the science, and especially that it will stimulate more recognition for the science of sensory evaluation.

References

Aaker, D.A., 1995. Developing Business Strategies, fourth ed. Wiley, New York.

Aaker, D.A., 1996. Building Strong Brands. Free Press, New York.

Aaker, D.A., 2010. Building Strong Brands. Simon & Schuster, New York.

Alder, H.L., Roessler, E.B., 1977. Introduction to Probability and Statistics, sixth ed. Freeman, San Francisco.

American Meat Science Association (AMSA), 1978. Guidelines for Cookery and Sensory Evaluation of Meat. AMSA, Chicago.

American Society for Testing and Materials (ASTM), 1996. Manual on Sensory Testing Methods, MNL 26, second ed. ASTM International, West Conshohocken, PA.

American Society for Testing and Materials (ASTM), 1998. Standard guide for sensory claims substantiation, E-1958-98. In: ASTM Book of Standards, vol. 15.08 (November 2003). ASTM International, West Conshohoken, PA.

American Society for Testing and Materials (ASTM), 2002. Standard test method for sensory analysis—Triangle test, E1985–97. In: Annual Book of ASTM Standards, vol. 15.08. ASTM International, West Conshohoken, PA.

American Society for Testing and Materials (ASTM), 2007. Standard Guide for Sensory Claims Substantiation, ASTM E1958-07. ASTM International, West Conshohocken, PA.

American Society for Testing and Materials (ASTM), 2010a. Standard practice for determining odor and taste thresholds by a forced-choice ascending concentration series of methods of limits, E679-04. In: Annual Book of Standards, vol. 15.08. ASTM International, West Conshohocken, PA.

American Society for Testing and Materials (ASTM), 2010b. Standard practice for defining and calculating individual and group sensory thresholds from forced-choice data sets of intermediate size, E1432-04. In: Annual Book of Standards, vol. 15.08. ASTM International, West Conshohocken, PA.

American Society for Testing and Materials (ASTM), 2011. Standard Practice for Descriptive Skinfeel Analysis of Creams and Lotions, ASTM E1490-11. ASTM International, West Conshohocken, PA.

Amerine, M.A., Pangborn, R.M., Roessler, E.B., 1965. Principles of Sensory Evaluation of Food. Academic Press, New York.

Amerine, M.A., Roessler, E.B., 1983. Wines: Their Evaluation. Freeman, San Francisco.

Amerine, M.A., Roessler, E.B., Filipello, F., 1959. Modern sensory methods of evaluating wine. Hilgardia 28, 477–565.

Andani, Z., Jaeger, S.R., Wakeling, I., MacFie, H.J.H., 2001. Mealiness in apples: towards a multilingual consumer vocabulary. J. Food Sci. 66 (6), 872–879.

Anderson, N.H., 1970. Functional measurement and psychological judgment. Psychol. Rev. 77, 153–170.

Anderson, N.H., 1974. Algebraic models in perception In: Carterette, E.C., Friedman, M.P. (Eds.), Handbook of Perception, vol. 2. Academic Press, New York, pp. 215–298.

Anonymous, 1968. Basic Principles of Sensory Evaluation, ASTM STP 433. American Society for Testing and Materials, Philadelphia.

Anonymous, 1975. Minutes of Division Business Meeting. Institute of Food Technologists–Sensory Evaluation Division, IFT, Chicago.

Anonymous, 1981a. Guidelines for the Selection and Training of Sensory Panel Members, STP 758. American Society for Testing and Materials, Philadelphia.

Anonymous, 1981b. Sensory evaluation guide for testing food and beverage products. Food Technol. 35 (11), 50–59.

Anonymous, 1984. Overview: outstanding symposia in Food Science & Technology: use of computers in the sensory lab. Food Technol. 38 (9), 66–88.

Anonymous, 1989, March 20. Playing it safe with new products. Food Bus., 40–42.

Aust, L.B., 1984. Computers as an aid in discrimination testing. Food Technol. 38 (9), 71–73.

Aust, L.B., Gacula Jr., M.C., Beard, S.A., Washam II, R.W., 1985. Degree of difference test method in sensory evaluation of heterogeneous product types. J. Food Sci. 50, 511–513.

Baker, G.A., Amerine, M.A., Roessler, E.B., 1954. Errors of the second kind in organoleptic difference testing. Food Res. 19, 205–210.

Baker, R.C., Hahn, P.W., Robbins, K.R., 1988. Fundamentals of New Food Product Development. Elsevier, New York.

Bartlett, P.N., Elliot, J.M., Gardner, J.W., 1997. Electronic noses and their application in the food industry. Food Technol. 51 (12), 44–48.

Baten, W.D., 1946. Organoleptic tests pertaining to apples and pears. Food Res. 11, 84–94.

Baumgardner, M., Tatham, R., 1988, June/July. No preference in paired-preference testing. Quirk's Market. Res. Rev., 18–20.

Bendig, A.W., Hughes, J.B., 1953. Effect of amount of verbal anchoring and number of rating-scale categories upon transmitted information. J. Exp. Psychol. 46, 87–90.

Bengtsson, K., Helm, E., 1946. Principles of taste testing. Waller. Lab. Commun. 9 (28), 171–181.

Bennet, G.B., Spahr, M., Dodds, M.L., 1956. The value of training a sensory test panel. Food Technol. 10, 205–208.

Bi, J., 2005. Similarity testing in sensory and consumer research. Food Qual. Prefer. 16, 139–149.

Bi, J., 2011. Similarity tests using forced-choice methods in terms of Thurstonian discriminal distance, d. J. Sensory Stud. 26, 151–157.

Bi, J., Ennis, D.M., 2001a. Statistical models for the A-not A method. J. Sensory Stud. 16, 215–237.

Bi, J., Ennis, D.M., 2001b. Exact beta-binomial tables for small experiments. J. Sensory Stud. 16, 319–325.

Bi, J., Ennis, D.M., O'Mahony, M., 1997. How to estimate and use the variance of d' from difference tests. J. Sensory Stud. 12, 87–104.

Bi, J., Lee, H.S., O'Mahony, M., 2010. d' and variance of d' for four-alternative forced choice (4-AFC). J. Sensory Stud. 25, 740–750.

Birnbaum, M.H., 1982. Problems with so-called direct scaling. In: Kuznicki, J.T., Johnson, R.A., Rutkiewick, A.F. (Eds.), Problems and Approaches to Hedonics. American Society for Testing and Materials, Philadelphia.

Blake, S.P., 1978. Managing for Responsive Research and Development. Freeman, San Francisco.

Bleibaum, R.N., Barnewolt, D. Willis, S., 2011. Exploration and innovation using qualitative sensory immersion. Oral presentation, 2011 Pangborn Symposium, Toronto, Canada.

Bleibaum, R.N., McDermott, B., 2008, October/November. Gaining sensory preferences. World Food Ingredients, 58–63.

Bleibaum, R.N., Robichaud, E.J., 2007. Using consumer's sensory experience to achieve strategic market segmentation. Cosmetics Toiletries 122 (11), 75–80.

Bleibaum, R.N., Robichaud, J., Tao, C., 2009. Exploring potential to increase descriptive analysis capacity by creating alternate data collection environments. Oral presentation, 2009 Pangborn Symposium, Florence, Italy.

Bleibaum, R.N., Schutz, H.C., 2010, March. Stimulating sensory. World Food Ingredients 36–37.

Bleibaum, R.N., Stone, H., Tan, T., Labreche, S., Saint-Martin, E., Isz, S., 2002. Comparison of sensory and consumer results with electronic tongue sensors for apple juices. Food Qual. Prefer. 13, 409–422.

Bleibaum, R.N., Walker, L., 2011. Global understanding of consumer perceptions of products. Poster presentation, 2011 Pangborn Symposium, Toronto, Canada.

Blom, G., 1955. How many taste testers? Wallerstein Lab. Commun. 18 (62), 173–177.

Bock, R.D., Jones, L.V., 1968. The Measurement and Prediction of Judgment and Choice. Holden-Day, San Francisco.

Boelens, M., Boelens, H., 2002. Main verbal responses during human olfaction. Perfumer Flavorist 27, 34–43.

Boggs, M.M., Hansen, H.L., 1949. Analysis of foods by sensory difference test. Adv. Food Res. 2, 219–258.

Bomio, M., 1998. Neural networks and the future of sensory evaluation. Food Technol. 52 (8), 62–63.

Bone, R., 1987. The importance of consumer language in developing product concepts. Food Technol. 41 (11), 58–60, 86.

Boring, E.G., 1950. A History of Experimental Psychology, second ed. Appleton, New York.

Box, G.E.P., 1954. The exploration and exploitation of response surfaces: some general considerations and examples. Biometrics 10, 16–61.

Box, G.E.P., Draper, N.R., 1987. Empirical Model-Building and Response Surfaces. Wiley, New York.

Box, G.E., Hunter, W.G., Hunter, J.S., 1978. Statistics for Experimenters. Wiley, New York.

Bradley, R.A., 1953. Some statistical methods in taste testing and quality evaluation. Biometrics 9, 22–38.

Bradley, R.A., 1963. Some relationships among sensory difference tests. Biometrics 19, 385–397.

Bradley, R.A., 1975. Science, Statistics, and Paired Comparisons, ONR Tech. Rep. No. 92 (Contract N00014-67-A-0235-0006). Office of Naval Research, Washington, DC.

Bradley, R.A., Harmon, T.J., 1964. The modified triangle test. Biometrics 20, 608–625.

Bradley, R.A., Terry, M.E., 1952. The rank analysis of incomplete block designs: I. The method of paired comparison. Biometrika 39, 324–345.

Brady, L.P., 1984. Computers in sensory research. Food Technol. 38 (9), 81–83.

Brady, L.P., Ketelsen, S.M., Ketelsen, L.J.P., 1985. Computerized system for collection and analysis of sensory data. Food Technol. 39 (5) 82, 84, 86, 88.

Brandt, F.I., Arnold, R.G., 1977. Sensory tests used in food product development. Prod. Dev. 8, 56.

Brandt, M.A., Skinner, E., Coleman, J., 1963. Texture profile method. J. Food Sci. 28, 404–410.

Bressan, L.P., Behling, R.W., 1977. The selection and training of judges for discrimination testing. Food Technol. 31 (11), 62–67.

Brockhoff, P.B., Christensen, R.H.B., 2010. Thurstonian models for sensory discrimination tests as generalized linear models. Food Qual. Prefer. 21 (3), 330–338.

Bruning, J.L., Kintz, B.L., 1987. Computational Handbook of Statistics, third ed. Scott Foresman, Glenview, IL.

Buchanan, B., Givon, M., Goldman, A., 1987. Measurement of discrimination ability in taste tests: an empirical investigation. J. Market. Res. 24, 154–163.

Buchanan, B., Henderson, P., 1992. Assessing the bias of preference, detection, and identification measures of discrimination ability in product design. Market. Sci. 11, 64–75.

Byer, A.J., Abrams, D., 1953. A comparison of the triangular and two-sample taste-test methods. Food Technol. 7, 185–187.

Cain, W.S., 1977. Differential sensitivity for smell: "noise" at the nose. Science 195, 796–798.

Cain, W.S., Marks, L.E., 1971. Stimulus and Sensation: Readings in Sensory Psychol. Little, Brown, Boston.

Cairncross, W.E., Sjöström, L.B., 1950. Flavor profile—A new approach to flavor problems. Food Technol. 4, 308–311.

Cardello, A., Lawless, H.T., Schutz, H., 2008. Effects of extreme anchors and interior label spacing on labeled magnitude scales. Food Qual. Prefer. 21, 232–334.

Cardello, A.V., Maller, D., Kapsalis, J.G., Segars, R.A., Sawyer, F.M., Murphy, C., et al. 1982. Perception of texture by trained and consumer panels. J. Food Sci. 47, 1186–1197.

Cardello, A.V., Schutz, H.G., 1996. Food appropriateness measures as an adjunct to consumer preference/acceptability evaluation. Food Qual. Prefer. 7, 239–249.

Carlson, K.D., 1977, September. New products—Problems or opportunities? Food Prod. Dev. 11 (34), 36–37.

Carpenter, R.P., Lyon, D.H., Hasdell, T.A., 2000. Guidelines for Sensory Analysis in Food Product Development and Quality Control, second ed. C.H.I.P.S., Weimar, TX.

Carter, C., Riskey, D., 1990. The roles of sensory research and marketing research in bringing a product to market. Food Technol. 44 (11), 160–162.

Carterette, E.C., Friedman, M.P., 1974. Handbook of Perception, vol. 2. Academic Press, New York.

Castura, J.C., 2010. Equivalence testing: a brief review. Food Qual. Prefer. 21, 257–258.

Caul, J.F., 1957. The profile method of flavor analysis. Adv. Food Res. 7, 1–40.

Chen, A.W., Resurreccion, A.V.A., Paguio, L.P., 1996. Age appropriate hedonic scales to measure food preferences of young children. J. Sensory Stud. 11, 141–163.

Christensen, C.M., Vickers, Z., Fahrenholtz, S.K., Gengler, I.M., 1988, June 19–22. Effect of questionnaire design on hedonic ratings, Paper No. 388. Presented at the 48th annual meeting of the Institute of Food Technologists, New Orleans, LA.

Civille, G.V., Lawless, H.T., 1986. The importance of language in describing perceptions. J. Sensory Stud. 1, 203–215.

Clapperton, J.R., Dalgliesh, C.E., Meilgaard, M.C., 1975. Progress towards an international system of beer flavor terminology. MBAA Tech. Q. 12, 273–280.

Clark, R., 1991. Sensory-texture profile analysis correlation in model gels. In: Frontiers in Carbohydrate Research, vol. 2. Elsevier, New York.

Clarke, R.J., Bakker, J., 2004. Wine Flavour Chemistry. Blackwell, Oxford.

Cloninger, M.R., Baldwin, R.E., Krause, G.F., 1976. Analysis of sensory rating scales. J. Food Sci. 41, 1225–1228.

Clurman, A., 1989, June. Kids are consumers. Market. Res., 70–71.

Clute, M., 2008. Food Industry Quality Control Systems. CRC Press, Boca Raton, FL.

Cochran, W.G., Cox, G.M., 1957. Experimental Designs, second ed. Wiley, New York.

Cohen, E., 1993. TURF analysis. Quirk's Market. Res. Rev. 7 (6), 10.

Cohen, J., 1977. Statistical Power Analysis for the Behavioral Sciences, rev. ed. Academic Press, New York.

Cohen, J., Cohen, P., 1983. Applied Multiple Regression/Correlation Analysis for the Behavioral Sciences, second ed. Erlbaum, Hillsdale, NJ.

Coombs, C.H., 1964. A Theory of Data. Wiley, New York.

Cooper, R.G., 1993. Winning at New Products: Accelerating the Process from Idea to Launch, second ed. Addison-Wesley, Cambridge, MA.

Cover, S., 1936. A new subjective method of testing tenderness in meat—The paired-eating method. Food Res. 1, 287–295.

Cox III, E.P., 1980. The optimal number of response alternatives for a scale: a review. J. Market. Res. 17, 407–422.

Crawford, C.M., 1977. Market research and the new product failure rate. J. Market. 21 (2), 51–61.

Cross, H.R., Moen, R., Stanfield, M.S., 1978. Training and testing of judges for sensory analysis of meat quality. Food Technol. 32, 48–54.

Daniel, W.W., 1978. Applied Nonparametric Statistics. Houghton Mifflin, Boston.

Davenport, K., 1994. NBC's television advertising review procedures and guidelines. Food Technol. 48 (8), 83–84.

Dawson, E.H., Brogdon, J.L., McManus, S., 1963a. Sensory testing of differences in taste: I. Methods. Food Technol. 17 (9), 45–48. 51.

Dawson, E.H., Brogdon, J.L., McManus, S., 1963b. Sensory testing of differences in taste: II. Selection of panel members. Food Technol. 17 (10), 39–41, 43–44.

Dawson, E.H., Dochterman, E.F., 1951. A comparison of sensory methods of measuring differences in food qualities. Food Technol. 5, 79–81.

Day, R.L., 1974. Measuring preferences. In: Ferber, R. (Ed.), Handbook of Marketing Research (Part A). McGraw-Hill, New York, pp. 3-101–3-125.

Delwiche, J., O'Mahony, M., 1996. Flavour discrimination—An extension of Thurstonian paradoxes to the tetrad test. Food Qual. Prefer. 7, 1–5.

Dember, W.N., 1963. The Psychology of Perception. Holt, New York.

Dorland, W.E., Rogers Jr., J.A., 1977. The Flavor and Fragrance Industry. Dorland, Mendham, NJ.

Downs, P.A., Anderson, E.O., Babcock, C.T., Herzer, F.H., Trout, G.M., 1954. Evaluation of collegiate student dairy products judging since World War II. J. Dairy Sci. 34, 1021–1026.

Draper, N., Smith, H., 1981. Applied Regression Analysis, second ed. Wiley, New York.

Duncan, A.J., 1974. Quality Control and Industrial Statistics, fourth ed. Irwin, Homewood, IL.

Eggert, J., 1989. Sensory strategy for success in the food industry. J. Sensory Stud. 3, 161–167.

Eggert, J., Zook, K., 1986. Physical Requirement Guidelines for Sensory Evaluation Laboratories, ASTM STP 913. American Society for Testing and Materials, Philadelphia.

Eisler, H., 1963a. How prothetic is the continuum of smell? Scand. J. Psychol. 4, 29–32.

Eisler, H., 1963b. Magnitude scales, category scales, and Fechnerian integration. Psychol. Rev. 70, 243–253.

Ekman, G., Sjöberg, L., 1965. Scaling. Annu. Rev. Psychol. 16, 451–474.

Ellis, B.H., 1966. Guide Book for Sensory Testing, second ed. Continental Can Company, Chicago.

Engel, J.F., Blackwell, R.D., Miniard, P.W., 1995. Consumer Behavior, eighth ed. Harcourt Brace, Orlando, FL.

Engen, T., 1960. Effect of practice and instruction on olfactory thresholds. Perceptual Motor Skills 10, 195–198.

Engen, T., 1971. Psychophysics: II. In: Kling, J.W., Rigg, L.A. (Eds.), Scaling Methods in Experimental Psychology (third ed.). Holt, New York, pp. 47–86.

Ennis, D.M., 1993. The power of sensory discrimination methods. J. Sensory Stud. 8, 353–370.

Ennis, D.M., Bi, J., 1998. The beta-binomial model: accounting for inter-trial variation in replicated difference and preference tests. J. Sensory Stud. 13, 389–412.

Ennis, D.M., Mullen, K., 1986. Theoretical aspects of sensory discrimination. Chem. Senses 11, 513–522.

Ennis, J.M., Jesionka, V., 2011. The power of sensory discrimination methods revisited. J. Sensory Stud. 26 (5), 371–382.

Farnbach, J.S., Kuesten, C., 2012. Eliminating waste in the front end of innovation. World Food Sci. (in press).

Ferdinandus, A., Oosterom-Kleijngeld, I., Runneboom, A.J.M., 1970. Taste testing. MBAA Tech. Q. 7, 210–227.

Fern, E.F., 1982. The use of focus groups for idea generation: the effects of group size, acquaintanceship and moderator on response quantity and quality. J. Market. Res. 19, 1–13.

Ferrier, J.G., Block, D.E., 2001. Neural-network-assisted optimization of wine blending based on sensory analysis. Am. J. Enol. Vitic. 52 (4), 386–395.

Fisher, C., Scott, T.R., 1997. Food Flavours: Biology and Chemistry. Royal Society of Chemistry, Cambridge, UK.

Fleiss, J.L., 1981. Statistical Methods for Rates and Proportions, second ed. Wiley, New York.

Friedman, M., 1990. Twenty-five years later and 98,900 new products later. Prepared Foods New Products Annu. 159 (8), 23–25.

Frijters, J.E.R., 1979. Variations of the triangular method and the relationship of its unidimensional probabilistic model to three-alternate forced-choice signal detection theories. Br. J. Math. Stat. Psychol. 32, 229–241.

Frijters, J.E.R., 1980. Three-stimulus procedures in olfactory psychophysics: an experimental comparison of Thurstone–Ura and three-alternative forced-choice models of signal detection theory. Percept. Psychophys. 28, 390–397.

Frijters, J.E.R., 1988. Sensory difference testing and the measurement of sensory discriminability. In: Piggott, J.R. (Ed.), Sensory Analysis of Foods (second ed.). Elsevier, London, pp. 131–153.

Frijters, J.E.R., Blauw, Y.H., Vermoaat, S.H., 1982. Incidental training in the triangular method. Chem. Senses 7, 63–69.

Fuller, G.W., 1994. New Food Product Development: From Concept to Marketplace. CRC Press, Boca Raton, FL.

Gacula Jr., M.C., 1975. The design of experiments for shelf life study. J. Food Sci. 40, 99–403.

Gacula Jr., M.C., 1978. Analysis of incomplete block designs with reference samples in every block. J. Food Sci. 43, 1461–1466.

Gacula Jr., M.C., 1993. Design and Analysis of Sensory Optimization. Food and Nutrition Press, Trumbull, CT.

Gacula Jr., M.C., Kubala, J.J., 1975. Statistical models for shelf life failures. J. Food Sci. 40, 404–409.

Gacula Jr., M.C., Singh, J., 1984. Statistical Methods in Food and Consumer Research. Academic Press, Orlando, FL.

Garber Jr., L.L., Hyatta, E.M., Starr Jr., R.G., 2003. Measuring consumer response to food products. Food Qual. Prefer. 14 (1), 3–15.

Garner, W.R., 1960. Rating scales, discriminability, and information transmission. Psychol. Rev. 67, 343–352.

Garner, W.R., Hake, H.W., 1951. The amount of information in absolute judgments. Psychol. Rev. 58, 446–459.

Geldard, F.A., 1972. The Human Senses, second ed. Wiley, New York.

Giovanni, M.E., Pangborn, R.M., 1983. Measurement of taste intensity and degree of liking of beverages by graphic scales and magnitude estimation. J. Food Sci. 48, 1175–1182.

Giradot, N.E., Peryam, D.R., Shapiro, R., 1952. Selection of sensory testing panels. Food Technol. 6, 140–143.

Goldman, A.E., McDonald, S.S., 1987. The Group Depth Interview. Prentice-Hall, Englewood Cliffs, NJ.

Gordin, H.H., 1987. Intensity variation descriptive methodology: development and application of a new sensory evaluation technique. J. Sensory Stud. 2, 187–198.

Gorman, B., 1990. New products for a new century. Prepared Foods New Products Annu. 159 (8), 16–18.

Graf, E., Saguy, I.S., (Eds.), 1991. Food Product Development: From Concept to the Marketplace. Van Nostrand, New York.

Granit, R., 1955. Receptors and Sensory Perception. Yale University Press, New Haven, CT.

Green, B.G., Shaffer, G.S., Gilmore, M.M., 1993. Derivation and evaluation of a semantic scale of oral sensation magnitude with apparent ratio properties. Chem. Sen. 18, 683–702.

Green, D.M., Swets, J.A., 1966. Signal Detection Theory and Psychophysics. Wiley, New York.

Greene, P.E., Carmone, F.J., 1970. Multidimensional Scaling and Related Techniques in Marketing Analysis. Allyn & Bacon, Boston.

Gridgeman, N.T., 1955a. The Bradley–Terry probability model and preference testing. Biometrics 11, 335–343.

Gridgeman, N.T., 1955b. Taste comparisons: two samples or three. Food Technol. 9, 148–150.

Gridgeman, N.T., 1958. Psychophysical bias in taste testing by pair comparison, with special reference to position and temperature. Food Res. 23, 217–220.

Gridgeman, N.T., 1959. Pair comparison, with and without ties. Biometrics 15, 382–388.

Gridgeman, N.T., 1964. Sensory comparisons: the 2-stage triangle test with sample variability. J. Food Sci. 29, 112–117.

Gridgeman, N.T., 1984. Tasting panels: sensory assessment in quality control. In: Herschdeorfer, S.M. (Ed.), Quality Control in the Food Industry. Academic Press, London, pp. 299–349.

Groopman, J., 2009, December 17. In: New York Rev. Books 56 (20), 22, 24.

Gruenwald, G., 1992. New Product Development: Responding to Market Demand, second ed. NTC Business Books, Lincolnwood, IL.

Guilford, J.P., 1954. Psychometric Methods, second ed. McGraw-Hill, New York.

Guinard, J.-X., Pangborn, R.M., Shoemaker, C.F., 1985. Computerized procedure for time-intensity sensory measurements. J. Food Sci. 50, 543–544, 546.

Gunst, R.F., Mason, R.L., 1980. Regression Analysis and Its Application: A Data-Oriented Approach. Dekker, New York.

Hall, R.L., 1958. Flavor study approaches at McCormick and Co., Inc. In: Little, A.D. (Ed.), Flavor Research and Food Acceptance. Reinhold–Van Nostrand, Princeton, NJ, pp. 224–240.

Hanson, J.E., Kendall, D.A., Smith, N.F., Hess, A.P., 1983, November. The missing link: correlation of consumer and professional sensory descriptions. Beverage World, 108–115.

Harper, R.M., 1950. Assessment of food products. Food 19, 371–375.

Harper, R.M., 1972. Human Senses in Action. Churchill-Livingstone, Edinburgh, UK.

Harper, R.M., 1982a. Art and science in the understanding of flavour. Food Flavour Ingredients Packag. Process. 4 (1), 13–15, 17, 19, 21.

Harper, R.M., 1982b. Art and science in the understanding of flavour. Food Flavour Ingredients Packag. Process. 4 (2), 17–19, 21, 23.

Harries, J.M., 1956. Positional bias in sensory assessments. Food Technol. 10, 86–90.

Harrison, S., Elder, L.W., 1950. Some applications of statistics to laboratory taste testing. Food Technol. 4, 434–439.

Hauser, J.R., Clausing, D., 1988, May 1. House of quality. Harvard Business Rev.

Hays, W.L., 1973. Statistics for the Social Sciences, second ed. Holt, New York.

Helm, E., Trolle, B., 1946. Selection of a taste panel. Wallerstein Lab. Commun. 9 (28), 181–194.

Henika, R.G., 1982. Use of response-surface methodology in sensory evaluation. Food Technol. 36 (11), 96–101.

Henry, J., Walker, D., 1994. Managing Innovation. Sage, Thousand Oaks, CA.

Herschdoerfer, S.M., 1984. Quality Control in the Food Industry, vol. 1, second ed. Academic Press, London.

Highsmith, J., 2009. Agile Project Management: Creating Innovative Products, second ed. Addison-Wesley, Boston.

Hinreiner, E.H., 1956. Organoleptic evaluation by industry panels—The cutting bee. Food Technol. 31 (11), 62–67.

Hollander, M., Wolfe, D.A., 1973. Nonparametric Statistical Methods. Wiley, New York.

Hopkins, J.W., 1954. Some observations on sensitivity and repeatability of triad taste difference tests. Biometrics 10, 521–530.

Hopkins, J.W., Gridgeman, N.T., 1955. Comparative sensitivity of pair and triad flavor intensity difference tests. Biometrics 11, 63–68.

Horsfield, S., Taylor, L.J., 1976. Exploring the relationship between sensory data and acceptability of meat. J. Sci. Food Agric. 27, 1044–1053.

Hubbard, M.R., 1996. Statistical Quality Control for the Food Industry, second ed. Van Nostrand Reinhold, New York. pp. 161–173.

Hubbard, M.R., 2003. Statistical Quality Control for the Food Industry, third ed. Kluwer/Plenum, New York.

Hughes, G.D., 1974. The measurement of beliefs and attitudes. In: Ferber, R. (Ed.), Handbook of Marketing Research. McGraw-Hill, New York, pp. 3–16.

Huitson, A., 1989. Problems with Procrustes analysis. J. Appl. Stat. 16, 39–45.

Inglett, G.E., 1974. Symposium: Sweeteners. Avi, Westport, CT.

Ishii, R., O'Mahony, M., 1990. Group taste concept measurement: verbal and physical definition of the umami taste concept for Japanese and Americans. J. Sensory Stud. 4, 215–227.

Ishii, R., O'Mahony, M., 1991. Use of multiple standards to define sensory characteristics for descriptive analysis: aspects of concept formation. J. Food Sci. 56, 838–842.

Jackson, R.A., 2002. Wine Tasting: A Professional Handbook. Academic Press, San Diego.

Joanes, D., 1985. On a rank sum test due to Kramer. J. Food Sci. 50, 1442–1444.

Johnson, P.B., Civille, G.V., 1986. A standard lexicon of meat WOF descriptors. J. Sensory Stud. 1, 99–104.

Johnson, R.M., 1973. Simultaneous Measurement of Discrimination and Preference. Market Facts, Chicago.

Jones, F.N., 1958. Prerequisites for test environment. In: Little, A.D. (Ed.), Flavor Research and Food Acceptance. Van Nostrand-Reinhold, Princeton, NJ, pp. 107–111.

Jones, L.V., Peryam, D.R., Thurstone, L.L., 1955. Development of a scale for measuring soldiers' food preferences. Food Res. 20, 512–520.

Kamen, J.M., Peryam, D.R., Peryam, D.B., Kroll, B.J., 1969. Hedonic differences as a function of number of samples evaluated. J. Food Sci. 34, 475–479.

Kamenetzky, J., 1959. Contrast and convergence effects in ratings of foods. J. Appl. Psychol. 43, 47–52.

Kimmel, S.A., Sigman-Grant, M., Guinard, J.-X., 1994. Sensory testing with young children. Food Technol. 48 (3), 92–94, 96–99.

Kline, P., 1994. An Easy Guide to Factor Analysis. Routledge, London.

Korth, B., 1982. Use of regression in sensory evaluation. Food Technol. 36 (11), 91–95.

Kramer, A., 1960. A rapid method for determining significance of differences from rank sums. Food Technol. 10, 576–581.

Kramer, A., 1963. Revised tables for determining significance differences. Food Technol. 17, 124–125.

Kramer, A., Twigg, B.A., 1970., Quality Control for the Food Industry, vol. 1, third ed. Avi, Westport, CT.

Kroll, B.J., 1990. Evaluating rating scales for sensory testing with children. Food Technol. 44 (11), 78–86.

Krzanowski, W.J., 1988. Principles of Multivariate Analysis. Oxford University Press, Oxford.

Kuesten, C.L., Kruse, L., 2008. Physical Requirement for Guidelines for Sensory Evaluation Laboratories, second ed. (No. MNL60). ASTM International, West Conshohocken, PA.

Labovitz, S., 1970. The assignment of numbers to rank order categories. Am. Soc. Rev. 36, 515–524.

Laming, D, 1986. Sensory Analysis. Academic Press, Orlando, FL.

Land, D.G., Shepherd, R., 1988. Scaling and ranking methods. In: Piggott, J.R. (Ed.), Sensory Analysis of Foods (second ed.). Elsevier Applied Science, London, pp. 155–185.

Larson-Powers, N., Pangborn, R.M., 1978. Descriptive analysis of the sensory properties of beverages and gelatins containing sucrose or synthetic sweeteners. J. Food Sci. 43, 42–51.

Laue, E.A., Ishler, N.H., Baltman, G.A., 1954. Reliability of taste testing and consumer testing methods. Food Technol. 8, 387–388.

Lawless, H.T., 1989. Logarithmic transformation of magnitude estimation data and comparisons of scaling methods. J. Sensory Stud. 4, 75–86.

Lawless, H.T., Heymann, H., 1999. Sensory Evaluation of Food Principles and Practices. Aspen, Gaithersburg, MD.

Lawless, H.T., Heymann, H., 2010. Sensory Evaluation of Food Principles and Practices. Springer, New York.

Lawless, H.T., Horne, J., Spiers, W., 2000. Contrast and range effects for category, magnitude and labeled magnitude scales in judgements of sweetness intensity. Chem. Senses 25, 85–92.

Lawless, H.T., Malone, G.J., 1986a. The discriminative efficiency of common scaling methods. J. Sensory Stud. 1, 85–98.

Lawless, H.T., Malone, G.J., 1986b. A comparison of rating scales: sensitivity, replicates and relative measurement. J. Sensory Stud. 1, 155–174.

Lawless, H.T., Popper, R., Kroll, B.J., 2010. A comparison of the labeled magnitude (LAM) scale, an 11-point category sale and the traditional nine-point hedonic scale. Food Qual. Prefer. 2, 4–12.

Lieb, M.E., 1989, March 20. Playing it safe with new products. Food Business 2, 40–42.

Lim, J., 2011. Hedonic scaling: a review of methods and theory. Food Qual. Prefer. 22 (8), 733–747.

Lindquist, E.F., 1953. Design and Analysis of Experiments in Psychology and Education. Houghton Mifflin, Boston.

Lockhart, E.E., 1951. Binomial systems and organoleptic analysis. Food Technol. 5, 428–431.

Lotong, V., Chambers, D.H., Dus, C., Chambers IV., E., Civille, G.V., 2001. Matching results of two independent highly trained panels using different descriptive analysis methods. Paper presented at The 4th Pangborn Sensory Science Symposium 2001: A Sense Odyssey, Dijon, France.

Luce, R.D., Edwards, W., 1958. The derivation of subjective scales from just noticeable differences. Psychol. Rev. 65, 222–237.

Lyon, B.G., 1987. Development of chicken flavor descriptive attribute terms aided by multivariate statistical procedures. J. Sensory Stud. 2, 55–67.

Lyon, C.E., Lyon, G.B., Townsend, W.E., Wilson, R.L., 1978. Effect of level of structured protein fiber on quality of mechanically deboned chicken meat patties. J. Food Sci. 43, 1524–1527.

Lyon, D.H., Francombe, M.A., Hasdell, T.A., Lawson, K., 1992. Guidelines for Sensory Analysis in Food Product Development and Quality Control. Chapman & Hall, London.

MacFie, H.J.H., Thomson, D.M.H., 1988. Preference mapping and multidimensional scaling. In: Piggot, J.R. (Ed.), Sensory Analysis of Foods, second ed. Elsevier, London, pp. 381–409.

Mackey, A.O., Jones, P., 1954. Selection of members of a food tasting panel: discernment of primary tastes in water solution compared with judging ability for foods. Food Technol. 8, 527–530.

Mahoney, C.H., Stier, H.L., Crosby, E.A., 1957. Evaluating flavor differences in canned foods: I. Genesis of the simplified procedure for making flavor difference tests. Food Technol. 11, 29–41.

Manly, B.F.J., 1986. Multivariate Statistical Methods: A Primer. Chapman & Hall, New York.

Manoski, P., Gantwerker, S., 2002a. Managing the new product development process: Part 1. Food Process 64 (5), 48, 50–51.

Manoski, P., Gantwerker, S., 2002b. Managing the new product development process: Part 2. Food Processing 64 (6), 40, 42, 44, 46.

Marks, L.E., 1974. On scales of sensation: prolegomena to any future psychophysics that will be able to come forth as science. Percept. Psychophys. 16, 358–376.

Marshall, R.J., Kirby, S.P.J., 1988. Sensory measurement of food texture by free-choice profiling. J. Sensory Stud. 3, 63–80.

Martens, M., Martens, H., 1986. Partial least squares regression. In: Piggott, J.R. (Ed.), Statistical Procedures in Food Research. Elsevier, London.

Marx, M.H., Hillix, W.A., 1963. Systems and Theories in Psychology. McGraw-Hill, New York.

Masuoka, S., Hatjopoulos, D., O'Mahony, M., 1995. Beer bitterness detection: testing Thurstonian and sequential sensitivity analysis models for triad and tetrad methods. J. Sensory Stud. 10, 295–306.

May, R.M., 2004. Uses and abuses of mathematics in biology. Science 303 (5659), 790–793.

McBride, R.L., 1983. A JND-scale/category-scale convergence in taste. Percept. Psychophys. 34, 77–83.

McBride, R.L., 1990. The Bliss Point Factor. Macmillan, South Melbourne.

McBride, R.L., Hall, C., 1979. Cheese grading versus consumer acceptability: an inevitable discrepancy. Aust. J. Dairy Technol. 34, 66–68.

McBride, R.L., Laing, D.G., 1979. Threshold determination by triangle testing: effect of judgmental procedure, positional bias and incidental training. Chem. Senses Flav. 4, 319–326.

McBride, R.L., MacFie, H.J.H. (Eds.), 1990. Psychological Basis of Sensory Evaluation. Elsevier, New York.

McCall, R.B., 1975. Fundamental Statistics for Psychology, second ed. Harcourt Brace Jovanovich, New York.

McClure, S., Lawless, H.T., 2010. Comparison of the triangle and a self-defined two alternative forced choice test. Food Qual. Prefer. 21, 547–552.

McDaniel, M.R., Sawyer, F.M., 1981. Preference testing of whiskey sour formulation: magnitude estimation versus the 9-point hedonic. J. Food Sci. 46, 182–185.

McDermott, B.J., 1990. Identifying consumers and consumer test subjects. Food Technol. 44 (11), 154–158.

McGrath, M.E., 1996. Setting the PACE in Product Development: A Guide to Product and Cycle-time Excellence. Butterworth-Heinemann, Boston.

McNemar, Q., 1969. Psychological Statistics, fourth ed. Wiley, New York.

Meilgaard, M., Civille, G.V., Carr, B.T., 1999. Sensory Evaluation Techniques, third ed. CRC Press, Boca Raton, FL.

Meilgaard, M., Civille, G.V., Carr, B.T., 2006. Sensory Evaluation Techniques, fourth ed. CRC Press, Boca Raton, FL.

Meiselman, H.L., MacFie, H.J.H., 1996. Food Choice, Acceptance and Consumption. Chapman & Hall, New York.

Meiselman, H.L, Schutz, H., 2003. History of food acceptance research in the U.S. Army. Appetite 40 (3), 199–216.

Meiselman, H.L., Waterman, D., Symington, L.E., 1974. Armed Forces Food Preferences, (Tech. Rep. 75-63-FSL). U.S. Army Natick Dev. Cent., Natick, MA.

Mela, D.J., 1989. A research note: a comparison of single and concurrent evaluations of sensory and hedonic attributes. J. Food Sci. 54, 1098–1100.

Mermelstein, N.H., 2012a. Coffee quality testing. Food Technol. 66 (1), 68–72.

Mermelstein, N.H., 2012b. Chocolate quality testing. Food Technol. 66 (2), 66–70.

Meyer, R.S., 1984. Eleven stages of successful new product development. Food Technol. 38 (7), 71–78, 98.

Miller, I., 1978. Statistical treatment of flavor data. In: Apt, C.M. (Ed.), Flavor: Its Chemical, Behavioral and Commercial Aspects. Westview, Boulder, CO., pp. 149–159.

Moncrieff, R.W., 1951. The Chemical Senses. Leonard Hill, London.

Morgan, C.T., Stellar, E, 1950. Physiological Psychology, second ed. McGraw-Hill, New York.

Morrison, D.F., 1990. Multivariate Statistical Methods, third ed. McGraw-Hill, New York.

Morrison, D.G., 1981. Triangle taste tests: are the subjects who respond correctly lucky or good? J. Market. 45, 111–119.

Mosher, H.A., Dutton, H.J., Evens, C.D., Cowan, J.C., 1950. Conducting a taste panel for the evaluation of edible oils. Food Technol. 4, 105–109.

Moskowitz, H.R., 1972. Subjective ideals and sensory optimization in evaluating perceptual dimensions in food. J. Appl. Psychol. 56, 60–66.

Moskowitz, H.R., 1975. Applications of sensory measurement to food evaluations: II. Methods of ratio scaling. Lebens. Wiss. Technol. 8, 249–254.

Moskowitz, H.R., Jacobs, B., Firtle, N., 1980. Discrimination testing and product decisions. J. Market Res. 17 (2), 35–43.

Moskowitz, H.R., Sidel, J.L., 1971. Magnitude and hedonic scales of food acceptability. J. Food Sci. 36, 677–680.

Mullen, K., Ennis, D.M., 1979. Rotatable designs in product development. Food Technol. 33 (7), 74–75, 78–80

Mullet, G.M., 1988. Applications of multivariate methods in strategic approaches to product marketing and promotion. Food Technol. 42 (11), 145, 152, 153, 155, 156.

Muñoz, A.M., 1986. Development and application of texture reference scales. J. Sensory Stud. 1, 55–83.

Myers, J.L., 1979. Fundamentals of Experimental Design, third ed. Allyn & Bacon, Boston.

Nakayama, M., Wessman, C., 1979. Application of sensory evaluation to the routine maintenance of product quality. Food Technol. 33 (9), 38–39.

Newell, G.J., MacFarlane, J.D., 1987. Expanded tables for multiple comparison procedures in the analysis of ranked data. J. Food Sci. 52, 1721–1725.

Noble, A., 1975. Instrumental analysis of the sensory properties of food. Food Technol. 29 (11), 56–60.

Noble, A.C., Arnold, R.A., Buechsenstein, J., Leach, E.J., Schmidt, J.O., Stern, P.M., 1987. Modification of a standardized system of wine aroma terminology. Am. J. Enol. Vitric. 38, 143–151.

Nunnally, J.C., 1978. Psychometric Theory, second ed. McGraw-Hill, New York.

O'Mahony, M., 1979. Short-cut signal detection measures for sensory analysis. J. Food Sci. 44, 302–303.

O'Mahony, M., 1982. Some assumptions and difficulties with common statistics for sensory analysis. Food Technol. 36 (11), 75–82.

O'Mahony, M., 1986. Sensory Evaluation of Food. Dekker, New York.

O'Mahony, M., Garske, S., Klapman, K., 1980. Rating and ranking procedures for short-cut signal detection multiple difference tests. J. Food Sci. 45, 392–393.

O'Mahony, M., Kulp, J., Wheeler, L., 1979. Sensory detection of off-flavors in milk incorporating short-cut signal detection measures. J. Dairy Sci. 62, 1857–1864.

O'Mahony, M., Rathman, L., Ellison, T., Shaw, D., Buteau, L., 1990. Taste descriptive analysis: concept formation, alignment and appropriateness. J. Sensory Stud. 5, 71–103.

O'Mahony, M., Thieme, V., Goldstein, L.R., 1988. The warm-up effect as a means of increasing the discriminability of sensory difference tests. J. Food Sci. 53, 1848–1850.

Orne, M.T., 1981. The significance of unwitting cues for experimental outcomes: toward a pragmatic approach. Ann. N. Y. Acad. Sci. 364, 152–159.

Pangborn, R.M., 1964. Sensory evaluation of foods: a look backward and forward. Food Technol. 18 (9), 63–67.

Pangborn, R.M., 1979. Physiological and psychological misadventures in sensory measurement or the crocodiles are coming. In: Johnson, M.R. (Ed.), Sensory Evaluation Methods for the Practicing Food Technologists. Institute of Food Technologists, Chicago, pp. 2-1-2-22.

Pangborn, R.M., 1980. Sensory science today. Cereal Foods World 25 (10), 637–640.

Pangborn, R.M., Dunkley, W.L., 1964. Laboratory procedures for evaluating the sensory properties of milk. Dairy Sci. Abstr. 26, 55–62.

Pangborn, R.M., Guinard, J.X., Meiselman, H.L., 1989. Evaluation of bitterness of caffeine in hot chocolate drink by category, graphic, and ratio scaling. J. Sensory Stud. 4, 31–53.

Passman, N., 1994. Support advertising superiority claims with taste tests. Food Technol. 48 (8), 71–72. 74

Payne, S.L., 1965. Are open-ended questions worth the effort? J. Market. Res. 10, 417–419.

Pearce, J.H., Korth, B., Warren, C.B., 1986. Evaluation of three scaling methods for hedonics. J. Sensory Stud. 1, 27–46.

Pecore, S., Stoer, N., Hooge, S., Holschuh, N., Hulting, F., Case, F., 2006. Degree of difference testing: a new approach incorporating control lot variability. Food Qual. Prefer. 17 (7–8), 552–555.

Pecore, S.D., 1984. Computer-assisted consumer testing. Food Technol. 38 (9), 78–80.

Perfetti, T.A., Gordin, H.H., 1985. Just noticeable difference studies of mentholated cigarette products. Tob. Sci. 57, 20–29.

Perrin, L., Symoneaux, R., Maître, I., Asselin, C., Jourjon, F., Pagès, J., 2008. Comparison of three sensory methods for use with the Napping procedure: case of ten wines from Loire valley. Food Qual. Prefer. 19 (1), 1–11.

Peryam, D.R., 1991. A history of ASTM Committee E-18. ASTM Standard. News 19 (3), 28–35.

Peryam, D.R., Haynes, J.H., 1957. Prediction of soldiers' food preferences by laboratory methods. J. Appl. Psychol. 41, 2–6.

Peryam, D.R., Pilgrim, F.J., 1957. Hedonic scale method of measuring food preferences. Food Technol. 11 (9), 9–14.

Peryam, D.R., Pilgrim, F.J., Peterson, M.S., (Eds.), 1954. Food Acceptance Testing Methodology. National Academy of Sciences/National Research Council, Washington, DC.

Peryam, D.R., Polemis, B.W., Kamen, J.M., Eindhoven, J., Pilgrim, F.J., 1960. Food Preferences of Men in the Armed Forces. Quartermaster Food and Container Institute of the Armed Forces, Chicago.

Peryam, D.R., Swartz, V.W., 1950. Measurement of sensory differences. Food Technol. 4, 390–395.

Petty, M.F., Scriven, F.M., 1991. The use of Fourier analysis to compare the sensory profiles of products. J. Sensory Stud. 6, 17–23.

Pfaffman, C., Schlosberg, H., Cornsweet, J., 1954. Variables affecting difference tests. In: Peryam, D.R., Pilgrim, F.J., Peterson, M.S. (Eds.), Food Acceptance Testing Methodology. National Academy of Sciences/National Research Council, Washington, DC, pp. 4–17.

Phillips Jr., J.L., 1973. Statistical Thinking. Freeman, San Francisco.

Piggot, J.R. (Ed.), 1988. Sensory Analysis of Foods. Second ed. Elsevier, London.

Pilgrim, F.J., Wood, K.R., 1955. Comparative sensitivity of rating scale and paired comparison methods for measuring consumer preference. Food Technol. 9, 385–387.

Pokorný, J., Marcin, A., Davidek, J., 1981. Comparison of the efficiency of triangle and tetrad tests for discriminatory sensory analysis of food. Nahrung 25, 561–564.

Popper, R., Kroll, J.J., 2003. Conducting sensory research with children. Food Technol. 57 (5), 60–65.

Powers, J.J., 1988. Current practices and applications of descriptive methods. In: Piggot, J.R. (Ed.), Sensory Analysis of Foods, second ed. Elsevier, London, pp. 187–266.

Powers, J.J., Warren, C.B., Masurat, T., 1981. Collaborative trials involving three methods of normalizing magnitude estimations. Lebensm. Wiss. Technol. 14, 86–93.

Poynder, T.M. (Ed.), 1974. Transduction Mechanisms in Chemoreception. Information Retrieval, London.

Prince, G.M., 1970. The Practice of Creativity. Harper & Row, New York.

Radkins, A.P., 1957. Some statistical considerations in organoleptic research: triangle, paired, duo–trio tests. Food Res. 22, 259–265.

Radkins, A.P., 1958. Sequential analysis in organoleptic research: triangle, paired, duo–trio tests. Food Res. 23, 225–234.

Rainey, B.A., 1986. Importance of reference standards in training panelists. J. Sensory Stud. 1, 149–154.

Read, J., 1994. Role of marketing research in claims testing. Food Technol. 48 (8), 75–77.

Resurreccion, A.V.A., 1988. Applications of multivariate methods in food quality evaluation. Food Technol. 42 (11), 128, 130, 132–134, 136.

Resurreccion, A.V.A., 1998. Consumer Sensory Testing for Product Development. Aspen, Gaithersburg, MD.

Robichaud, J., Bleibaum, R.N., Thomas, H., 2007. Cracking the consumer code—Linking winemakers to consumers to increase brand loyalty. In: Proceedings of the Thirteenth Australian Wine Industry Technical Conference, Adelaide, South Australia, Australian Wine Industry Technical Conference, Adelaide, South Australia, pp. 151–156.

Roessler, E.B., Baker, G.A., Amerine, M.A., 1953. Corrected normal and chi-square approximations to the binomial distribution in organoleptic tests. Food Res. 18, 625–627.

Roessler, E.B., Baker, G.A., Amerine, M.A., 1956. One-tailed and two-tailed tests in organoleptic comparisons. Food Res. 21, 117–121.

Roessler, E.B., Pangborn, R.M., Sidel, J.L., Stone, H., 1978. Expanded statistical tables for estimating significance in paired-preference, paired-difference, duo–trio and triangle tests. J. Food Sci. 43, 940–943.

Roessler, E.B., Warren, J., Guymon, J.F., 1948. Significance in triangular taste tests. Food Res. 13, 503–505.

Roper, G., 1989, June. Research with marketing's paradoxical subjects: children. Market. Res., 16–23.

Rousseau, B., Ennis, D.M., 2001. A Thurstonian model for the dual pair (4IAX) discrimination method. Percept. Psychophys. 63, 1083–1090.

Rousseau, B., Meyer, A., O'Mahony, M., 1998. Power and sensitivity of the same–different test: comparison with triangle and duo–trio methods. J. Sensory Stud. 13, 149–173.

Rousseau, B., Stroh, S., O'Mahony, M., 2002. Investigating more powerful discrimination tests with consumers: effects of memory and response bias. Food Qual. Prefer. 13, 39–45.

Russell, G.F., 1984. Some basic considerations in computerizing the sensory laboratory. Food Technol. 38 (9), 67–70, 77.

Rutledge, K.P., Hudson, J.M., 1990. Sensory evaluation: method for establishing and training a descriptive flavor analysis panel. Food Technol. 44 (12), 78–84.

Ryan, T.A., 1959. Multiple comparisons in psychological research. Psychol. Bull. 56, 26–47.

Ryan, T.A., 1960. Significance tests for multiple comparisons of proportions, variances, and other statistics. Psychol. Bull. 57, 318–328.

Savoca, M.R., 1984. Computer applications in descriptive testing. Food Technol. 38 (9), 74–77.

Sawyer, F.M., Stone, H., Abplanalp, H., Stewart, G.F., 1962. Repeatability estimates in sensory-panel selection. J. Food Sci. 27, 386–393.

Scheffé, H., 1952. An analysis of variance for paired comparisons. J. Am Stat. Assoc. 47, 381–400.

Schiffman, S.S., Reynolds, M.L., Young, F.W., 1981. Introduction to Multidimensional Scaling: Theory, Methods, and Applications. Academic Press, New York.

Schlich, P., 1993. Risk tables for discrimination tests. J. Food Qual. Prefer. 4, 141–151.

Schlosberg, H., Pfaffmann, C., Cornsweet, J., Pierrel, R., 1954. Selection and training of panels. In: Peryam, D.R., Pilgrim, J.J., Peterson, M.S. (Eds.), Food Acceptance Testing

Methodology. National Academy of Sciences/National Research Council, Washington, DC, pp. 45–54.

Schutz, H.G., 1954. Effect of bias on preference in the difference-preference test. In: Peryam, D.R., Pilgrim, F.J., Peterson, M.S. (Eds.), Food Acceptance Testing Methodology. National Academy of Sciences/National Research Council, Washington, DC, pp. 85–91.

Schutz, H.G., 1965. A food action rating scale for measuring food acceptance. J. Food Sci. 30, 365–374.

Schutz, H.G., 1983. Multiple regression approach to optimization. Food Technol. 37 (11), 46–48, 62.

Schutz, H.G., 1988a. Multivariate analyses and the measurement of consumer attitudes and perceptions. Food Technol. 42 (11), 141–144, 156.

Schutz, H.G., 1988b. Beyond preference: appropriateness as a measure of contextual acceptance of food. In: Thomson, D.M.H. (Ed.), Food Acceptability. Elsevier, London, pp. 115–134.

Schutz, H.G., 1998. Evolution of the sensory science discipline. Food Technol. 52 (8), 42–46.

Schutz, H.G., Cardello, A.V., 2001. A labeled affective magnitude (LAM) scale for assessing food liking/disliking. J. Sensory Stud. 16, 117–159.

Schutz, H.G., Cardello, A.V., 2003. Sensory science II: Consumer acceptance. In: Rahman, M.S. (Ed.), Handbook of Food Science. Dekker, New York.

Schutz, H.G., Damrell, J.D., Locke, B.H., 1972. Predicting hedonic ratings of raw carrot texture by sensory analysis. J. Text. Stud. 3, 227–232.

Schutz, H.G., Pilgrim, F.J., 1957. Differential sensitivity in gustation. J. Exp. Psychol. 54, 41–48.

Seaton, R., 1974. Why ratings are better than comparisons. J. Adv. Res. 14, 45–48.

Shepard, R.N., 1966. Metric structure in ordinal data. J. Math. Psychol. 3, 287–315.

Sherman, P., 1969. A texture profile of foodstuffs based upon well-defined rheological properties. J. Food Sci. 34, 458–462.

Sidel, J.L., Stone, H., 1976. Experimental design and analysis of sensory tests. Food Technol. 30 (11), 32–38.

Sidel, J.L., Stone, H., 1983. Introduction to optimization research—Definitions and objectives. Food Technol. 37 (11), 36–38.

Sidel, J.L., Stone, H., 2006. Sensory science: methodology. In: Hui, Y.H. (Ed.) Handbook of Food Science, Technology, and Engineering, vol. 2. Taylor & Francis, London, pp. 57-3–57-24.

Sidel, J.L., Stone, H., Bloomquist, J., 1981. Use and misuse of sensory evaluation in research and quality control. J. Dairy Sci. 64, 2296–2302.

Sidel, J.L., Stone, H., Bloomquist, J., 1983. Industrial approaches to defining quality. In: Williams, A.A., Atkin, R.K. (Eds.), Sensory Quality in Foods and Beverages: Its Definitions, Measurement and Control. Horwood, Chichester, UK, pp. 48–57.

Sidel, J.L., Stone, H., Thomas, H.A., 1994. Hitting the target: sensory and product optimization. Cereal Foods World 39 (11), 826–830.

Sidel, J.L., Woolsey, A.L., Stone, H., 1975. Sensory analysis: theory, methodology, and evaluation. In: Inglett, G.E. (Ed.), Fabricated Foods. Avi, Westport, CT, pp. 109–126.

Silver, D., 2003. The seven deadly sins of product development. Food Product Design 12 (2), 93, 95, 97, 99

Sjöström, L.B., Cairncross, S.E., 1954. The descriptive analysis of flavor. In: Peryam, D.R., Pilgrim, F.J., Peterson, M.S. (Eds.), Food Acceptance Testing Methodology. National Academy of Sciences/National Research Council, Washington, DC, pp. 25–30.

Smith, G.C., Savell, J.W., Cross, H.R., Carpenter, Z.L., 1983. The relationship of USDA quality grade to beef flavor. Food Technol. 37 (5), 233–238.

Smith, G.L., 1981. Statistical properties of simple sensory difference tests: confidence limits and significance tests. J. Sci. Food Agric. 32, 513–520.

Smith, G.L., 1988. Statistical analysis of sensory data. In: Piggot, J.R. (Ed.), Sensory Analysis of Foods, second ed. Elsevier, London, pp. 335–379.

Smith, P.G., 2007. Flexible Product Development. Jossey-Bass, San Francisco.

Smith, P.G., Oltmann, J., 2010. Flexible Project Management: Extending Agile Techniques Beyond Software Projects. PMI Global Congress Proceedings, Washington, DC.

Smith, P.G., Reinertsen, D.G., 1998. Developing Products in Half the Time: New Rules, New Tools. Wiley, New York.

Smithies, R.H., 1994. Resolving advertising disputes between competitors. Food Technol. 48 (8), 68, 70.

Stagner, R., Osgood, C.E., 1946. Impact of war on a nationalistic frame of reference: I. Changes in general approval and qualitative patterning of certain stereo types. J. Soc. Psychol. 24, 187–215.

Stampanoni, C.R., 1993. The quantitative flavor profiling technique. Perfumer Flavorist 18, 19–24.

Steiner, E.H., 1966. Sequential procedures for triangular and paired comparison tasting tests. J. Food Technol. 1, 41–53.

Stevens, S.S., 1951. Mathematics, measurement and psychophysics. In: Stevens, S.S. (Ed.), Handbook of Experimental Psychology. Wiley, New York, pp. 1–49.

Stevens, S.S., 1957. On the psychophysical law. Psychol. Rev. 64, 153–181.

Stevens, S.S., 1962. The surprising simplicity of sensory metrics. Am. Psychol. 17, 29–38.

Stevens, S.S., Galanter, E.H., 1957. Ratio scales and category scales for a dozen perceptual continua. J. Exp. Psychol. 54, 377–411.

Stewart, R.A., 1971. Sensory evaluation and quality assurance. Food Technol. 25 (4), 103–106.

Stone, H., 1963. Determination of odor differences limens for three compounds. J. Exp. Psychol. 66, 466–473.

Stone, H., 1972. Food Products and Processes. SRI International, Menlo Park, CA.

Stone, H., 1999. Sensory evaluation: science and mythology. Food Technol. 53 (10), 124.

Stone, H., 2002. Sensory evaluation and the consumer in the 21st century. Presentation at the New Products Conference, Phoenix, AZ, October 13–16.

Stone, H., Bleibaum, R.N., 2009. Sensory evaluation. In: Campbell-Platt, G. (Ed.), Food Science and Technology. Wiley Blackwell, Oxford.

Stone, H., Bosley, J.J., 1965. Olfactory discrimination and Weber's law. Percept. Mot. Skills 20, 657–665.

Stone, H., Drexhage, K.A., 1985. Use of computers in sensory evaluation. Tragon Newslett. 1, 1–5.

Stone, H., McDermott, B.J., Sidel, J.L., 1991. The importance of sensory analysis for the evaluation of quality. Food Technol. 45 (6), 88, 90, 92–95.

Stone, H., Oliver, S., 1969. Measurement of the relative sweetness of selected sweeteners and sweet mixtures. J. Food Sci. 34, 215–222.

Stone, H., Pangborn, R.M., 1968. Intercorrelations of the senses. In: Basic Principles of Sensory Evaluation, ASTM STP, 433. American Society for Testing and Materials, Philadelphia. pp. 30–46.

Stone, H., Sidel, J.L., 1978. Computing exact probabilities in sensory discrimination tests. J. Food Sci. 43, 1028–1029.

Stone, H., Sidel, J.L., 1981. Quantitative descriptive analysis in optimization of consumer acceptance. Paper presented at the Eastern Food Science and Technology Conference on Strategies of Food Product Development, Lancaster, PA.

Stone, H., Sidel, J.L., 1995. Strategic applications for sensory evaluation in a global market. Food Technol. 49 (2), 80, 85–89.

Stone, H., Sidel, J.L., 1998. Quantitative descriptive analysis: developments, applications, and the future. Food Technol. 52 (8), 48–52.

Stone, H., Sidel, J.L., 2003. Descriptive analysis In: Encyclopedia of Food Science, second ed. Academic Press, London. pp. 5152–5161.

Stone, H., Sidel, J.L., 2007. Sensory research and consumer-led food product development. In: MacFie, H. (Ed.), Consumer-Led Food Product Development. Woodhead, Cambridge, UK.

Stone, H., Sidel, J.L., 2009. Sensory science and consumer behavior. In: Barbosa-Cánovas, G., Mortimer, A., Lineback, D., Spiess, W., Buckle, K., Colonna, P. (Eds.), Global Issues in Food Science and Technology. Elsevier, New York, pp. 67–77.

Stone, H., Sidel, J.L., Bloomquist, J., 1980. Quantitative descriptive analysis. Cereal Foods World 25 (10), 642–644.

Stone, H., Sidel, J.L., Oliver, S., Woolsey, A., Singleton, R.C., 1974. Sensory evaluation by quantitative descriptive analysis. Food Technol. 28 (11), 24, 26, 28, 29, 32, 34.

Swartz, M.L., Furia, T.E., 1977. Special sensory panels for screening new synthetic sweeteners. Food Technol. 31 (11), 51–55, 67.

Szczesniak, A.S., 1963. Classification of textural characteristics. J. Food Sci. 28, 385–389.

Szczesniak, A.S., Brandt, M.A., Friedman, H.H., 1963. Development of standard rating scales for mechanical parameters of texture and correlation between the objective and the sensory methods of texture evaluation. J. Food Sci. 28, 397–403.

Taguchi, G., 1987. System of Experimental Design, English ed. vols. 1 and 2. Unipub/Kraus International, White Plains, NY.

Tamar, H., 1972. Principles of Sensory Physiology. Thomas, Springfield, IL.

Thorngate III., J.H., 1995. Power analysis of sensory taste panels. Paper presented at the 2nd International Pangborn Sensory Symposium, Davis, CA.

Thurstone, L.L., 1927. The law of comparative judgement. Psychol. Rev. 34, 273–286.

Thurstone, L.L., 1959. The Measurement of Values. University of Chicago Press, Chicago.

Tilgner, D.J., 1962. Dilution tests for odor and flavor analysis. Food Technol. 16, 47–50.

Tilgner, D.J., 1965. Flavor dilution profilograms. Food Technol. 19 (12), 25–29.

Tobias, R.D., 1999. An Introduction to Partial Least Squares Regression. SAS Institute, Cary, NC.

Trant, A.S., Pangborn, R.M., Little, A.C., 1981. Potential fallacy of correlating hedonic responses with physical and chemical measurements. J. Food Sci. 46, 583–588.

Tushman, M.L., Anderson, P., 1997. Managing Strategic Innovation and Change. Oxford University Press, New York.

Vickers, Z.M., 1983. Magnitude estimation vs. category scaling of the hedonic quality of food sounds. J. Food Sci. 48, 1183–1186.

Vie, A., Gulli, D., O'Mahony, M., 1991. Alternative hedonic measures. J. Food Sci. 56, 1–5.

Warren, C., Pearce, J., Korth, B., 1982, March. Magnitude estimation and category scaling. ASTM Stand. News 10, 15–16.

Weiss, D.J., 1972. Averaging: an empirical validity criterion for magnitude estimation. Percept. Psychophys. 12, 385–388.

Wenzel, B.M., 1949. Differential sensitivity in olfaction. J. Exp. Psychol. 39, 129–143.

Wiley, R.C., Briant, A.M., Fagerson, I.S., Murphy, E.F., Sabry, J.H., 1957. The Northeast regional approach to collaborative panel testing. Food Technol. 11, 43–49.

Williams, A.A., Arnold, G., 1985. A comparison of the aromas of six coffees characterized by conventional profiling, free-choice profiling and similarity scaling methods. J. Sci. Food Agric. 36, 204–214.

Williams, A.A., Langron, S.P., 1984. The use of free-choice profiling for the evaluation of commercial ports. J. Sci. Food Agric. 35, 558–568.

Winer, R.J., 1971. Statistical Principles in Experimental Design, second ed. McGraw-Hill, New York.

Winn, R.L., 1988. Touch screen system for sensory evaluation. Food Technol. 42 (11), 68–70.

Wolfe, K.A., 1979. Use of reference standards for sensory evaluation of product quality. Food Technol. 33 (9), 43–44.

Woodward, W.A., Schucany, W.R., 1977. Combination of a preference pattern with the triangle taste test. Biometrics 33, 31–39.

Yandell, B.S., 1997. Practical Data Analysis for Designed Experiments. Chapman & Hall, London.

Yankelovich, D., 1964, March. New criteria for market segmentation. Harvard Business Rev.

Young, P.T., 1961. Motivation and Emotion. Wiley, New York.

Young, T., Pecore, S., Stoer, N., Hulting, F., Holschuh, N., Case, F., 2008. Incorporating test and control product variability in degree of difference tests. Food Qual. Prefer. 19, 734–736.

Zelek Jr., E.F., 1990. Legal aspects of sensory analysis. Paper presented at the IFT Annual Meeting and Food Expo, June 16–20, Anaheim, California.

Zook, K., Wessman, C., 1977. The selection and use of judges for descriptive panels. Food Technol. 31 (11), 56–60.

Index

Note: Page references followed by *t* and *f* denote tables and figures, respectively.

Food Science and Technology
International Series

Amerine, M. A., Pangborn, R. M., and Roessler, E. B., *Principles of Sensory Evaluation of Food*. 1965.

Glicksman, M., *Gum Technology in the Food Industry*. 1970.

Joslyn, M. A., *Methods in Food Analysis*, 2nd ed. 1970.

Stumbo, C. R., *Thermobacteriology in Food Processing*, 2nd ed. 1973.

Altschul, A. M. (Ed.), *New Protein Foods*: Volume 1, Technology, Part A—1974. Volume 2, Technology, *Part B*—1976. Volume 3, *Animal Protein Supplies, Part A*—1978. Volume 4, *Animal Protein Supplies, Part B*—1981. Volume 5, *Seed Storage Proteins*—1985.

Goldblith, S. A., Rey, L., and Rothmayr, W.W., *Freeze Drying and Advanced Food Technology*. 1975.

Bender, A. E., *Food Processing and Nutrition*. 1975.

Troller, J. A., and Christian, J. H. B., *Water Activity and Food*. 1978.

Osborne, D. R., and Voogt, P., *The Analysis of Nutrients in Foods*. 1978.

Loncin, M., and Merson, R. L., *Food Engineering: Principles and Selected Applications*. 1979.

Vaughan, J. G. (Ed.), *Food Microscopy*. 1979.

Pollock, J. R. A. (Ed.), *Brewing Science*, Volume 1—1979. Volume 2—1980. Volume 3—1987.

Christopher Bauernfeind, J. (Ed.), *Carotenoids as Colorants and Vitamin A Precursors: Technological and Nutritional Applications*. 1981.

Markakis, P. (Ed.), *Anthocyanins as Food Colors*. 1982.

Stewart, G. G., and Amerine, M. A. (Eds.), *Introduction to Food Science and Technology*, 2nd ed. 1982.

Iglesias, H. A., and Chirife, J., *Handbook of Food Isotherms: Water Sorption Parameters for Food and Food Components*. 1982.

Dennis, C. (Ed.), *Post-Harvest Pathology of Fruits and Vegetables*. 1983.

Barnes, P. J. (Ed.), *Lipids in Cereal Technology*. 1983.

Pimentel, D., and Hall, C. W. (Eds.), *Food and Energy Resources*. 1984.

Regenstein, J. M., and Regenstein, C. E., *Food Protein Chemistry: An Introduction for Food Scientists*. 1984.

Gacula, M. C., Jr., and Singh, J., *Statistical Methods in Food and Consumer Research*. 1984.

Clydesdale, F. M., and Wiemer, K. L. (Eds.), *Iron Fortification of Foods*. 1985.

Decareau, R. V., *Microwaves in the Food Processing Industry*. 1985.

Herschdoerfer, S. M. (Ed.), *Quality Control in the Food Industry*, 2nd ed. Volume 1—1985. Volume 2—1985. Volume 3—1986. Volume 4—1987.

Cunningham, F. E., and Cox, N. A. (Eds.), *Microbiology of Poultry Meat Products*. 1987.

Urbain, W. M., *Food Irradiation*. 1986.

Bechtel, P. J., *Muscle as Food*. 1986.

Chan, H. W.-S., *Autoxidation of Unsaturated Lipids*. 1986.

McCorkle, C. O., Jr., *Economics of Food Processing in the United States*. 1987.

Japtiani, J., Chan, H. T., Jr., and Sakai, W. S., *Tropical Fruit Processing*. 1987.

Solms, J., Booth, D. A., Dangborn, R. M., and Raunhardt, O., *Food Acceptance and Nutrition*. 1987.

Macrae, R., *HPLC in Food Analysis*, 2nd ed. 1988.

Pearson, A. M., and Young, R. B., *Muscle and Meat Biochemistry*. 1989.

Penfield, M. P., and Campbell, A. M., *Experimental Food Science*, 3rd ed. 1990.

Blankenship, L. C., *Colonization Control of Human Bacterial Enteropathogens in Poultry*. 1991.

Pomeranz, Y., *Functional Properties of Food Components*, 2nd ed. 1991.

Walter, R. H., *The Chemistry and Technology of Pectin*. 1991.

Stone, H., and Sidel, J. L., *Sensory Evaluation Practices*, 2nd ed. 1993.

Shewfelt, R. L., and Prussia, S. E., *Postharvest Handling: A Systems Approach*. 1993.

Nagodawithana, T., and Reed, G., *Enzymes in Food Processing*, 3rd ed. 1993.

Hoover, D. G., and Steenson, L. R., *Bacteriocins*. 1993.

Shibamoto, T., and Bjeldanes, L., *Introduction to Food Toxicology*. 1993.

Troller, J. A., *Sanitation in Food Processing*, 2nd ed. 1993.

Hafs, D., and Zimbelman, R. G., *Low-Fat Meats*. 1994.

Phillips, L. G., Whitehead, D. M., and Kinsella, J., *Structure–Function Properties of Food Proteins*. 1994.

Jensen, R. G., *Handbook of Milk Composition*. 1995.

Roos, Y. H., *Phase Transitions in Foods*. 1995.

Walter, R. H., *Polysaccharide Dispersions*. 1997.

Barbosa-Cánovas, G. V., Marcela Góngora-Nieto, M., Pothakamury, U. R., and Swanson, B. G., *Preservation of Foods with Pulsed Electric Fields*. 1999.

Jackson, R. S., *Wine Tasting: A Professional Handbook*. 2002.

Bourne, M. C., *Food Texture and Viscosity: Concept and Measurement*, 2nd ed. 2002.

Caballero, B., and Popkin, B. M. (Eds.), *The Nutrition Transition: Diet and Disease in the Developing World*. 2002.

Cliver, D. O., and Riemann, H. P. (Eds.), *Foodborne Diseases*, 2nd ed. 2002.

Kohlmeier, M., *Nutrient Metabolism*. 2003.

Stone, H., and Sidel, J. L., *Sensory Evaluation Practices*, 3rd ed. 2004.

Han, J. H., *Innovations in Food Packaging*. 2005.

Sun, D.-W. (Ed.), *Emerging Technologies for Food Processing*. 2005.

Riemann, H. P., and Cliver, D. O. (Eds.), *Foodborne Infections and Intoxications*, 3rd ed. 2006.

Arvanitoyannis, I. S., *Waste Management for the Food Industries*. 2008.

Jackson, R. S., *Wine Science: Principles and Applications*, 3rd ed. 2008.

Sun, D.-W. (Ed.), *Computer Vision Technology for Food Quality Evaluation*. 2008.

David, K., and Thompson, P., (Eds.), *What Can Nanotechnology Learn from Biotechnology?* 2008.

Arendt, E. K., and Bello, F. D. (Eds.), *Gluten-Free Cereal Products and Beverages*. 2008.

Bagchi, D. (Ed.), *Nutraceutical and Functional Food Regulations in the United States and around the World*. 2008.

Singh, R. P., and Heldman, D. R., *Introduction to Food Engineering*, 4th ed. 2008.

Berk, Z., *Food Process Engineering and Technology*. 2009.

Thompson, A., Boland, M., and Singh, H. (Eds.), *Milk Proteins: From Expression to Food*. 2009.

Florkowski, W. J., Prussia, S. E., Shewfelt, R. L. and Brueckner, B. (Eds.), *Postharvest Handling*, 2nd ed. 2009.

Gacula, M., Jr., Singh, J., Bi, J., and Altan, S., *Statistical Methods in Food and Consumer Research*, 2nd ed. 2009.

Shibamoto, T., and Bjeldanes, L., *Introduction to Food Toxicology*, 2nd ed. 2009.

BeMiller, J., and Whistler, R. (Eds.), *Starch: Chemistry and Technology*, 3rd ed. 2009.

Jackson, R. S., *Wine Tasting: A Professional Handbook*, 2nd ed. 2009.

Sapers, G. M., Solomon, E. B., and Matthews, K. R. (Eds.), *The Produce Contamination Problem: Causes and Solutions*. 2009.

Heldman, D. R., *Food Preservation Process Design*. 2011.

Tiwari, B. K., Gowen, A., and McKenna, B., *Pulse Foods: Processing, Quality and Nutraceutical Applications*. 2011.

Cullen, P. J., Tiwari, B. K., and Valdramidis, V. P., *Novel Thermal and Non-Thermal Technologies for Fluid Foods*. 2012.

Printed and bound by CPI Group (UK) Ltd, Croydon, CR0 4YY

03/10/2024

01040317-0001